D0085066

# A Guide to Lasers in Chemistry

# A Guide to Lasers in Chemistry

**Gerald R. Van Hecke**

**Kerry K. Karukstis**

*Harvey Mudd College*
*Claremont, California*

**Jones and Bartlett Publishers**
Boston London Singapore

*Editorial, Sales, and Customer Service Offices*
Jones and Bartlett Publishers
40 Tall Pine Drive
Sudbury, MA 01776
978-443-5000

Jones and Bartlett Publishers International
Barb House, Barb Mews
London W6 7PA
UK



**Library of Congress Cataloging-in-Publication Data**
Van Hecke, Gerald R.
A guide to lasers in chemistry / Gerald R. Van Hecke, Kerry K. Karukstis.
p. cm.
Includes bibliographical references (p. – ) and index.
ISBN 0-7637-0412-1
1. Lasers in chemistry. 2. Laser spectroscopy. 3. Laser photochemistry. I. Karukstis, Kerry K. II. Title.
QD63.L3V36 1998
542′.8—dc21 97-13440
CIP

*Sponsoring Editor:* Christopher W. Hyde
*Manufacturing Buyer:* Jenna Sturgis
*Editorial-Production Service:* WordCrafters Editorial Services, Inc.
*Typesetting:* Publishers' Design and Production Services, Inc.
*Cover Design:* Hannus Design Associates
*Printing and Binding:* Courier Companies, Inc.
*Cover Printing:* Coral Graphic Services, Inc.

Printed in the United States of America

01 00 99 98 97 10 9 8 7 6 5 4 3 2 1

# Contents

## Part II The Laser as a Probe 69

# Preface

Chemists today creatively use laser-generated photons as probes of molecular structure and as reagents to effect chemical transformations. This book will help you appreciate the laser revolution in chemistry and, perhaps once intrigued, entice you to contribute to this expanding frontier.

*A Guide to Lasers in Chemistry* is written for you, the first-time student, to introduce you to general principles and to illustrate those principles through specific case studies. The book is also intended to be a reference and a resource, allowing you to quickly review terms and concepts by using the glossary and appendices. Whether you come to read this book because of self-interest, in preparation for research, or as a text, we believe you will find it informative yet light in style, for it was written to be both. Following introductory chapters on light, lasing action, and laser components, we offer you a menu of applications illustrated by case studies and organized to highlight the laser in its role as a structural probe or chemical reagent.

To enable you to appreciate the contribution of lasers to the study of chemistry, the four chapters of Part I review the general principles of laser operation and the specific features of various laser types. The properties of light and the unique attributes of laser light are detailed in Chapter 1. The basis for the vast array of laser types is described in Chapter 2, with a discussion of the principles of lasing action and the features of laser construction. Phenomena discussed include optical and electrical pumping, generation of population inversions, stimulated emission, and light amplification. Central properties of laser cavities and lasing media are also examined. Chapter 3 delineates several modification schemes of laser light to obtain new characteristics of lasing emission such as polarized light, pulsed output, or enhanced power. Modifications of lasing output include production of polarized light, $Q$-switching, cavity dumping, mode-locking, synchronous pumping, pulse compression, and continuum generation. Finally, the principal features and current chemical applications of six major categories of lasers are characterized in Chapter 4: monatomic gas lasers, molecular gas lasers, metal vapor lasers, solid-state lasers, semiconductor lasers, and dye lasers. This fundamental discussion of laser principles and types constitutes a solid foundation for understanding the critical role of the selected laser apparatus in the chemical applications subsequently presented in the text.

The seven chapters of Part II feature the laser as a structural probe. In Chapter 5 the laser is featured as a tool to explore chemical reactions in the challenging fast and ultrafast time regimes. Chemical processes that exhibit rates on these timescales span the range from simple unimolecular dissociations to complex protein-ligand binding reactions with concurrent conformational transitions. We survey the dynamical mapping of several chemical systems to illustrate ultrafast laser techniques and highlight the potential directions for lasers in the study of fast chemical events. In Chapter 6 we examine the relatively new field of multiphoton spectroscopy, a field that is based on the extraordinary condition where target molecules absorb multiple photons. The sheer intensity of light that can be obtained from lasers has made multiphoton experiments possible.

Chapters 7 and 8 highlight the use of laser-induced fluorescence as a powerful optical technique to characterize chemical systems. The variety of fluorescence applications is apparent from the case studies selected in Chapter 7: time-resolved studies of photosynthetic organisms on the picosecond timescale and airborne remote sensing of laser-induced fluorescence in terrestrial and oceanographic targets. The coupling of the separation and analysis technique of capillary electrophoresis with the detection method of laser-induced fluorescence is described in Chapter 8 with two case studies highlighting the extensive capabilities of this combination of approaches. Chapter 9 discusses two novel chemical applications of the scattering of laser light to probe the existence of intermolecular forces between two component molecules in a liquid mixture and to detect the existence of specific molecules in hostile or otherwise unattainable environments. Recent advances using tunable lasers in mass spectral analysis are presented in Chapter 10 to reveal the applicability of the technique to two classes of molecules previously precluded from study due to their low volatility and tendency to decompose upon ionization—high-molecular-weight biomolecules and molecular adsorbates on solid surfaces. Our final example of the laser as a chemical probe examines the use of pulsed or modulated laser light to characterize the thermodynamic properties of a chemical substance through the induction of the photoacoustic effect.

The four chapters in Part III survey several exciting applications of laser photons as chemical reagents. Chapters 12 and 13 feature the laser as a photochemical tool in both research and industrial settings. Several properties of laser light—monochromaticity, tunability, intensity, and mode of operation (continuous vs. pulsed)—assist the synthetic chemist in achieving enhanced product yields and selectivity in product distribution. The growing applications of lasers in medicine are discussed in Chapter 14. After an initial discussion of the interaction of light with biological tissue and a brief survey of the current medical applications of lasers, we examine the light-activated technique of photodynamic therapy. This promising therapeutic procedure uses laser illumination of photosensitizers preferentially localized in cancerous tissues to initiate chemical reactions that ultimately destroy the cancerous tissue. As our final illustration of the laser as a chemical reagent, Chapter 15 describes the particular control afforded chemists by the selective excitation of a specific chemical bond with laser light. This bond-selective chemistry is currently limited to those chemical systems with "localized" bonds, but is anticipated to be an ever-increasing occurrence in more complex chemical systems.

*A Guide to Lasers in Chemistry* is not intended to be encyclopedic, but is meant to provide you with enough fundamentals and background to appreciate exciting applications of lasers. We hope to set you on your own course of discovery in the use of lasers in chemistry.

Gerald R. Van Hecke
Kerry K. Karukstis
Claremont, CA

# Acknowledgments

The genesis of this monograph was a Pew Foundation Grant to the authors that allowed Professor Richard N. Zare to spend a week on the Harvey Mudd campus lecturing on applications of lasers in chemistry. With Professor Zare's encouragement and inspiration, we developed a course to illustrate to our undergraduates the exciting and expanding uses of lasers in chemical investigations. We wrote this monograph for use as a text in the course. We thank the students of Chemistry 168: Special Topics in Chemistry for their patience and constructive feedback as our writing efforts progressed. We also thank Lisa Gann (class of '94) for her editorial assistance. We particularly wish to single out for acknowledgment and thanks J. Michael Underhill (class of '95) for his insightful editorial and graphics contributions.

# A Guide to Lasers in Chemistry

PART I

# Introduction to the Laser Itself

An astounding array of laser devices confronts the chemist today. While the laser in general is a remarkable and versatile tool, not all lasers afford the same advantages to every chemical study. The success of a specific laser application may be critically dependent on the optimal choice of laser source. The key to selecting the most appropriate laser for a particular chemical application lies in understanding the general principles of laser operation and knowing the unique features of various laser types. The following four chapters aim to provide you with the principles essential for understanding the crucial role of lasers in chemistry.

Chapter 1 defines the properties of laser light that distinguish it from ordinary light and radiation and contribute to the laser's utility. Chapter 2 is an overview of the essential concepts underlying laser action. Chapter 3 describes the various techniques to modify and control laser output that enhance the flexibility of lasers. These increased performance capabilities enable chemists to expand the repertoire of laser applications and to envision even more exciting directions for lasers in chemistry. Finally, Chapter 4 highlights the principal features of the primary laser types.

CHAPTER 1

# Properties of Light and Laser Light

## Chapter Overview

Chemists are presented with a vast and impressive array of experimental approaches to address fundamental questions, characterize chemical systems, and perform quantitative analyses. How does one sort through the myriad of options to select an appropriate technique? A clear understanding of the central principles of a potential technique and the parameters governing its advantages and limitations is vital to the decision-making process. Such understanding is also essential to design new applications for existing methods. In this chapter we review the properties of light in order to understand how photons enable characterizations of molecular structure, as well as effect chemical transformations. We further review the specific and unique attributes of laser light to recognize the potential uses of this modern light source.

## Light

### Light and Light Regimes

A laser is a light source. To begin understanding this incredible source of light, or photons if you prefer, let's review some properties of light. Today we appreciate the dualistic particle/wave nature of light: as a particle, we discuss the photon; as a wave, we discuss a transverse electromagnetic wave which propagates as a vector with an electric field and a magnetic field at right angles to each other. Most of the time we neglect the magnetic field because its interactions with matter are much weaker than those of the electric field. The electric field can be further thought of as the oscillation of an electric dipole, the displacement in time of a charge. For virtually all of our discussion, we will think of light as a wave. For a wave, we can define a wavelength as the distance between successive maxima or successive minima. Moreover, we can define a frequency of the wave as the number of wavelengths which pass a fixed point per second. Propagating light is illustrated in Figure 1-1.

Over time, we have come to characterize light by dividing the entire electromagnetic spectrum into wavelength regimes. We all know of radio waves, X rays, and of course, visible light, which by definition is what we can see. The various scales and energy regimes are "defined" in Figure 1-2 and discussed in greater detail in Appendix I. What is the regime of visible light? Generally this is accepted to be from 400 to 700 nm, and as you look at the total electromagnetic spectrum, you will discover the visible range to be only a very small portion.

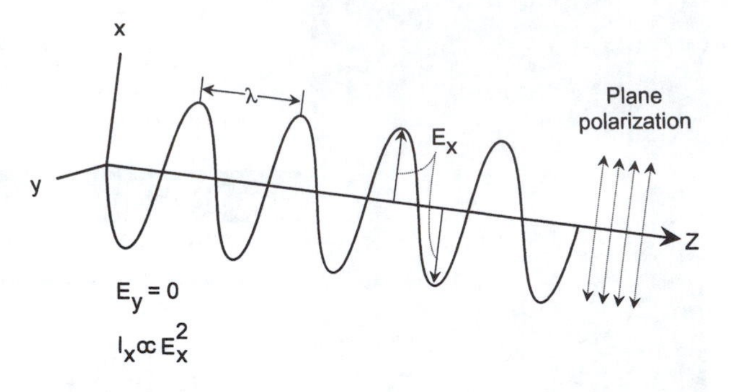


**Figure 1-1**
Plane polarized electromagnetic radiation of wavelength $\lambda$ propagating along the $z$ axis. The plane of polarization is the $xz$ plane. Only the oscillating electric field is depicted. The perpendicular magnetic field is omitted, as warranted by the weak interactions of the magnetic field with matter in comparison with the interactions of the electric field.

### Frequency and Wavelength

Perhaps one of the most amazing characteristics of light is that no matter what its color or frequency, $\nu$,

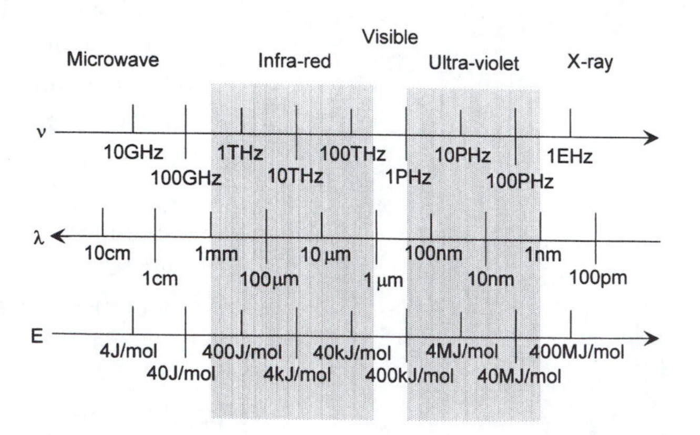


**Figure 1-2**
Frequency, wavelength, and energy regimes of the electromagnetic spectrum. Both frequency and energy increase from the microwave to the X-ray region; correspondingly, the wavelength is longest for microwave radiation, shortest for X rays. For nomenclature, the prefixes are: **k** = **kilo** = $\times 10^3$, **M** = **mega** = $\times 10^6$, **G** = **giga** = $\times 10^9$, **T** = **tera** = $\times 10^{12}$, **P** = **peta** = $\times 10^{15}$, and **E** = **exa** = $\times 10^{18}$.

the wavelength ($\lambda$) of light times the frequency is a constant we call the speed of light, $c$. The speed of light in a vacuum is now known to high precision as $2.99792458 \times 10^8$ m/s, or more commonly as $3 \times 10^8$ m/s.

We have all seen the gas flame in the burner of a kitchen gas stove periodically exhibit a yellow color. Moreover, you probably have seen yellow light from a Bunsen burner in the laboratory. This yellow light comes from a visible light emission of atomic sodium, called the sodium-D line, whose wavelength $\lambda$ is about 589 nm. The frequency of 589 nm sodium light is $5 \times 10^{14}$ Hz as shown by the calculation below:

$$\nu = \frac{c}{\lambda}$$

$$\nu = \frac{3 \times 10^8 \text{ms}^{-1}}{589 \text{ nm} \times 10^{-9} \text{ m nm}^{-1}}$$

$$= 5 \times 10^{14} \text{ s}^{-1} = 5 \times 10^{14} \text{ Hz} \qquad (1)$$

We can describe this frequency in other terms as $5 \times 10^8$ MHz, $5 \times 10^5$ GHz, or even 500 THz, where we have taken advantage of the nomenclature for powers of ten: M = mega = $10^6$, G = giga = $10^9$, T = tera = $10^{12}$.

## Light Energy

Knowing about wavelength and frequency of light, can we comprehend the energy of light? To answer this query, we will need to appreciate some aspects of the wave/particle duality. As a result of his study of blackbody radiation, Planck made the remarkable assertion that the energy of a given wavelength of light is directly proportional to its frequency. We now call his assertion Planck's law, which is:

$$E = h\,\nu \qquad (2)$$

Moreover, light should not be viewed just as a wave with a frequency, but as particles—each with the energy given by Equation 2. Thus, photons come in energy quanta $h\ \nu$, where $h$ is Planck's constant and has a value of $6.626 \times 10^{-34}$ J s photon$^{-1}$. A common laser is made from neodymium doped yttrium/aluminum oxide crystals and is often known as the Nd:YAG or YAG laser. You might ask: How many photons are there in a typical 1 joule output pulse from an Nd:YAG laser whose output light has a wavelength of 1.06 microns?

$$1\text{J} = (E \text{ per photon})(\#\ \text{photons})$$

$$1\text{J} = (h\nu)(\#\ \text{photons})$$

$$1\text{J} = h\frac{c}{\lambda}(\#\ \text{photons})$$

$$1\text{J} = 6.626 \times 10^{-34}\ \text{J s photon}^{-1}\ \frac{3 \times 10^8\ \text{ms}^{-1}}{1.06 \times 10^{-6}\ \text{m}}$$

$$\times (\#\ \text{photons})$$

$$\#\ \text{photons} \approx 5 \times 10^{18} \qquad (3)$$

The very large number of photons necessary to total 1 joule suggests that each individual photon actually represents very little energy. An Einstein is Avogadro's number of photons, $6.02 \times 10^{23}$. Thus, while a photon itself is only a small number of joules for 1.06 micron light, an Einstein of such light represents over 1.1 MJ of energy. Many types of energy units are used to describe photons. The origin of the many types is historical and based on the user's preferred energy scale. Comparisons of these scales are detailed in Appendix I.

## Wave or Particle?

What experimental evidence supports the dual nature of light as a wave and as a particle? Consider for a moment a dramatic illustration of the wave nature of light. If you put two of your fingers close together and look through them at a lamp, you can see lines of light and dark parallel to your fingers. Moving your fingers closer and farther apart will dramatically affect the number of lines and the ease with which you can see them. The lines you see result from diffraction—the constructive and destructive interference of light waves when they pass through slits or pinholes or reflect off surfaces. Diffraction is associated with waves, not particles.

However, a phenomenon known as the photoelectric effect is more consistent with light having the characteristics of particles. When you detect light with a device known as a photocell, photons collide with the photoemissive metal surface of the photocell. When the photon energy exceeds the binding energy of an electron on the metal surface, the collision of the photon with the surface causes an electron to be ejected. You can measure the rate of flow of the electrons from the surface as a current. Collision is a process associated with particles. In fact, the photoelectric effect occurring in the simple photocell is one of the most compelling experiments available to demonstrate that light comes in particle packages of energy called photons.

## Light Waves

### Light Goes Which Way?

How does light travel in a given direction? First consider light as if it were a plane wave. In a perfect plane wave, all the waves are travelling in exactly the same direction and all are in phase such that all of the maxima are maxima at the same time; all of the minima are minima at the same time. Thus, a plane wave consists of surfaces of constant phase or wave front as shown in Figure 1-3. Such waves can be described as travelling in perfect spatial and temporal coherence. Coherence here means that, as the waves travel, they will be exactly superimposable on each other at any position or at any time.

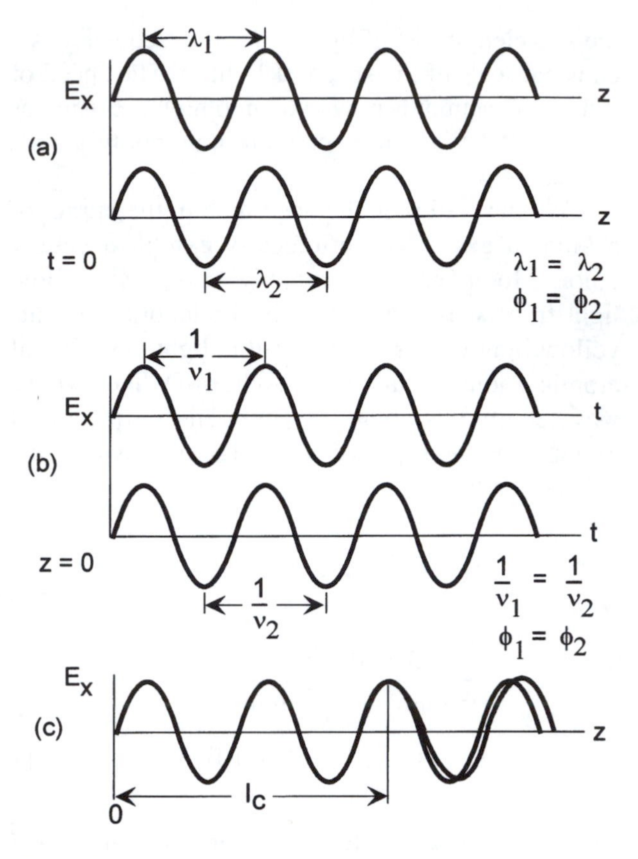


**Figure 1-3**
Perfect plane waves with constant phase or wave front. The waves depicted in *(a)* illustrate spatial coherence, where waves with equal phase and wavelength are superimposable at any distance along the axis of propagation. The waves illustrated in *(b)* are characterized by temporal coherence and are superimposable at any time. Loss of spatial coherence is indicated in *(c)*: The wave fronts remain coherent as they travel over the distance $l_c$.

### Plane Waves

The idealization of a perfect plane wave with infinite extent—no beginning and no end—is wonderful, but it does not exist in nature. Consider, however, what happens if you toss a pebble into a puddle. Waves are created from the point of pebble entry. These waves, now called spherical waves, travel outward from the origin in ever-widening circles, decreasing in curvature until they reach a point far enough away from the origin to look like plane waves. What does this imply about spherical waves far from their origin? One could infer that any plane wave is simply a spherical wave far away from its origin. This concept is important for diffraction to explain the origin of constructive and destructive interference.

### A Mathematical Wave

What mathematical equation can we write to describe a travelling wave such as those shown in Figures 1-1 and 1-3? A general equation for wave motion in one direction, say along a $z$ axis, is:

$$E(z, t) = A \cos\left(\frac{2\pi v}{\lambda} t - \frac{2\pi}{\lambda} z + \phi\right) \tag{4}$$

While this equation looks somewhat daunting, a little explanation should make it tractable. Here $A$ (the amplitude) is the maximum value of $E(z,t)$, v is the velocity of the wave contour, $t$ is time, $z$ is the distance travelled from some originating point, $\lambda$ is the wavelength, and $\phi$ is called the phase angle, which is useful when comparing one wave to another. The units of the argument of the cosine are radians.

Now some substitutions can be made. Recall that the frequency $\nu$ of a wave is the number of maxima that pass a point in space per unit of time. If the distance between maxima is the wavelength $\lambda$, then how fast the wave contour moves past a point in space should be the velocity given by $v = \nu \lambda$. The expression $2\pi/\lambda$ is the number of waves in the linear distance $2\pi$ meters (if $\lambda$ is measured in meters) and is often referred to as the magnitude of the wave vector **k** (in this case along the $z$ direction exclusively). Rewriting the expression for $E(z,t)$ in terms of these new definitions gives:

$$E(z,t) = A\cos(2\pi\nu t - kz + \phi) \quad (5)$$

which now describes a wave contour travelling in the positive $z$ direction whose oscillation frequency is $\nu$. Note this expression places no restriction on the magnitude of $z$ or $t$. The range of $z$ could be infinite, as it must in fact be to define the cosine function. Such waves are the infinite plane waves pictured in Figure 1-3. In Figure 1-3(a), a wave motion is described for $t = 0$ and defines the wavelength. The wave motion in Figure 1-3(b) illustrates wave motion for $z = 0$ and defines the period of a wave, which is $\nu^{-1}$.

## Wave Vectors and Phases

An elaborate picture of light waves travelling in three dimensions is given in Figure 1-4. Mathematically such light waves can be described by:

$$E(x,y,z,t) = A\cos(2\pi\nu t - \mathbf{k}\cdot\mathbf{r} + \phi) \quad (6)$$

In Figure 1-4, **r** represents the distance from our reference origin to the point at which we look to follow the amplitude of the wave as a function of time. The wave vector **k** defines the direction of the wave as it travels spatially with a wavelength of $\lambda$. The waves shown in Figure 1-4 are travelling in the $xz$ plane.

If, as in Figure 1-3, each wave had its maxima and minima in the same place at the same time, then each wave would have the same phase. Such waves are also called coherent. When the phase of each wave is the same, we can set the phase angle equal to zero with no loss of information. Suppose the phase angle is not the same for each wave, however. Over time the waves would lose coherence and begin to interfere with each other as illustrated in Figure 1-3(c). Coherence can be described as either spatial or temporal. Spatial coherence describes how long a distance a wave travels spatially before coherence is lost; for example, $l_c$ is the coherence length shown in Figure 1-3(c). Temporal coherence describes the time over which the waves are coherent.

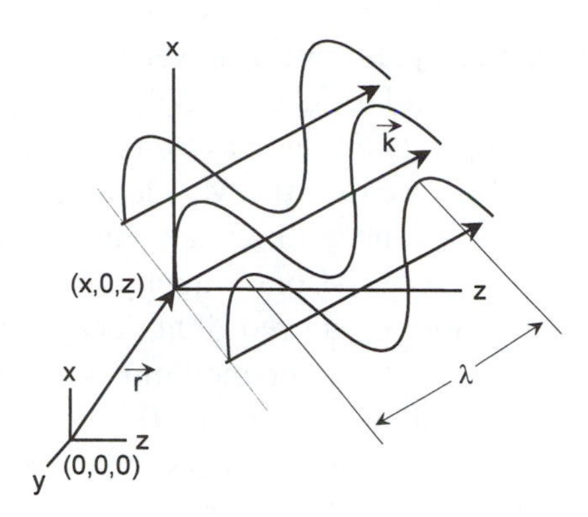


**Figure 1-4**
A representation of the three-dimensional travel of a wave of wavelength $\lambda$. The direction of spatial travel is represented by the wave vector **k** as the waves travel in the $xz$ plane. The distance from the reference origin to the point where the amplitude of the wave will be monitored as a function of time is given by the vector **r**.

## Light in Condensed Phases

How does light travel? You have seen that light travels in straight lines in the direction defined by the wave vector **k**. So far, we have assumed that light has been travelling in a vacuum. In various other media, since light is an electromagnetic wave, it creates an electromagnetic disturbance in the medium. The light slows down depending on how much it interacts with the medium. Even though a medium is transparent to light, the light slows down because the electric field of the light interacts with the electrons in the material. The decrease in speed is even more dramatic the closer the medium is to being nontransparent, that is, absorbing.

The ratio of the velocity of light in a vacuum, $c$, to that in the medium, v, is called the refractive index: $n = c/v$. For a vacuum, the refractive index is unity. Since light always slows down in any media other than a vacuum, for all other media $n$ must be greater than unity. For air, $n$ is very near unity. For water, $n$ is 1.33 and for glass, $n$ is 1.5. The refractive index also depends on the wavelength of light. The bluer the light, the closer the wavelength is to the region where substances like water and glass really absorb light energy. Consequently, blue light is slowed much more than red light. The variation in the speed of light in a given medium as a function of wavelength is called dispersion. This effect is used all the time to separate the colors of white light, for example in a prism. Moreover, the phenomenon of dispersion is the origin of rainbows. The fact that the frequency of light does not change when travelling through media but the wavelength does is very important. Let's try to make this clear with a very simple picture.

## Wavelength or Frequency?

Suppose your dog Spot jumps up and down in the back seat of your car, and you somehow keep track of your dog's nose as you drive along the freeway. Let Spot jump up and down at a constant rate, that is, a constant frequency. Follow the path of Spot's nose, first for your car travelling at 45 km/hr, then at 90 km/hr. Spot bounces at the same frequency, but the velocity of the car changes the horizontal distance travelled by Spot's nose—a larger distance is traversed during one complete bounce at 90 km/hr than at 45 km/hr. Thus, the wavelength changes even though the frequency does not. This phenomenon is illustrated in Figure 1-5.

To further emphasize this point, suppose you were swimming underwater and observed scattered "red" helium-neon laser light. What would you see? Red light! Suppose you saw the same laser light on land when you came out of the water. You detect the same color red. What, however, is the wavelength of the light in the water and in the air? For water, the actual $\lambda$ is really $\lambda$ in a vacuum divided by $n_{\text{water}}$. The "red" He/Ne output in a vacuum is 632.8 nm. Since the red light slows down in water, the wavelength in water is 475.8 nm.

$$\lambda_{\text{water}} = \frac{\mathrm{v}}{\nu} = \frac{c}{n\nu} = \frac{\lambda_{\text{vac}}}{n_{\text{water}}}$$

$$= \frac{632.8\ \text{nm}}{1.33} = 475.8\ \text{nm} \qquad (7)$$

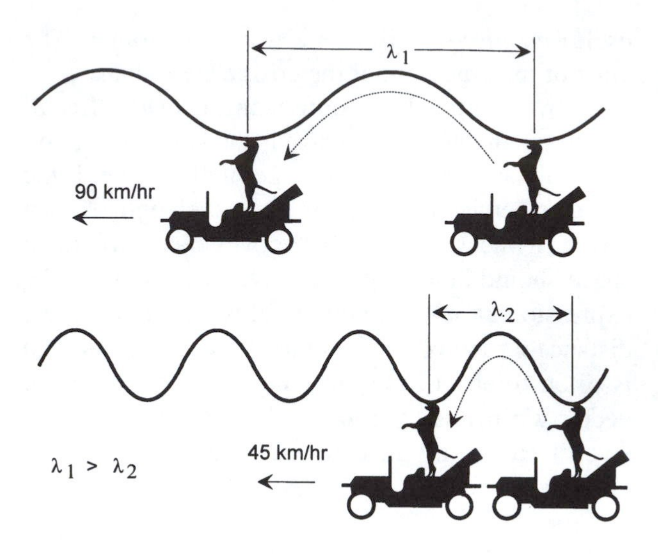


**Figure 1-5**
An illustration of the effect of a medium on the wavelength of light and not on its frequency. The dog Spot jumps up and down at a constant frequency, but the differing speeds of the car cause the distance travelled during one complete bounce to vary.

However, when you looked at the laser emission in water you noticed no difference in color. Does the eye respond to wavelength or frequency? We have the answer. The eye is not a dispersive device, a diffractive grating, or a prism. The eye senses only frequency and hence only energy.

# Characteristics of Laser Ligtht

## What is Unique About Laser Light?

We began this chapter by noting that a laser is a light source and discussed some general properties of light. What are some characteristics of laser light? Several important features that characterize a laser are listed in Table 1-1.

We will discuss these laser features in some detail.

## Monochromaticity

Laser light is monochromatic, which means it is a very pure color. However, even laser light cannot be made a perfectly pure color since that would require the existence of a perfect plane wave. However, to better appreciate monochromaticity, we have to un-

**Table 1-1**
Characteristics of Laser Light

| |
|---|
| High monochromaticity |
| Low divergence (directionality) |
| Extreme brightness |
| Spatial and temporal coherence |
| Speckle |
| Controllable polarization |

derstand plane waves, which we will attempt to do by first discussing divergence.

## Divergence

Closely related to directionality is divergence, which is a measure of how much the diameter of a light beam spreads out as the beam travels. A laser beam has a very small divergence which means the beam can travel great distances and remain a narrow beam. An automobile headlamp has a much greater divergence. We all have seen how the light from a headlamp spreads out or diverges in fog. There is a simple reason for the laser's small divergence. Inside the laser, more specifically inside what will be called the laser cavity, are two mirrors facing each other. Light in the cavity bounces back and forth between these two mirrors, as illustrated in Figure 1-7. An excellent analogy is that of looking into two mirrors facing each other and watching your image go off seemingly to infinity in both directions. The light in the laser cavity bounces off these mirrors between 50 and 100 times before exiting from the laser. These internal reflections make the effective length of the laser cavity 50 to 100 times longer than its nominal length. Moreover, the mirrors of the laser cavity force the light to travel in a straight line with small divergence, thus collimating the beam. Ordinary light sources such as the auto headlamp do not have this "collimation" and as a result have a much larger divergence.

Lasers cannot have zero divergence, however, because of diffraction. To illustrate this point, con-

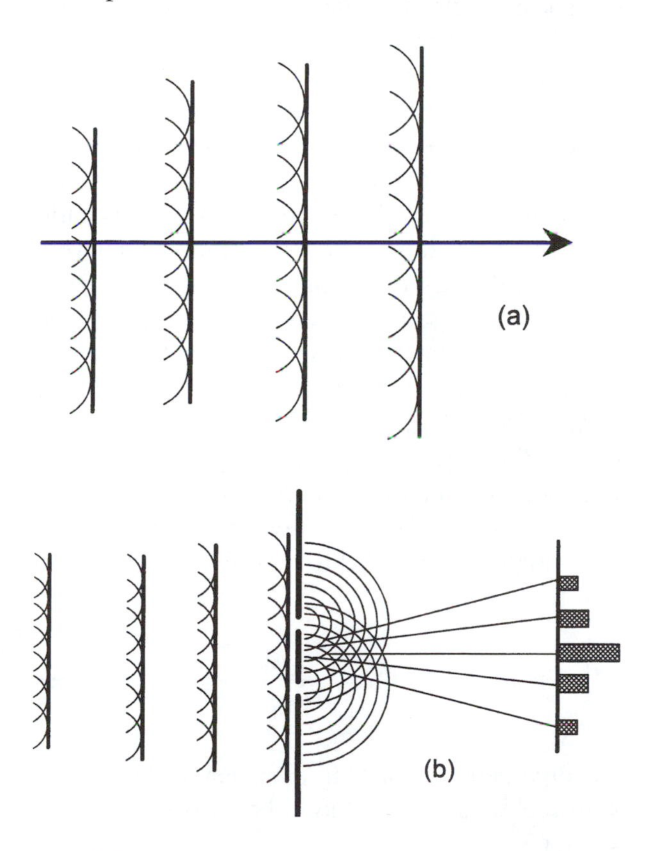


**Figure 1-6**
*(a)* Huygens' construction of the propagation of a plane wave in free space. The ray which is the plane wave is represented by the tangent to the spherical wave fronts. *(b)* The isolation of a ray is impossible because the plane waves incident on ever-decreasing slits are diffracted into their component spherical waves. The diffraction becomes more noticeable as the slit size approaches zero.

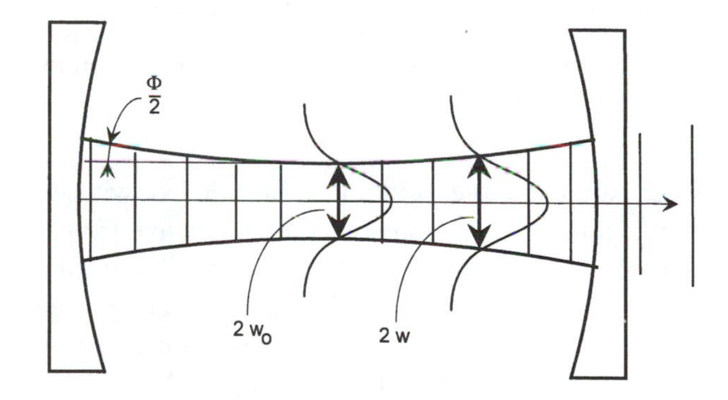


**Figure 1-7**
A typical helium-neon laser with a laser cavity defined by the two reflecting mirrors. A laser beam with a Gaussian cross-sectional profile is presented, where the beam intensity decreases according to a Gaussian distribution as distance from the beam center increases. The beam radius **w** is measured from the center of the beam to the point where the intensity has fallen to $1/e \times 100\%$ of its original intensity. At each point along the axis of the laser cavity, the intensity of the Gaussian profile varies. The beam radius at the point where the Gaussian profile is at a maximum is called $\mathbf{w}_o$. This beam radius is also known as the beam waist, which occurs at the minimum beam radius, $\mathbf{w}_{min} = \mathbf{w}_o$. The divergence of the beam, how much it spreads out as it travels, is characterized by the angle $\Phi/2$.

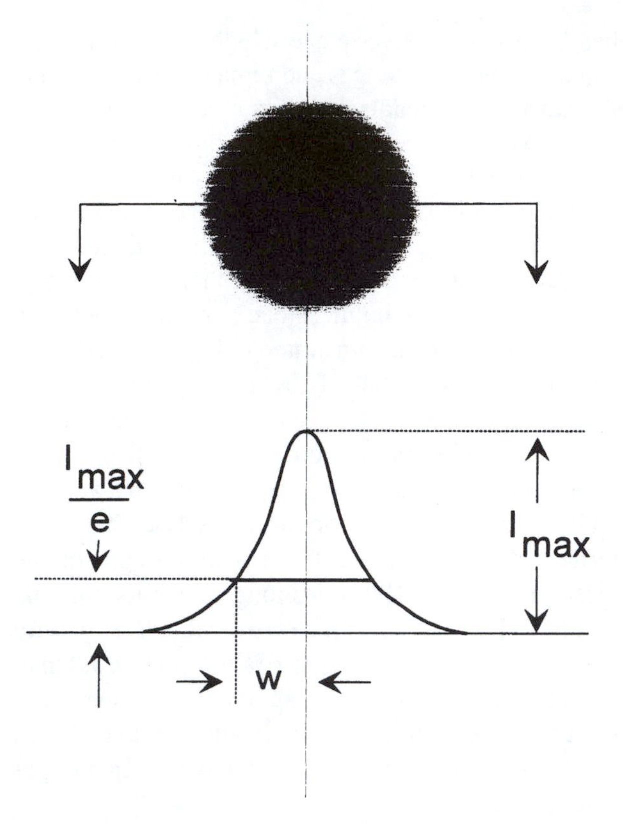


**Figure 1-8**
The variation in the laser beam intensity as viewed down the beam axis. The beam is most intense in the center, falling off in intensity with a Gaussian dependence as distance from the center increases. The Gaussian cross-sectional profile of the laser beam intensity is plotted, demarcating the beam radius where the intensity has fallen to 37% of the maximum intensity at the center of the beam.

sider some simple principles introduced without proof. Suppose one constructs a plane wave from a line of equally spaced point sources, each emitting circular waves all in phase. The development of such a plane wave is called a Huygens' construction, which is illustrated in Figure 1-6(a). When light reaches a slit, it spreads or diverges on the other side of the aperture, as in Figure 1-6(b). Constructive and destructive interference give rise to the diffraction pattern of light and dark regions shown in Figure 1-6(b). Because any plane wave can be constructed from a number of spherical waves travelling essentially but not exactly parallel, the beam must spread or diverge as it propagates. Divergence can be quantified by determining how much the beam spreads out per length of travel.

Let us explore this for a typical helium-neon laser whose simplified construction is illustrated in Figure 1-7. Light is reflected back and forth by the two mirrors defining the laser cavity. Generally these mirrors are curved slightly to focus or narrow the beam to a minimum in the middle of the cavity. Moreover, one of the mirrors is made more transmissive (not as reflective) as the other, which means a fraction of the light is transmitted ("leaks") through that window. Without such cavity leaks, the laser would not be a light source! The light propagated out of the cavity forms spherical waves that diverge. The question here is: How great is the divergence? If you were to look directly at a laser beam, but with a piece of paper between you and the beam, the intensity would appear as shown in Figure 1-8. *A NOTE OF CAUTION: Never, never do this for any real laser beam—paper or not.* Studies of such beam intensities have shown them to have a Gaussian distribution in cross section. Many lasers are very close to pure Gaussian beams. The cross section of such a beam is a spot with intensity falling off from the center with a Gaussian dependence, as shown in Figure 1-8. This cross-sectional intensity is called a $TEM_{00}$ transverse mode of laser output. The cross-sectional view of the output laser beam can have many shapes which depend on what is called the transverse mode of operation. (Transverse modes are discussed in Chapter 2.)

We can define a beam radius $w$ to be the distance from the center of the beam to where the intensity has fallen to 37% of the maximum intensity (this is $100\% \times 1/e$; $e = 2.78$). Note in Figure 1-7 that the beam radius varies along the cavity axis. The beam waist occurs when the beam radius is at a minimum and is denoted by $2w_o$. The divergence is quantified as the angle $\Phi$ illustrated in Figure 1-7 and is calculated from:

$$\Phi = \frac{1.27\lambda}{2w_0} \tag{8}$$

The divergence of a Gaussian beam He/Ne laser (with red emission) having a beam waist of 1 mm ($= 2w_o$) is:

$$\Phi = \frac{(1.27)(6.328 \times 10^{-7}\text{ m})}{(10^{-3}\text{ m})} = 0.8 \text{ millirad} \tag{9}$$

A beam characterized by $\Phi = 0.8$ millirad increases its size 0.8 mm for each meter travelled.

Compared to a conventional light source, the laser's directionality is far superior because of the small divergence. The laser's mirrors cause the wave fronts to be planar in the region of the beam waist, and study of Figure 1-7 will reveal that the beam waist occurs where the wave fronts are planar. Most modern He/Ne lasers are designed through mirror control to place the beam waist just outside of the exit mirror (window) to the laser cavity. Conventional light sources suffer from large divergence because they have no control over the output wave fronts.

## Brightness Is Radiance

As a result of having a compact, collimated, spectrally pure light source, the laser offers unexcelled brightness. Brightness seems an easily understood concept, but it is worthwhile to introduce some terms from the field of radiometry (see Appendix III for further discussion of brightness, luminance, and irradiance). Brightness is properly called radiance and is defined as radiant intensity per unit area of the source of radiation. Radiant intensity is the power of the light source per unit solid angle. Usually we think of intensity as the power (energy per unit of time) per unit area. We must note then that radiant intensity quantifies the power per solid angle, or watts/steradian. Thus:

$$\text{brightness} = \text{radiance} = \frac{\left(\dfrac{\text{power}}{\text{unit solid angle}}\right)}{\text{area of the source}} \qquad (10)$$

Recall that a radian measures the ratio of the length of an arc $s$ on a circle of radius $r$ to the radius ($s/r$) and a solid angle $\Omega$ measures the ratio of the area $A$ on the surface of a sphere of radius $r$ to the square of the radius ($A/r^2$). The radiance at some area can then be calculated using Equation 11:

$$\text{brightness} = B = \frac{P/\Omega}{A} = \frac{P}{A\Omega}$$

$$\frac{\Phi}{2} = \frac{s}{r} \Rightarrow r = \frac{2s}{\Phi}$$

$$\Omega = \frac{A}{r^2} = \frac{\pi s^2}{\left(\dfrac{2s}{\Phi}\right)^2} = \frac{\pi\Phi^2}{4} \qquad (11)$$

$$B = \frac{P}{(\pi s^2)\left(\dfrac{\pi\Phi^2}{4}\right)} = \frac{4P}{(\pi s^2)\pi\Phi^2}$$

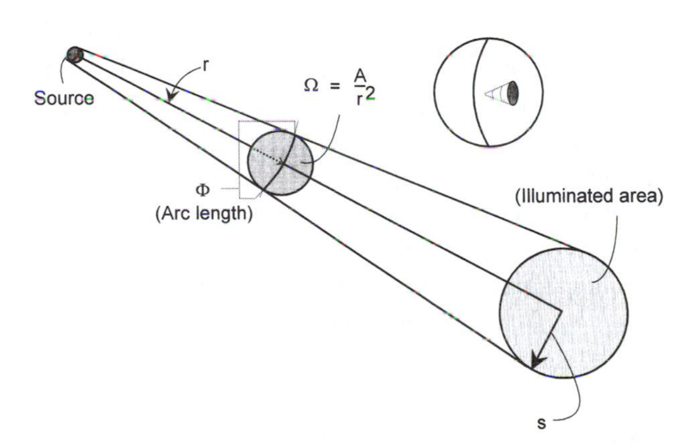


**Figure 1-9**
Brightness of a source on a given area. Brightness, or radiance, is defined as the radiant intensity (power $P$ of the light source per unit solid angle $\Omega$) per unit area of the source. The illuminated region is estimated to be circular, and the area of illumination ($\pi s^2$) is used as an approximation of the source area. Photons emanating from the source radiate energy into a certain sector defined by the solid angle $\Omega$. The solid angle $\Omega$ is the angle of a cone which subtends an area $A$ within the surface of a sphere of radius $r$ divided by the square of the sphere's radius: $\Omega = A/r^2$. The area $A$ or the sphere's surface, defined by the solid angle and the planar area of illumination, are approximated as equal. The plane angle $\Phi/2$, expressed in radians, is the ratio of the length of an arc on a circle of radius $r$ to the radius. The length of the arc is approximated as the radius of the illustrated area; thus, $\Phi/2 = s/r$. Thus, by substitution, brightness $B = (P/\Omega)/A = 4P/\{(A)(\pi\Phi^2)\} = 4P/\{(\pi s^2)(\pi\Phi^2)\}$.

Figure 1-9 illustrates the pertinent parameters.

So, how bright is a He/Ne laser compared to the sun? The intensity of the sun is about 0.1 W per cm$^2$ at the earth's surface. Given the distance of the earth from the sun and the diameter of the earth, $\Phi/2$ is about $4 \times 10^{-3}$ radians. The sun's radiance is calculated as:

$$B_{\text{sun}} = \frac{P}{A\Omega} = \frac{4P}{A\pi\Phi^2}$$

$$B_{\text{sun}} = \frac{4(0.1\ \text{W})}{1\ \text{cm}^2\pi(8 \times 10^{-3})^2\text{steradian}} \qquad (12)$$

$$= \frac{2 \times 10^2\ \text{W}}{\text{cm}^2\ \text{sterad}}$$

How bright is a 1 mW He/Ne laser whose divergence ($\Phi$) is 0.8 mradian? This laser will illuminate a 1 $cm^2$ area 7 meters away (using Equation 11). The brightness is:

$$B_{\mathrm{He/Ne}} = \frac{4(0.001\ \mathrm{W})}{1\ \mathrm{cm}^2 \pi (0.8 \times 10^{-3})^2 \mathrm{steradian}} \tag{13}$$

$$= \frac{2 \times 10^3\ \mathrm{W}}{\mathrm{cm}^2\ \mathrm{sterad}}$$

The humble He/Ne laser is brighter than the sun. We should note this brightness is the result of the low divergence of the laser, that is, the fact that all of its light energy is essentially going in the same direction.

While the brightness comparison above is dramatic, the comparison is really impressive when you ask the question: What is the brightness per unit of frequency? This measure is called the *spectral brightness*. To make the comparison on the basis per unit of frequency we need to appreciate *spectral distribution* or *spectral width*.

Spectral or frequency widths are often denoted by $g(\nu)$, and Figure 1-10 illustrates three frequency distributions. Basically the spectral distribution describes how many colors are output by the source of interest. Clearly, if the power of a source comes from many colors, the power per color, that is, the power per unit frequency, must be smaller than the same quantity calculated for a source of the same power but from a single color. The spectral distribution of the sun can be thought of, for these purposes, as extending from 400 to 700 nm while that of a 633 nm He/Ne laser is about 0.01 nm. Thus per unit of frequency, the spectral brightness of the He/Ne laser compared to the sun is greater by an additional factor of 50,000. When brightness is normalized in this manner, the spectral width of the laser will always enhance its output compared to other sources.

## Coherence

Perhaps the most unique aspect of lasers is coherence. What is coherent light? Light whose waves of the same wavelength go in the same direction and have the same phase or polarization of the electric vectors in space are coherent. This is illustrated in Figure 1-3. Even a laser does not have perfect coherence because it is not perfectly monochromatic and because it diverges. How do we find out whether a light source is coherent or not? The absence of interference effects implies coherence. The most famous experimental realization of interference is Young's double slit experiment which is illustrated in Figure 1-6 along with Huygens' construction. As discussed earlier, when a plane wave is incident on a screen with two slits, spherical wave fronts emerge from both of the slits and may strike a screen some distance away from the slits. Sometimes the spherical waves arrive at a screen in phase and other times they arrive out of phase. This is a two-dimensional phenomenon and is illustrated in Figure 1-6(b). Bragg or X-ray diffraction is the same concept, but in

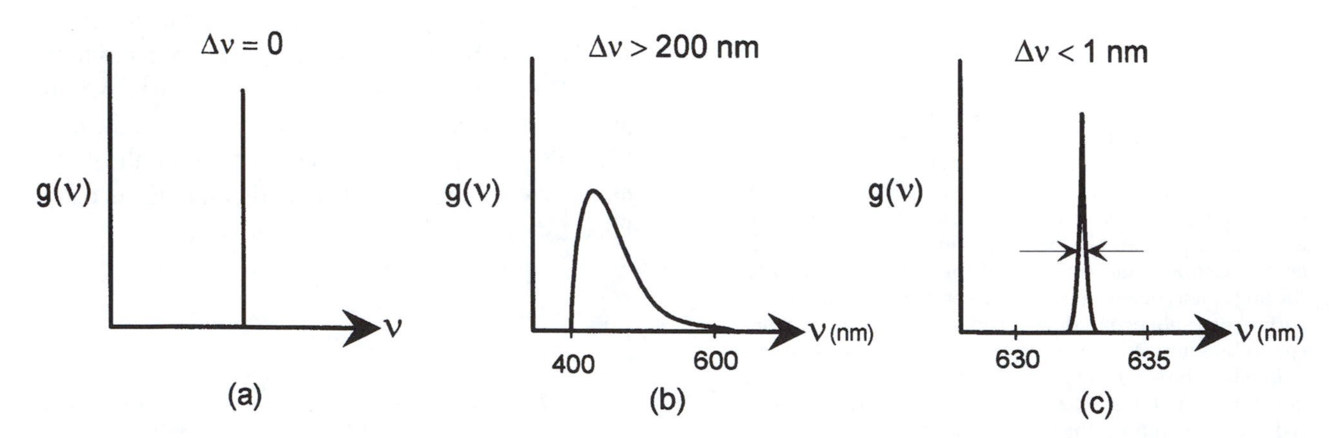


**Figure 1-10**
Three examples of spectral (frequency) distributions, $g(\nu)$. In *(a)* all of the intensity of the source is output at exactly one frequency. The frequency distribution in *(b)* illustrates the spectral width of a typical blackbody source. The frequency distribution in *(c)* is characteristic of a typical laser distribution.

three dimensions. Passage through slits establishes a relative phase difference and a path length difference which lead to constructive and destructive interference, that is, bright and dark spots. Such interferences are very readily seen with a laser source.

## Speckle

An almost unique consequence of the extreme coherence possessed by laser light is the so-called laser speckle. Each point on a rough surface illuminated by laser light scatters light as a spherical wave. The roughness of a surface causes a distribution of phases which are constant in time since the surface is still. Your eye's retina will detect points of light caused by destructive and constructive interference from neighboring points. These points are speckle interference patterns. Speckle is characteristic of a very coherent source. Try to see speckle with a flashlight!

## Coherence Length and Time

There are various ways of measuring coherence. Probably the best way to measure coherence is to use an interferometer. Figure 1-11 shows a very simple variety. By passing a light source through a pinhole and then through a beam splitter, some of the beam travels straight to a screen and some travels at right angles to a mirror, which reflects the light to yet another mirror. The light reflected off the second mirror is also beam split causing part of the split beam to travel in the same direction as the original beam. Both the unsplit and split beams reach the screen. If you move the two mirrors back and forth a distance $X$, you will observe patterns on the screen caused by interference. One path takes $Z/c$ to reach the screen, where $Z$ is the distance from the pinhole to the screen. Now $Z/c$ is the transit time for light to travel the distance $Z$. The transit time of the other path to the screen is $(Z + 2X)/c$. As $X$ is changed, the interference pattern on the screen changes. The $X$ for which the "best" or maximum pattern is observed is half the coherence length and is a quantitative measure of spatial coherence. The coherence time which is a quantitative measure of temporal coherence is given by:

$$\text{coherence time} \equiv \frac{\text{coherence length}}{c}$$

$$t_c = \frac{l_c}{c} \tag{14}$$

Recall that monochromaticity measures how close to a single color is the output of a given light source. A perfectly monochromatic source yields exactly one color. Using the ideas introduced in Figure 1-10, a pure-color, exactly monochromatic source would have a frequency distribution $g(\nu)$, as illustrated by Figure 1-10(a). Real sources have a frequency distribution $g(\nu)$ which covers some finite region of frequency $\nu$. The degree of monochromaticity depends on the range of the frequencies $\nu$ which characterize the source output energy. A common way to quantify monochromaticity is by bandwidth, which has many synonyms (e.g., bandwidth at half height, FWHM =

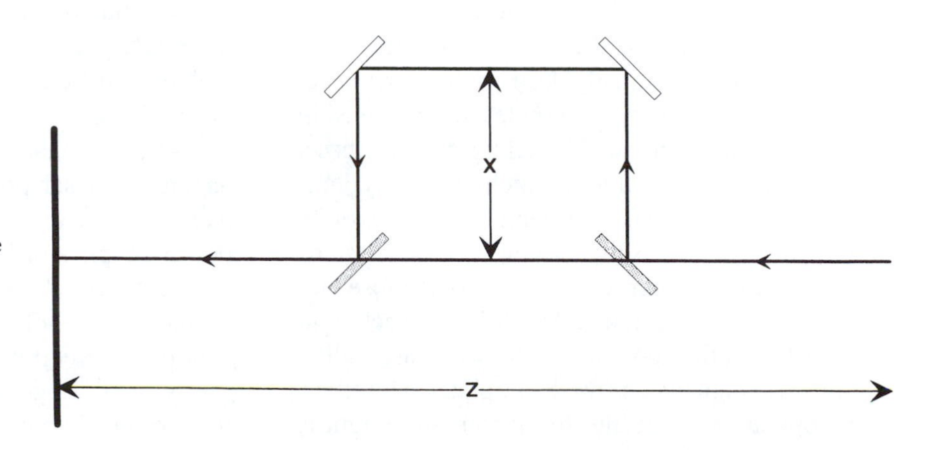


**Figure 1-11**
A simple Michelson interferometer used to measure the coherence length, $l_c$, of a source. Light from the source is passed through a pinhole separated a distance $Z$ from a screen. With beam splitters and mirrors, the light path can be extended to a distance $Z + 2X$. The time for the light to travel directly to the screen, the transit time, is $Z/c$. The transit time for the indirect path is $(Z + 2X)/c$. The interference pattern on the screen of the two beams varies with $X$, with the maximum pattern observed for $X = 0.5\ l_c$. The coherence time can also be calculated as $l_c/c$.

**Table 1.2**
Examples of Coherence Length and Time

| *Light Source* | *Coherence Length/m* | *Coherence Time/s* |
|---|---|---|
| Interference filter 500 nm $\Delta\nu =$ 10 nm | $25\times10^{-6}$ | $10^{-13}$ (0.1 picosecond) |
| Hg lamp 546 nm | 0.08 | $10^{-10}$ (100 picoseconds) |
| He/Ne laser 633 nm | 50 | $1.7\times10^{-7}$ (0.17 microsecond) |

full width at half maximum, spectral bandwidth, and linewidth). Basically, the bandwidth measures the "thickness" of the $g(\nu)$ function at some value of $g(\nu)$. A very common bandwidth is at 1/2 $g(\nu_o)$, often denoted $\Delta\nu$, the full width at half maximum when centered at $\nu_o$. This is illustrated in Figure 1-10(c).

A laser's degree of monochromaticity is also related to its temporal coherence and spatial coherence:

$$\Delta\nu = \text{bandwidth of spectral source} = \text{degree of monochromaticity}$$

$$\text{coherence length } l_c = \frac{c}{\Delta\nu} = \text{measure of spatial coherence}$$

$$\text{coherence time} = t_c = \frac{1}{\Delta\nu} = \text{measure of temporal coherence} \quad (15)$$

The physical meaning of these concepts is that if all the waves were the same color, they would stay both spatially and temporally coherent forever. In fact, $\Delta\nu$ never equals zero and neither $l_c$ or $t_c$ are infinite because interference eventually destroys the coherence.

Consider some simple examples, as presented in Table 1-2. A conventional filtered light source has a 10 nm bandwidth at 500 nm, which means a coherence time $t_c$ of $10^{-13}$ s and a coherence length $l_c$ of 25 microns. Physically this means that two waves, starting out as shown in Figure 1-3(a), become those pictured in Figure 1-3(c) in just 0.000025 m. Such light is monochromatic enough to see what are called Newton's rings between two glass plates. However, for the operation of an interferometer, such light is woefully incoherent. Other comparisons are presented in Table 1-2. The Hg lamp has a coherence length of 8 cm. However, compare the Hg lamp coherence of 8 cm to a single mode He/Ne laser with a coherence length of 50 m. Such laser light travels 50 m before the light illustrated in Figure 1-3(a) becomes like that pictured in Figure 1-3(c).

As an application of spatial coherence, the LIGO (laser interferometric gravitational-wave observatory) project seeks to demonstrate the existence of gravitational waves using lasers with very long coherence lengths which are built into a Michelson interferometer. A gravity wave would be detected by a change in the interference fringe pattern established by the lasers.

## Laser Safety

If a laser is so bright, it must be dangerous. Let's consider some aspects of laser safety. Common sense always is the rule when using lasers. All high-power laser beams are dangerous. Keep in mind the most dangerous wavelengths may be those which you cannot see, UV and IR. You cannot blink to avoid light you cannot see but you can still be injured by such light! Even for visible light you generally cannot blink fast enough to avoid injury should a beam somehow be directed toward your eye. Moreover, IR and UV lasers are very dangerous because they can cause damage to biological tissue beneath the skin. Of course you always need to worry about reflections from any laser and not just those you think might cause damage. Also remember that while working in a laser laboratory co-workers might generate a reflection in your direction and not even know it, but you sure might! You *must* wear proper laser safety glasses all of the time when working with or near IR and UV lasers, and indeed, even with visible lasers.

As unfortunate as laser-induced optical or tissue damage is, laser power supplies can be even more dangerous. More people have been killed by the power supplies of lasers than the lasers themselves. Be careful out there! Many good, practical, and important aspects of laser safety are discussed in Appendix IV. Keep in mind that many points of laser safety are just good common sense and have the force of law behind them.

## SUMMARY

This introduction has illustrated some features of light found in the laser tools we have available to us. The high degree of coherence exhibited by laser light is intimately related to the laser's characteristics of monochromaticity, directionality, and brightness (polarization control is discussed in Chapter 3). The bandwidth of a laser is inversely proportional to the coherence length. Were it not for high spatial coherence, the laser beam could not consist of highly directional planar wave segments. All of these features work together. No single parameter describes a laser completely, and they are all consequences of one another. A white light source, directed through enough filters to make its bandwidth very narrow, would yield coherent light. A laser, however, has the advantage of much more power. The high spectral brightness results from both high directionality (spatial coherence) and small bandwidth (temporal coherence). The single greatest aspect that distinguishes laser light from all other sources of light is coherence. Coherence is caused by the great regularity, in time and in space, of the light waves. Somehow within the laser, light waves have been set up in time and in space to work together in this way. In Chapter 2 we explore how laser light is generated.

## GENERAL REFERENCES

Halliday, D., and Resnick, R., *Physics, Part II*, John Wiley & Sons, New York, 1968.

Jenkins, F. A., and White, H. E., *Fundamentals of Optics, Third Edition*, McGraw-Hill, New York, 1957.

O'Shea, D. C., Callen, W. R., and Rhodes, W. T., *Introduction to Lasers and Their Applications*, Addison-Wesley Publishing Company, Inc., Reading, MA, 1977.

CHAPTER 2

# What Makes the Laser Shine?

## Chapter Overview

The unique characteristics of laser light arise from the coupling of several distinctive phenomena—optical and electrical pumping, generation of population inversions, stimulated emission, and light amplification. As a consequence of the complex mechanism of laser action, the diverse nature of lasing media, and the variable design of laser cavities, a vast spectrum of lasers and a variety of modes of operation are possible. In this chapter we discuss the general principles of lasing action and the features of laser construction that combine to yield the range of laser types and the variable modes of laser operation.

## Light Absorbed, Light Emitted

### Absorption and Emission

A *laser* is a light source based on the phenomenon of Light Amplification of Stimulated Emission of Radiation. To understand the acronym *laser*, we need to understand stimulated emission and then how it can be amplified. You are no doubt familiar with the processes of absorption and emission of light by matter. To be more precise we should be calling absorption *stimulated* absorption and emission either *stimulated* or *spontaneous* emission. Thus, we need to understand three phenomena: stimulated absorption, stimulated emission, and spontaneous emission. The first and third of these processes are in fact what happens when molecules normally absorb or emit light, respectively. The adjective *stimulated* is added to the terms to note specifically that these processes must occur in the presence of photons. The photons are the necessary stimulus for either the absorption or emission process to occur. Spontaneous emission, on the other hand, results when, in the absence of any external influence, a molecule in an excited energy state returns to its ground state or a lower state with the emission of a photon. These three processes are illustrated in Figure 2-1.

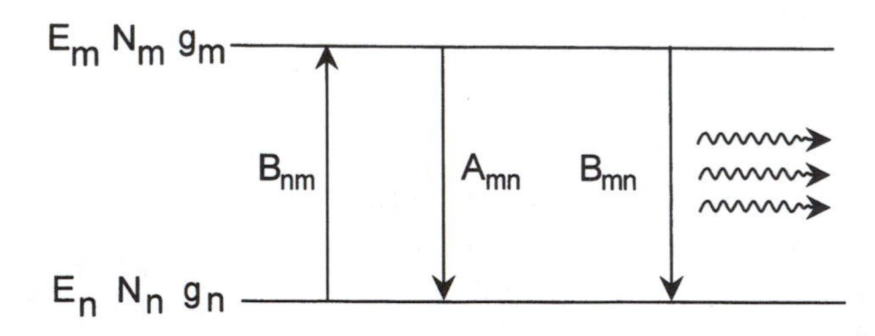


**Figure 2-1**
Energy level diagram differentiating between the processes of stimulated absorption, spontaneous emission, and stimulated emission with their rate constants $B_{nm}$, $A_{mn}$, and $B_{mn}$, respectively. Each of the energy levels $E_n < E_m$ is characterized by a *population*, $N_n$ and $N_m$, that indicates the number of molecules with that characteristic energy and a *degeneracy*, $g_n$ and $g_m$, that indicates the number of energy states with the same energy. Stimulated absorption from $E_n$ to $E_m$ occurs when a molecule absorbs a photon of light. Spontaneous emission returns a molecule in $E_m$ to the lower $E_n$ level via the emission of a photon without the presence of any external influence. Stimulated emission describes the photon emission process that is induced by the presence of additional photons.

Details of stimulated processes are our next topic for discussion. How the emitted light is amplified is then discussed later in the chapter.

### Stimulated Emission

Two aspects of stimulated emission are of critical importance. (1) The stimulation of a population of particles in an excited state with a photon of the appropriate energy, $h\nu = E_m - E_n$, yields emitted photons, the initial incident photon plus others. The fact that one photon can stimulate the emission of others is the origin of the amplification of the emitted light, and so the origin of laser light. (2) The direction of travel of the photons stimulated to emit is the same as the direction of the incident "stimulating" photon(s). This is also illustrated in Figure 2-1. The reason the emitted photon travels in the same direction lies in the details of the process, which we will not discuss here. The importance of this directionality of the stimulated emission is perhaps overlooked too often. In contrast, spontaneously emitted photons are emitted in all directions. The inherent unidirectionality of laser light contributes to the brightness of lasers.

### Rates of Absorption and Emission

It seems reasonable that the rate of absorption of incident photons is proportional to the number of particles (molecules or atoms) that have the initial energy $E_n$, that is, the population of the energy level $n$ which we will denote as $N_n$. To quantify this absorption rate, Einstein derived several expressions that now bear his name. These expressions are analogous to chemical reaction rate expressions. The pertinent rate constants, also called the Einstein coefficients, are shown diagrammatically in Figure 2-1 and are defined as:

$B_{nm}$—A second-order rate constant describing the stimulated absorption of a photon by the energy state $E_n$, expressed in units of $m^3\ J^{-1}\ s^{-2}$ or $m^3\ watt^{-1}\ s^{-3}$ or (light speed)$^3$ power$^{-1}$.

$B_{mn}$—A second-order rate constant describing the stimulated emission of a photon from the energy state $E_m$, expressed in units of $m^3\ J^{-1}\ s^{-2}$ or $m^3\ watt^{-1}\ s^{-3}$ or (light speed)$^3$ power$^{-1}$.

$A_{mn}$—A first-order rate constant describing the spontaneous emission of a photon from the energy state $E_m$, in units of $s^{-1}$.

The $B$ coefficients may be viewed as second-order rate constants because the process described depends in each case on the energy distribution of photons $\rho$ (expressed in units of $J\ s\ m^{-3}$ or the number of photons of energy $h\nu$ per unit volume per unit frequency) and the number of molecules in the initial energy state, $N_n$ or $N_m$. The $A$ coefficient is a first-order rate constant because it only depends on $N_m$ and not on the energy distribution of photons $\rho$. Clearly, at a steady state the rate of creating excited molecules must equal the rate of creating "de-excited" molecules so that, again with reference to Figure 2-1, the rate of excitation equals the total rate of de-excitation:

$$N_n B_{nm} \rho = N_m B_{mn} \rho + N_m A_{mn} \tag{16}$$

Einstein showed that under steady state conditions, the $B$ coefficients are related to each other by the degeneracy of each energy level, that is, by the number of times the states of the molecule have repeat magnitudes of a given energy level. The degeneracy of the energy level $E_m$ is represented by the variable $g_m$. If three independent energy states all have the energy $E_m$, then $g_m = 3$. Einstein also showed that the spontaneous emission and stimulated emission coefficients are related to each other through fundamental constants, and the frequency $\nu$ of the emitted photons $= (E_m - E_n)/h$. These ideas are summarized as:

$$g_m B_{mn} = g_n B_{nm}$$

$$A_{mn} = B_{mn} \frac{8\pi h \nu_{nm}^3}{c^3} \tag{17}$$

(The origin of the relationship between $A_{mn}$ and $B_{mn}$ requires many intermediate steps which we will not discuss here.)

Which is favored, spontaneous emission or stimulated emission? In other words, which emission process has the faster rate and therefore will lead to the depopulation of the excited state? The answer to this question largely depends on the magnitude of the frequency term in Equation 17:

$$\frac{8\pi h \nu^3_{nm}}{c^3} = 6.1 \times 10^{-58}\ \mathrm{J\ s^4\ m^{-3}\ photon^{-1}} \nu_{nm}^3$$

For a 100 Kev X ray:

$$\nu_{nm}^3 \approx 1.4 \times 10^{58}\ \mathrm{s}^{-3},\ A_{mn} = 7.3 B_{mn}$$

For a red 600 nm photon:

$$\nu_{nm}^3 \approx 1.3 \times 10^{44}\ \mathrm{s}^{-3},\ A_{mn} = 8 \times 10^{-14} B_{mn}$$

For visible light, the rate of spontaneous emission is orders of magnitude smaller than stimulated emission, and, hence, visible lasers function efficiently. For X-ray frequencies, however, the rate of spontaneous emission is greater or about the same order of magnitude as stimulated emission. Beating spontaneous emission is a major obstacle to the development of X-ray lasers.

## Lifetimes

What are the physical meanings of $A_{mn}$, $B_{mn}$, and $B_{nm}$? The reciprocals $1/A$ and $1/\rho B$ have units of time and, in fact, refer to lifetimes for the process under consideration.

$$\tau_{A_{mn}} = \frac{1}{A_{mn}} \Rightarrow$$

a time for spontaneous emission to occur

$$\tau_{B_{mn}} = \frac{1}{\rho B_{mn}} \Rightarrow$$

a time for stimulated emission to occur

Lifetimes of states become very important when we discuss the spectral distribution of the laser $g(\nu_{laser})$. Recall that the spectral distribution is related to the width of the laser line in frequency units. (Review the examples illustrated in Figure 1-10 to recall this relationship.)

The treatment above assumes that the density of radiant energy $\rho$ is approximately constant across the frequency range of interest. If $\rho$ is not uniform, then we need to introduce a spectral distribution $g(\nu)$ (also called a line shape function) that describes how often each frequency occurs over the frequency range of interest. We return to this important point later in this chapter.

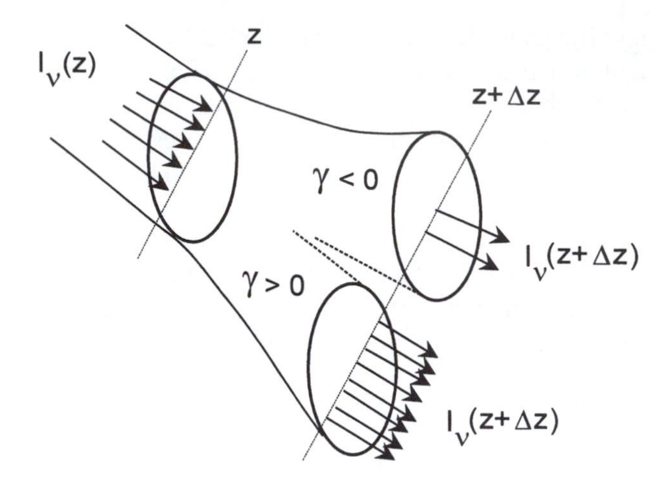


**Figure 2-2**
The fate of light passing through a medium. The intensity will change to either decrease, in the case of absorption by the medium, or increase, in the case of stimulated emission leading to lasing. The gain coefficient $\gamma$ is negative in the case of absorption and positive in the situation of stimulated emission.

## Excited States Are Useful

### Beer's Law in a New Light

Let's imagine that radiation passes through a material that can lase (the verb coined for the process of emitting light by light amplified stimulated emission). Let $I(\nu)$ be the intensity of light at the frequency $\nu$ that is travelling in the $+z$ direction. Recall that the units of intensity are power per unit area. As the light propagates through some medium, as diagrammed in Figure 2-2, the intensity as a function of distance is given by:

$$\left(\frac{I_{z+\Delta z}(\nu) - I_z(\nu)}{\Delta z}\right)_{\Delta z \to 0} \equiv \frac{dI_z(\nu)}{dz} = \gamma I_z(\nu) \quad (18)$$

where the derivative is proportional to the intensity via the proportionality coefficient $\gamma$. The parameter $\gamma$ is considered a gain or a loss coefficient depending on whether the light intensity decreases or increases on passing through the medium. A quiz: Will $\gamma$ be > 0 or < 0 for a lasing material?

For a generalized case of absorption according to Beer's Law, the solution to this differential equation is:

$$I_z(\nu) = I_0(\nu)\exp(\gamma z) \quad (19)$$

where $I_0(\nu)$ is the incident intensity at $z = 0$. For normal absorption, $\gamma$ is related to the absorptivity of the medium. Normally we use Beer's Law to describe the attenuation of light as it passes through a medium, and we expect $\gamma < 0$ if $I(\nu)$ decreases. However, if the population of the excited level is larger than the lower level, $N_m > N_n$ and stimulated emission is highly likely. This is consistent with $\gamma > 0$ and vice versa. Moreover, if $\gamma > 0$, there would be an exponential gain of intensity on passing through the medium. For these laser discussions we will call $\gamma$ the small gain coefficient. The small gain coefficient $\gamma$ is dependent on the populations of the two energy levels "connected" by $h\nu_{nm}$ energy. In the case of gain, however, the net increase in intensity of monochromatic radiation of frequency $dI(\nu)$ which results from travelling a distance $dz$ with a speed $c'$ is given (without derivation; for this see O'Shea et al. or Svanberg) by:

$$dI(\nu) = \frac{A_{mn}c'^2}{8\pi^2\nu^2}\left[N_m - \frac{g_m}{g_n}N_n\right]g(\nu)I(\nu)dz \quad (20)$$

Here $N_m$ and $N_n$ are the populations per unit volume weighted by energy state degeneracies and the speed $c'$ depends on the refractive index of the medium through $c = nc^1$. Included in this expression is the spectral distribution which accounts for the fact that no frequency is truly monochromatic. The nature and origin of $g(\nu)$ is of great interest particularly for a laser. We mention here, as a preview to more expansive discussion at the end of the chapter, the three common origins for $g(\nu_{laser})$: Doppler frequency shifts, collisional broadening, and natural linewidths reflecting the Heisenberg uncertainty principle.

## Population Inversions

Suppose we rearrange the left-hand side of Equation 20 and recognize that the rearrangement restates $\gamma(\nu)$ from Equation 18 as the middle term of Equation 20:

$$\gamma(\nu) = \frac{1}{I(\nu)}\frac{dI(\nu)}{dz} = \frac{A_{mn}c'^2}{8\pi^2\nu^2}\left[N_m - \frac{g_m}{g_n}N_n\right]g(\nu)$$

Now notice that $dI(\nu)/I(\nu)$ is the fraction of all of the intensity $I(\nu)$ observed in the region that is between $I(\nu)$ at $z$ and $I(\nu)$ at $z + dz$, and this fraction is in effect normalized to the region between $z$ and $z + dz$ by division by $dz$. A fraction of an observable in a region is called a distribution function, a concept extremely important in science. For example, just think about a Maxwell speed distribution or a Boltzmann distribution. What consequence has this interpretation for $\gamma(\nu)$? The units of $\gamma(\nu)$ are the fraction of light in the distance interval $dz$, or inverse length. Moreover, for $\gamma(\nu)$ to be positive, the population term [ ] must be positive:

$$\left[N_m - \frac{g_m}{g_n}N_n\right] > 0 \qquad (22)$$

This term is just the population of the upper level minus the population of the lower level multiplied by the degeneracy factors. Thus, for $\gamma(\nu) > 0$ there must be a population inversion, i.e., more effective population in the upper energy state than in the lower energy state (effective here referring to the influence of the degeneracy factors). Such inversions are not normal. Typically, substances absorb radiation because the populations of the ground and excited states are such that [ ] < 0. When we can put more population in an excited state than in the ground state, we have one of the necessary conditions for laser action by stimulated emission. Several other conditions must also be met, however, and we will discuss these requirements after further considering population inversions in more detail.

## Beating Boltzmann

To understand the importance of population inversions for laser operation, let's review some consequences of thermal equilibrium and the Boltzmann distribution. We describe the Boltzmann distribution as the condition where, in the absence of external influences, the lower energy levels are always more populated compared to the upper energy levels. Moreover, the populations in the upper levels always decrease at any fixed temperature according to a Boltzmann equation:

$$\frac{N_m}{N_n} = \frac{g_m}{g_n}e^{-\frac{(E_m - E_n)}{k_B T}} \qquad (23)$$

To promote light amplification by stimulated emission of radiation, we must achieve a sustainable nonequilibrium, nonthermal population distribution. We must first recognize that temperature changes alone cannot provide a population inversion. Even at infinite temperature, the exponential term goes to 1, and we do not get a population inversion but only a population ratio dependent on degeneracies. We have to do more than just change temperature to achieve a population inversion. Figuring out how to overcome the natural population distribution was accomplished by Townes, Prochorov, and Basov and recognized by the 1964 Nobel prize in physics.

The reader might find the brief history of laser development found in Appendix I to be a pleasant diversion at this point.

## Population Inversions, How?

We have just mathematically demonstrated that a simple thermal process cannot achieve the population of excited states needed to produce stimulated emission. Alternatively, what if we use photon energy to create excited states? Clearly we know that such a prospect is possible. However, suppose we only have a two-level system consisting of one upper and one lower energy state. Now shine light on the system to increase the number of molecules in the upper state relative to the lower state. Every absorbed photon of light decreases the population of the lower energy level by one while increasing the population of the upper energy level to the same extent. At best, light absorption achieves an equalization of the effective populations of the two states. This equalization is called "bleaching" the system, for as fast as the system's energy is increased by absorption, the system is stimulated to emit the same energy. At the condition of equalized populations, additional incident light is transmitted, not absorbed. Two-level systems, while simple, do not work.

Consider a system with three energy levels accessible by photon absorption or emission. The three-level system of the ruby laser is illustrated in Figure 2-3. A ruby laser is named for the fact that its laser cavity is a synthetic ruby crystal. Since a ruby is alumina $Al_2O_3$ with small amounts of chromium substituted for aluminum, the energy states of a ruby laser really depend on the energy states of atomic chromium. How can we achieve a population inversion in the chromium energy levels—that is, in the ruby laser? A possible mechanism first requires that the ground state absorb energy to reach the uppermost level. We note here that the process of supplying energy for absorption to reach upper energy states is most often called *pumping*. If some fast energy loss then occurs to reach the intermediate level, a stimulated emission—a laser transition—is possible from the intermediate level to ground state. The fast energy loss referred to here might occur by spontaneous emission or by nonradiative processes involving collisional energy losses (generally involving a transfer of translational energy to vibrational energy, etc.). Such losses are usually referred to as energy decay or just decay processes.

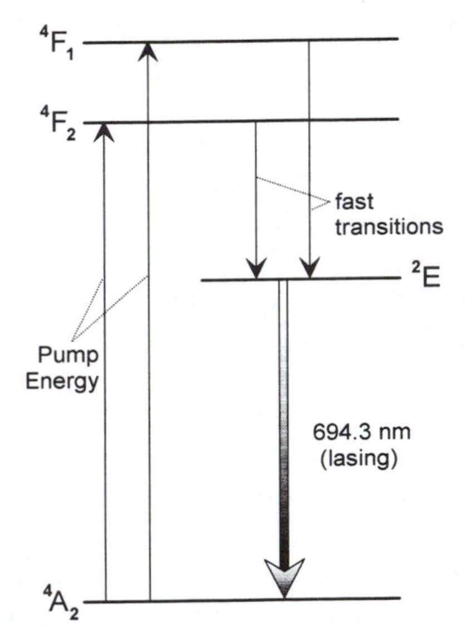


**Figure 2-3**
The energy levels of the three-level ruby laser system. The population achieved by optical pumping of the ground electronic state $^4A_2$ to the upper electronic states $^4F_1$ and $^4F_2$ is quickly depleted by fast transitions to the metastable state $^2E$. Lasing at 694.3 nm subsequently results when a population inversion is created between the intermediate state $^2E$ and the ground state $^4A_2$. For such a population inversion to be created, the lifetime of the intermediate state must be longer than the lifetime of the upper-level states. In addition, more than half of the ground-state population must be promoted to higher energy states.

Three-level systems require more than half of the ground-state population to be removed in order to function, however. If less than half of the ground-state population were removed, no population inversion would be created. (Can you see why this condition is a requirement? Consider the relative rates of the spontaneous absorption, spontaneous emission, and stimulated emission processes.) As a practical matter, then, a three-level system demands high pump power. The best known three-level system is the pulsed ruby laser first built by Ted Maiman in about 1960. After a pulse of pump light, a population inversion is created in the intermediate state on a time scale of 100 ns. Essentially all of the population in the intermediate state is simultaneously depleted by stimulated emission as a pulse of laser light. Another time period is then required to reestablish the population inversion. Ruby lasers and other three-level systems have been found to work only on a pulsed basis since a population inversion cannot be sustained on a continuous basis.

Consider an interesting calculation applicable to a pulsed ruby laser. How many photons are in a unit volume of a ruby laser while it is lasing with a peak intensity of $10^{16}$ W/m$^2$? The refractive index of the laser material is 1.6, its molar volume is approximately $6 \times 10^{-5}$ m$^3$/mol and its output frequency is $4.62 \times 10^{14}$ Hz (694 nm).

$$\text{Intensity} = I = \frac{\left(\dfrac{\text{Energy}}{\text{time}}\right)}{\text{area}} = \frac{\text{power}}{\text{area}}$$

$$\text{time in medium} = \frac{L}{c'} = L\left(\frac{n}{c}\right) \text{ with } c' = \frac{c}{n}$$

$$I = \frac{\dfrac{Nh\nu}{\left(L\left(\dfrac{n}{c}\right)\right)}}{L^2} = \frac{Nh\nu c}{nV} \qquad V = L^3$$

$$\frac{N}{V} = \frac{\text{\# photons}}{\text{vol of medium}} = \frac{I}{h\nu}\left(\frac{n}{c}\right)$$

$$N = \frac{10^{16}\ \dfrac{\text{J}}{\text{s m}^2}(1.6)\left(6 \times 10^{-5}\ \dfrac{\text{m}^3}{\text{mol}}\right)}{\left(6.6 \times 10^{-34}\ \dfrac{\text{J s}}{\text{part}}\right)\left(\dfrac{4.62 \times 10^{19}}{\text{s}}\right)\left(\dfrac{3 \times 10^{8}\text{m}}{\text{s}}\right)\left(\dfrac{6 \times 10^{23}}{\text{mol}}\right)}$$

$N \approx 0.02$ photon in the molar atomic vol.

or 100 atomic molar vol. per 2 photons

It may be surprising to realize that a power of $10^{16}$ W/m$^2$ only amounts to 2 photons per 100 molar volumes. We can appreciate this amazing number by recognizing two facts of nature. Taking wavelength as a measure of size, a 694 nm photon is large compared to atomic/molecular sizes measured in nm, and Avogadro's number is huge!

Suppose four energy levels could be arranged as shown in Figure 2-4. The four-level scheme is how the He/Ne laser and many other types of lasers work. Here, the pump energy populates the uppermost level which rapidly decays to the next lower level. This intermediate level undergoes stimulated emission and is the source of the laser light. The ending state for the laser light essentially stays unpopulated

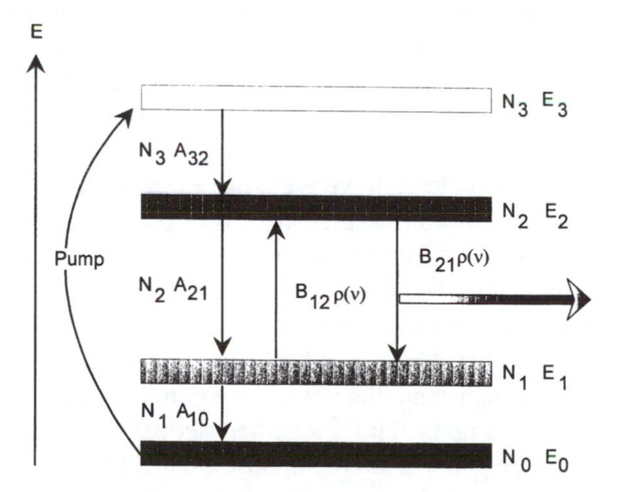


**Figure 2-4**
An energy diagram for a four-level laser system showing features important for efficient laser action. The energy states $E_0$, $E_1$, $E_2$, and $E_3$ are characterized by populations $N_0$, $N_1$, $N_2$, and $N_3$ whose shading reflects magnitudes, with the darkest being the most populated. Optical pumping promotes molecules in the ground state to the uppermost energy state. Rapid decay to the intermediate state $E_2$ leads to the condition for stimulated emission where $N_2 > N_1$ ($E_2 \rightarrow E_1$). Further deactivation of the excited state $E_1$ repopulates the ground state.

as it rapidly decays to the ground state by some means. In this case, we can continually excite the ground state with pump radiation and sustain a population inversion. This is referred to as continuous wave (CW) operation and was first achieved for the He/Ne laser by Javan, Bennett, and Herriott in 1961.

## Pumping

In the four-level scheme just described and illustrated in Figure 2-4, the phenomenon of lasing, the establishment of the population $N_2$ in energy $E_2$ suitable for stimulated emission, depends critically on the rates of spontaneous emission from $E_2$ and $E_1$, denoted $A_{21}$ and $A_{10}$, and the pump rate $P$ producing $N_3$ of the $E_3$ state. For a stable population inversion to be maintained, the pump rate must overcome the lasing and the spontaneous emissions according to:

$$\left[N_2 - \frac{g_2}{g_1}N_1\right] \geq \frac{P\left(1 - \dfrac{A_{21}}{A_{10}}\right)}{A_{21} + B_{21}\dfrac{I(\nu)}{4\pi c'}} \qquad (25)$$

where $I(\nu)$ is the intensity of the pump radiation and $c'$ is the speed of light in the laser medium ($c' = c/n$). If $A_{21} > A_{10}$, lasing is not possible, for the numerator of the right-hand side of Equation 25 is negative, implying that $[N_2 - (g_2/g_1)\,N_1] < 0$ and that no population inversion is possible. Recall the discussion of the inverse Einstein coefficients as a measure of a lifetime. The statement $A_{21} > A_{10}$ means $1/A_{21} < 1/A_{10}$ or $\tau_{21} < \tau_{10}$ which implies that the intermediate state $E_1$ builds up populations, instead of $E_2$, the desired lasing state. For practical operation of this four-level system, neither $E_3$ nor $E_1$ should build up populations. The pump rate primarily yields a highly populated $E_2$ ready to lase under the stimulation of the absorbed radiation of intensity $I(\nu)$.

Besides the lifetime requirements on $A_{21}$ and $A_{10}$, the pump intensity $I(\nu)$ plays a critical role in achieving a lasing condition. If $I(\nu)$ is too small, $B_{21}I(\nu)/4\pi c' < A_{21}$ and the population term takes on values given by:

$$\left[N_2 - \frac{g_2}{g_1}N_1\right] \geq P\,\frac{\left(1 - \dfrac{A_{21}}{A_{10}}\right)}{A_{21}} \qquad (26)$$

If this population inversion is too small, then $\gamma < 1$ and lasing cannot occur. The population inversion defined by the condition of equality for Equation 26 is called the threshold inversion, and its value determines the practical operating conditions of pump rate and intensity for the operation of a four-level laser.

## Sources of Pump Energy

There are various ways pump energy can be put into a system to excite the ground state to a stable excited state. Two of the most common ways are via electron discharge or via intense incident light. Intense light as a pump energy source is virtually self-explanatory. Electrical discharge may need some explanation, however. Highly energetic electrons produced by an electrical discharge excite the target molecules (atoms) by inelastic collisions with the electrons of the target. The He/Ne laser, which actually depends on an energy transition involving neon atoms for the laser light, can be made to operate without the presence of the helium using a strong electrical discharge. This is not done practically because too many energy levels are populated for efficient lasing. Far more practical is to excite helium by electrical discharge and then allow the excited helium to excite the neon to lasing energy levels via collisional energy transfer. We discuss aspects of the He/Ne laser in more detail later in this chapter and in Chapter 4.

## Preview of Laser Types

Before discussing further factors affecting the light output of lasers, let's digress a moment and survey the various types of lasers and their general style of operation. Even at this point it should be appreciated that every laser must be made of a material (or materials) that can develop a population inversion, either in a pulsed or continuous mode. Lasers could be classified by the means used to create the population inversion or by characteristics of the laser material. The more common classification is based on laser material. On this basis Table 2-1 presents the known laser types with a few examples of each type. In Chapter 4 we discuss in some detail the construction and principles of operation of the lasers in most common use today. Moreover, in Appendix II we summarize the principles of operation for nearly all currently known lasers.

**Table 2-1**
Types of Lasers Based on Lasing Medium

| *Type* | *Laser Medium* |
|---|---|
| Gas lasers | Monatomic gases: Ar, Kr, He/Ne, Cd/He |
| | Molecular gases: $CO_2$, $N_2$, $I_2$, KrF, HF |
| | Metal vapor: Cu, Au, Ba |
| Dye lasers | Fluorescent dye molecules |
| Solid state lasers | Host crystals doped with laser active metal ions |
| | Semiconductor *p*- or *n*-doped insulators |

## Amplification

Recall our earlier caution that the creation of a population inversion was only one requirement necessary to obtain laser light. It's time to discuss the other requirements. Our discussion so far has focused on developing a population inversion within a medium and then somehow stimulating light emission. If the stimulated emitted light were not modified further, the laser would be referred to as a superradiant system. The nitrogen gas laser is the most common example of a superradiant laser. A superradiant system is in effect a one-shot process with little opportunity for amplification. For power we need amplification and for amplification we need gain.

Suppose the gain were characterized by a magnitude of 3, meaning that, for every photon of the right color and frequency, three additional photons are created in the lasing medium. Each photon is capable of stimulating emission. Of course, only a gain greater than 1 is required for lasing, but an amplification greater than 3 is not an uncommon situation. Stringing "amplifiers" together sequentially enhances the gain possible. Suppose we had three photons to start the lasing process. How many stages would be required to achieve a joule of energy? Recalling our earlier calculation of an Einstein, one joule is about $10^{19}$ photons depending upon the frequency. If each stage were to provide a gain of 3, then 40 stages would be required to obtain a joule, for if $3^x = 10^{19}$, then $x \approx 40$. The idea that Schowlow and Townes came up with in 1958 to achieve multiple stages and high gains was to use facing mirrors.

If the lasing light reflects back and forth between two mirrors, a gain is achieved each time the stimulated light traverses the medium, thus immensely amplifying the initial stimulating photons.

The configuration of the lasing medium placed between two mirrors defines a laser cavity. A schematic laser cavity is illustrated in Figure 2-5. Since the laser light "oscillates" back and forth between the mirrors, this cavity with mirrors is also known as an oscillator. We will see next that properties of such oscillators play a large role in determining the nature of the light output by a laser. In order for a laser to be a useful light source and not just an energy storage device, some of the laser light in the cavity has to leave or "leak" out as light output. These losses are the laser's power output and thus central to laser operation. A desired power output dictates the value of a threshold gain coefficient $\gamma_{thres}$, in turn specifying a minimum population inversion defined by Equation 26.

To operate a laser continuously, the power lost to mirror transmission and to other sources, such as scattering and absorption, must be replaced via an adequate gain in the cavity. Imagine an initial flux of photons making one complete round trip of the laser cavity. If the round-trip gain is less than the round-trip loss, then the laser is below threshold and no laser action results. For the condition where round-trip gain > round-trip loss, a transient condition exists which is common in pulsed laser operation. For a continuously operating laser cavity oscillator the following condition must be satisfied:

Round-trip gain = Round-trip loss

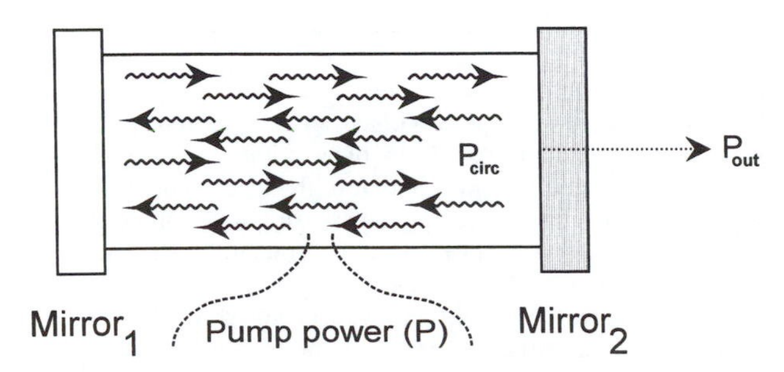


**Figure 2-5**
The elements of a laser cavity. The cavity is defined by the configuration of the lasing medium placed between two parallel mirrors.

As lasing occurs inside a laser cavity, the radiation density will deplete the population inversion to its threshold value until a surplus condition is reestablished. We can define a net gain as the ratio of cavity power at the end of one complete round trip of the cavity to the power at the start of the cavity traversal. The net gain $G$ can be quantified by Equation 27:

$$G = R_1 R_2 e^{(\gamma - \alpha)2L} \tag{27}$$

where $R_1$ and $R_2$ are the reflectivities of two mirrors, $L$ is the length of the cavity, $\gamma$ is the small gain coefficient introduced earlier (Equation 18), and $\alpha$ accounts for any losses, including the mirrors per unit length of the laser cavity. What is important about $G$, sometimes called the big gain coefficient, is that it is a figure of merit for the complete laser, for it incorporates properties of the lasing medium through $\gamma$ and the construction of the cavity through $R_1$, $R_2$, and $\alpha$, the loss coefficient.

When $\gamma = \alpha$, $G = 1$ and the oscillator is stable. Moreover, the condition $G = 1$ defines $\gamma_{thres}$:

$$\gamma_{thres} = \left[\frac{hB_{mn}g(\nu)}{\pi c'}\right]\left[N_m - \frac{g_m}{g_n}N_n\right]_{thres} = \left(\alpha - \frac{1nR_1R_2}{2L}\right) \tag{28}$$

The first bracketed term is intrinsic to the medium and is a function of frequency. The second bracketed term is the threshold population inversion. The most important aspect of this relationship is that the $\gamma_{thres}$ value depends not only on the intrinsic properties of the lasing medium but also on the design of the laser cavity itself via $\alpha$, $R_1$, $R_2$, and $L$.

## Laser Operation

How does one operate a laser? Generally we want as much power from the laser as is possible. It should seem obvious that increasing input pump power will increase laser output power. It can be shown that the laser's output power is given by:

$$P_{out} = T\left[\left(\frac{\gamma - \gamma_{thres}}{\gamma_{thres}}\right)b\right] \tag{29}$$

where $b$ is a construction parameter, $T$ is the transmission coefficient of the mirror that allows some of the laser cavity light to pass (typically 0.1% to 1%), and the term in the brackets is the laser light power circulating in the cavity. Since $\gamma_{thres}$ and $b$ are constants fixed by properties of the lasing medium and cavity construction, increasing the small signal gain by increasing incident pump power means $P_{out}$ must increase. As can be seen from Equation 29, increasing $T$ is another way to increase the output power. However, if $T$ becomes too large, too much of the power in the cavity leaks out and lasing stops. The construction and operation of a laser is a study in trade-offs.

As a brief aside, one of the technological "miracles" encouraged by the advent of the laser was the development of optical coatings to control reflectivity and transmission of lenses and mirrors.

### An Interesting Diversion: Why No Light in the Box?

We can consider the laser oscillator as an energy storage device, a container that stores light. A story is told that an ancient Greek philosopher tried to understand why if the lid of an open box in a lighted room were closed, he could not take the box into a dark room, open the lid, and let the light back out. If you were to do this, of course, you would find no light inside the box. Light travels about 30 cm in a nanosecond, 0.03 cm in a picosecond, and 0.00003 cm in a femtosecond. Even if the walls of the box were 99% reflective, after 500 bounces the intensity would be reduced to 0.0066 of initial and the time elapsed would only be 0.5 microseconds for a 30 cm box. The question was insightful but neither we today nor the ancient Greek philosopher could run fast enough to catch the light!

## Characteristics of Laser Light Output

### Energy Bundles: Transverse and Longitudinal Modes

The laser light energy stored in the cavity oscillator is not continuous in space but stored in what can be thought of as energy bundles. How do such bundles pack in the cavity space? The bundles could pack parallel to the long axis of the oscillator, which is known either as the longitudinal or axial direction. (Axial and longitudinal are terms that will be used interchangeably.) Such packings will be called axial modes. Another packing direction is perpendicular to the oscillator axis, which is known as the transverse direction. Such packings will be called transverse modes. The official name for these modes is transverse electromagnetic modes, or TEM. The study of such modes was greatly spurred on by the development of radar and microwave technology. Are all packings possible for laser oscillators? No, various restrictions limit the number and types of modes that will be observed in a given laser cavity. We will discuss some of those limitations now.

Transverse modes are easy to see experimentally. You can project these as patterns on a wall:

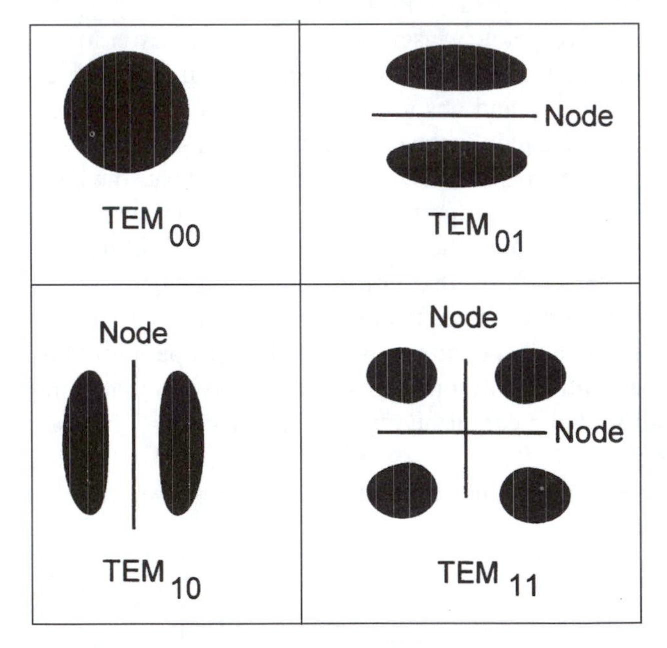


**Figure 2-6**
Examples of transverse modes of a laser cavity: TEM spots. Transverse modes are "packings" of laser light energy which pack in the direction perpendicular to the laser axis. Each mode leads to a characteristic cross-sectional distribution of laser intensity when the laser beam is projected on a screen or viewed end on. (*CAUTION: Never do this with your eye.*) These modes are designated by integral subscripts which describe for the projected pattern the number of nodes (those regions where no energy density occurs) along three mutually perpendicular directions, the $m$, $n$, and $q$ axes. The $q$ subscript denotes nodes along the cavity axis and is often not specified. The $TEM_{00}$ mode describes a Gaussian distribution of intensity, with a maximum in the center of the projection.

spots, rings, clover leaves, and others. Figure 2-6 illustrates several patterns with their $TEM_{mnq}$ designations. Modes are designated by integral subscripts which describe the number of nodes (regions of no energy density) along three mutually perpendicular axes, labeled in Figure 2-7 as the *m*, *n*, and *q* axes. The *q* subscript, which denotes modes along the axis of the cavity (axial), is usually not specified but such modes are very important. Thus *m*, *n*, or $TEM_{mn}$ describe transverse cavity modes. The $TEM_{00}$ is the Gaussian mode mentioned in Chapter 1. It is worth noting here that, unless the laser cavity is modified in some way, all polarizations of light are possible for each transverse mode.

How do these modes affect the output of the laser? Adjacent axial modes effectively define the spatial coherence of the laser; however, too many axial modes in a laser can affect the spatial coherence. As we see in Chapter 3, isolating a single axial mode of a laser is known as the technique of mode-locking and is quite important in many laser applications.

While transverse modes can be projected on a screen, axial or longitudinal modes are much harder to see. Nevertheless, longitudinal modes satisfy the

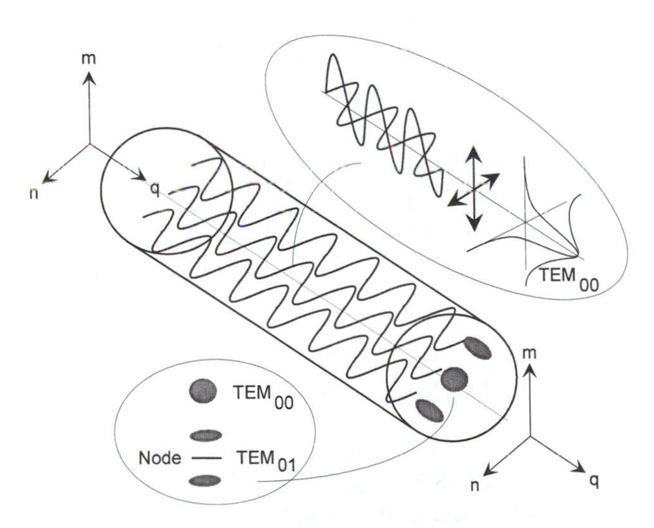


**Figure 2-7**
A view along a laser cavity showing the positions of the $TEM_{00}$ and $TEM_{01}$ modes. The designations for the $TEM_{mn}$ modes *m* and *n* denote the number of nodes in a given direction. The number of nodes in the *m* direction is noted by *m*, similarly for *n*. The insert shows two polarizations of the $TEM_{00}$ mode. Unless some polarization control is placed in the cavity, each mode is randomly polarized, which means the two polarizations shown in the insert are only two of an infinite number of polarizations possible.

condition that an integral number of wavelengths fit into the cavity length in order that the laser wavelength $\lambda$ has a node at each mirror. In effect, axial modes are standing waves in the oscillator and as such are not really visible.

The TEM are describable by theory. The various modes have frequencies (energies) given by Equation 30:

$$\nu_{mnq} = \left[q + (m + n + 1)K\right]\frac{c'}{2L} \qquad (30)$$

where *m*, *n*, and *q* are again integers that characterize the $TEM_{mnq}$ mode of interest. The mode energy is $h\nu_{mnq}$, $c'$ is the speed of light in the cavity medium, *L* is the cavity length, and *K* is a constant dependent on the geometry of the laser. Consider the frequency difference between two adjacent axial modes of a given transverse mode, that is, a mode of fixed *mn*. From Equation 30:

$$\Delta\nu_{\text{axial}} = \Delta\nu_{mn\Delta q} = (q+1)\frac{c'}{2L} - q\frac{c'}{2L} = \frac{c'}{2L} \qquad (31)$$

The axial modes of a given transverse mode are thus spaced in frequency by the value $c'/2L$ and independent of the actual transverse mode (*m*, *n* values), its nominal frequency, or its wavelength. This spacing is illustrated in Figure 2-8 showing bundles of axial modes in the longitudinal direction. Curiously, these laser cavity resonant frequencies are also just those frequencies of a Fabry-Perot interferometer of length *L*. Actually, this is not such curious behavior because a laser can be viewed as a Fabry-Perot interferometer filled with a lasing medium. The quantity $\Delta\nu = c'/2L$ is known as the *free spectral range* of a laser, and this value is independent of the wavelength of output laser light! The free spectral range is a figure of merit often quoted when comparing models or makes of lasers.

What is the axial mode spacing in an ordinary He/Ne laser that is about 50 cm long?

$$\Delta\nu = \frac{c'}{2L} = \frac{3 \times 10^8\,\text{m/s}}{2 \times 5 \times 10^{-1}\ \text{m}}$$

$$= 3 \times 10^8\,\text{Hz} = 300\ \text{MHz} \qquad (32)$$

So for every 300 MHz in frequency, there is another axial mode of the laser. Such spacings are noted in

Figure 2-8. If two adjacent axial modes are separated by 300 MHz, how different are the wavelengths of two adjacent mode frequencies? Consider a He/Ne laser 50 cm long with a mode wavelength of 633 nm and compute the wavelength of its adjacent mode spaced 300 MHz away. For a 633.0000 nm mode ($\nu = 4.73933649 \times 10^{14}$ Hz), the adjacent mode is 632.9997 nm ($\nu = 4.73933949 \times 10^{14}$ Hz), or 0.0003 nm apart. As just mentioned above, axial mode spacing is a physical parameter having to do with the resonator structure and would be the same for any laser 50 cm long. How many axial modes would be in the 50 cm laser cavity? If we first assume that $m$ and $n$ are small with respect to $q$, Equation 30 yields:

$$\nu_{mnq} \approx q\frac{c'}{2L} \Rightarrow q \approx \frac{2L\nu_{mnq}}{c'}$$

$$q \approx \frac{2 \times 0.5 \text{ m}}{3 \times 10^{8} \text{ m/s}} 4.7 \times 10^{14} \text{ Hz} \tag{33}$$

$$\approx 1.6 \times 10^{6}$$

Since $m$ and $n$ are typically on the order of simple integers, the assumption that they could be neglected in Equation 30 seems warranted. How should we interpret the results of this calculation? Simply stated, there are many axial modes in a laser cavity. Sometimes a great number of axial modes is counterproductive to a laser's spatial coherence, and we would like to eliminate all but one by some means. Selecting just one mode out of $10^6$ is a neat trick and we will say more about how this can be done at the end of this chapter when discussing etalons, and in Chapter 3 when discussing mode-locking.

As we have mentioned, transverse modes have different shapes. The $TEM_{11}$ looks like a clover leaf pattern. The Gaussian $TEM_{00}$ looks like a spot. A laser beam that consists only of the $TEM_{00}$ has the lowest divergence and can be focused to the smallest spot with a size on the order of the wavelength of the light. Normally, a laser oscillates in many transverse modes simultaneously. Multimode operation is especially desirable if high output power is wanted. For example, the $TEM_{01}$ mode shown in Figure 2-7 has two "spots" in projection, that is, two sets of energy bundles exist in the cavity with a single energy node between them. With two sets of energy bundles twice as much interaction of stimulating photons with the lasing medium is possible, yielding twice the power of a $TEM_{00}$ mode. Similar comments can be made about other $TEM_{mn}$ modes.

Often multimode operation is undesirable. Multimode operation can cause "hot spots" where energy bundles overlap. These "hot spots" have a spatial pattern and can lead to optical damage, or they can change the mode supported by the cavity to an unstable output. How can we make a laser oscillate only in the $TEM_{00}$ fundamental? The answer is to put a small aperture inside the cavity which has the effect of blocking higher order modes with their large spatial extent. In effect, only the peak of the Gaussian mode passes through the small aperture. The price which one pays for using the fundamental mode is lower power because the beam inside the active medium interacts with less of the population inversion.

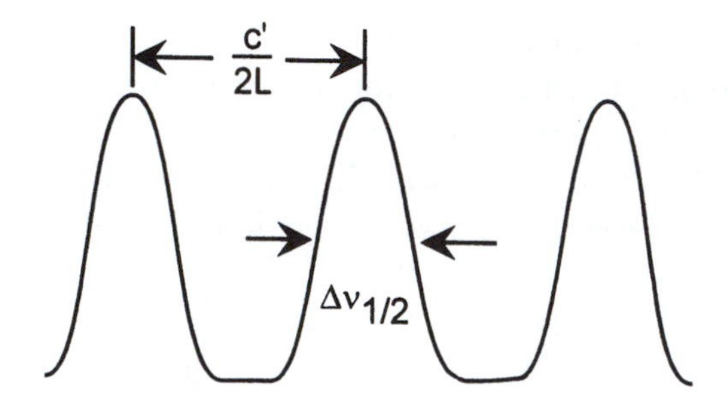


**Figure 2-8**
The spacing and linewidths of adjacent axial modes of a laser. The linewidth of an axial mode is denoted as $\Delta\nu_{1/2}$. The spacing of the axial modes is given by $c'/2L$ where $c'$ is the speed of light in the lasing medium.

## *Laser Cavity Bandwidth*

How narrow is the output of each laser frequency? This question is asking for the "thickness" of each laser axial mode. Consider again the $TEM_{00}$ output of a laser. The spacing of each frequency output is $c'/2L$, as illustrated in Figure 2-8. Of the many terms used to describe the width of the laser output line, we will use bandwidth for $\Delta\nu_{1/2}$. Recall from Chapter 1 that the bandwidth is related to the temporal coherence time, $t_c$, and that the laser cavity is an energy storage device. Starting at a time $t = 0$, suppose the stored energy leaks out mainly through the transmission mirror. Recalling the interpretation of the small signal gain as the fractional intensity loss per unit length (Equation 21), then the intensity loss expected while travelling a distance $2L$ for a laser held at threshold condition is:

$$2L\gamma_{\text{th}} = 2L\left(\alpha - \frac{\ln R_1 R_2}{2L}\right) \qquad (34)$$

The time it takes to lose this much intensity is the time it takes to travel $2L$, and that is given by:

$$t = \frac{2L}{c'} \qquad (35)$$

The decay constant $t_c^{-1}$ of the mode is the intensity loss divided by the time to lose the intensity:

$$t_c^{-1} = \frac{2L\left(\alpha - \dfrac{\ln R_1 R_2}{2L}\right)}{\dfrac{2L}{c'}} = \gamma_{\text{th}} c' \qquad (36)$$

Coherence time is given by Equation 15 in Chapter 1 which, on substitution into Equation 36, yields:

$$\Delta\nu_{1/2} = \frac{1}{t_c} = c'\left(\alpha - \frac{\ln R_1 R_2}{2L}\right) \qquad (37)$$

This expression estimates the inherent laser bandwidth $\Delta\nu_{1/2}$. There will always be such a bandwidth related to a decay time that is proportional to the temporal coherence time of the laser. How large is that width? Again consider a He/Ne laser 50 cm in length. Suppose $\alpha$ is negligible and $R_1R_2 = 0.99$. Equation 37 with these values yields 500 KHz or 0.5 MHz. This result, as well as axial mode spacings calculated earlier, is illustrated in Figure 2-8. Thus a He/Ne laser, if operated in a single mode, has a spectral purity of 1 part in $10^9$. A laser can be a very pure color!

### *Laser Medium Bandwidth*

The actual output of a laser depends on the bandwidth of the axial modes of the cavity and the bandwidth of the laser medium. The last section discussed the bandwidth of the axial modes which we will call the *laser cavity bandwidth*. We need to discuss the bandwidth of the lasing medium which arises from the gain of the laser as a function of frequency, $\gamma(\nu)$. The combination of these two bandwidths determines actual laser output. These bandwidths are depicted in Figure 2-9.

From Equation 21 we see that $\gamma(\nu) \propto g(\nu)$. It is time to discuss the factors that influence $g(\nu)$, for those factors also directly influence the laser medium's gain and bandwidth. The three most common factors affecting $g(\nu)$ are reviewed in Table 2-2. Practical lasers are most affected by collisional and Doppler broadening. Broadening from either of these sources leads to a larger laser medium's bandwidth. The He/Ne laser is primarily Doppler-broadened, and a typical value for the laser medium's bandwidth is about 1500 MHz. How many axial modes are contained in a 50 cm long He/Ne laser? (Recall that we calculated the axial mode spacing in

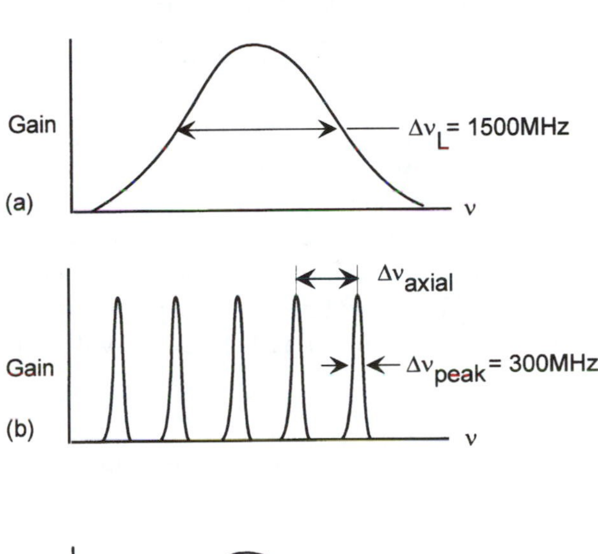


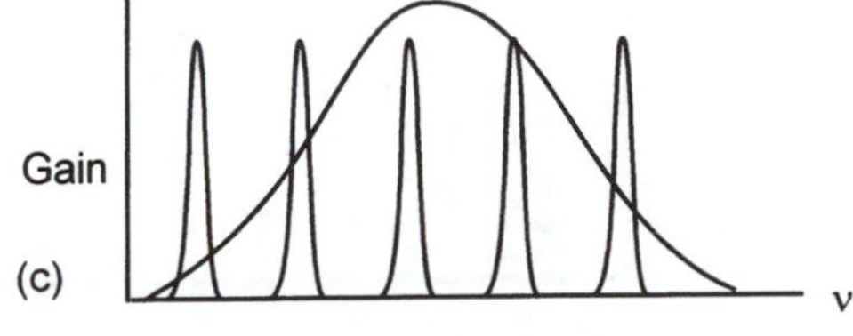


**Figure 2-9**
A laser medium gain curve, axial modes, and their combination on common frequency scales to illustrate relative magnitudes. These gain curves can be viewed as frequency distribution curves representing intensity as a function of frequency. *(a)* The gain curve of the lasing medium alone, which is relatively broad. *(b)* The axial modes due to the cavity construction and the speed of light in the lasing medium. *(c)* The super-position of the medium and axial mode gain curves.

**Table 2-2**
Factors Responsible for Laser Gain Bandwith

| *Effect* | *Features* |
|---|---|
| Natural linewidth | 1. Consequence of Uncertainty Principle $\Delta\nu\ \Delta t \approx 1$<br>2. Homogeneous process—all oscillators equally affected |
| Collisional broadening | 1. Collision alters phase of emitted photon<br>2. Homogeneous since collisions are random<br>3. Important in gases and *IR* frequencies |
| Doppler broadening | 1. Emitting photon travelling with or against the direction of the stimulated emission alters the frequency of emission by its speed<br>2. Broadening is inhomogeneous<br>3. Effect is most marked for UV/VIS frequencies |

an ordinary He/Ne laser as 300 MHz using Equation 32.) Thus, the answer is 1500 MHz/300 MHz = 5. These five axial modes are also illustrated as well in Figure 2-9. Note that while there are many axial modes, only a few will fit under the laser medium's gain curve. The output of a laser depends critically on this fitting. Appreciating Figure 2-10 will finally put all the pieces together to describe how a laser really works.

As pump energy is put into the laser medium, the appropriate population inversion is created and the medium's gain curve is high and its bandwidth large. This is shown in Figure 2-10(a). In this case the axial modes marked by *a*, *b*, and *c* are contained under the medium's gain curve and have intensities (= gains) greater than the indicated $\gamma_{thres}$. The laser outputs photons whose energies correspond to $\nu_a$, $\nu_b$, and $\nu_c$. As lasing continues, the population inversion decreases, approaching the threshold value defined by Equation 28. The medium's gain curve decreases in magnitude and bandwidth. The cavity modes remain the same but now fewer are contained under the medium's gain curve. Figure 2-10(b) illustrates a steady-state condition where now only the mode *b* can be output as laser frequency. This example features the condition whereby the laser outputs just one mode at steady state. Such is not the general case, however. It is worth noting that the combination of the axial modes and the medium's gain curve has greatly reduced the number of possible output axial modes from order $10^6$ to a small integer number.

Another way to further reduce the number of modes output by a laser is to use an etalon. Basically an etalon is a piece of quartz, several mm thick with especially flat, parallel optical surfaces. The etalon is placed in the laser cavity so that the laser beam strikes its surface at an angle. The laser beam travels

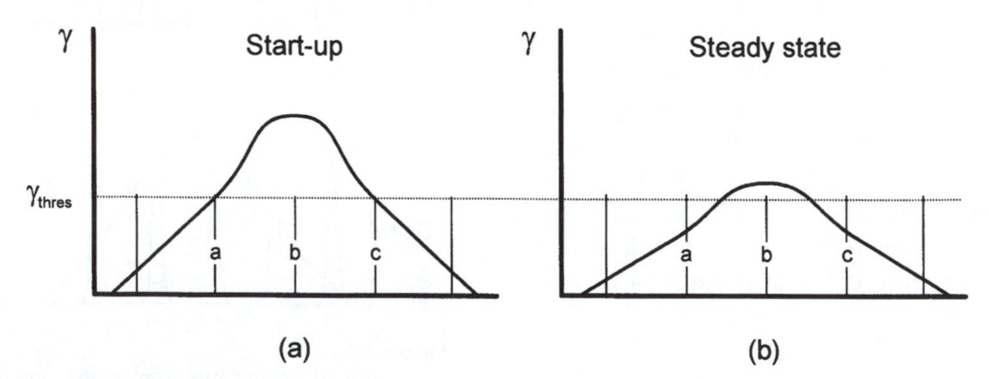


**Figure 2-10**
Lasing start-up and steady-state conditions. The laser medium threshold gain curve is taken to be a constant, independent of frequency. *(a)* After the pump has created a population inversion, the modes labelled *a*, *b*, and *c*, whose intensity and spacing are encompassed by the laser medium gain curve and which have at least the laser gain threshold magnitude, will be output by the laser. *(b)* As the population inversion is depleted by the lasing spontaneous emission process, the laser medium gain curve decreases in magnitude until only mode *b* is encompassed and has sufficient magnitude to satisfy the threshold condition. This describes a typical steady-state condition of operation.

through the etalon a certain distance, depending on its thickness and angle of refraction. Varying the angle of incidence varies the path length of the laser beam in the etalon. The etalon acts as an interferometer allowing only a small range of frequencies to pass through it. If a given laser cavity mode matches the range of frequencies passed by the etalon, then that mode is passed through the etalon and all other modes are reflected. The result is additional mode selection. Varying the angle of laser beam incidence will vary the mode(s) passed by the etalon.

An even more sophisticated, yet common way to reduce the number of axial modes output by a laser is mode-locking, which we detail in Chapter 3.

## SUMMARY

The output of a laser depends on many factors which can be classified either as construction parameters—for example, cavity length and mirror reflectivities—or as elements characterizing the laser medium—for example, the energy levels involved and the extent of Doppler or collisional broadening. The details of construction yield the cavity axial modes, while the properties of the laser medium yield the gain curve. It is the critical combination of these factors that allows the laser to shine.

## SUGGESTED EXPERIMENTS

Suggested references to experiments that illustrate the principles described in this chapter:

1. Grieneisen, H.P., "The Nitrogen Laser-Pumped Dye Laser: An Ideal Light Source for College Experiments; Experiment 1—The Nitrogen Laser-Pumped Dye Laser," Laser Science, Inc., Cambridge, MA. An experiment to demonstrate the phenomenon of superradiance.
2. Grieneisen, H.P., "The Nitrogen Laser-Pumped Dye Laser: An Ideal Light Source for College Experiments; Experiment 2—Wavelength Tunability," Laser Science, Inc., Cambridge, MA. An experiment to demonstrate the property of wavelength tunability.
3. Pollock, W.F., "A 'Hands-On' Helium-Neon Laser for Teaching the Principles of Laser Operation," *Physical Chemistry: Developing a Dynamic Curriculum*, R. Moore, and R. Schwenz, Eds., American Chemical Society Books, Washington, D.C., 1993.

## GENERAL REFERENCES

Bernath, P. F., *Spectra of Atoms and Molecules*, Oxford University Press, New York, 1995.

O'Shea, D. C., Callen, W. R., and Rhodes, W. T., *Introduction to Lasers and Their Applications*, Addison-Wesley Publishing Company, Inc., Reading, MA, 1977.

Silfast, W. T., *Laser Fundamentals,* Cambridge University Press, Cambridge, 1996.

Svanberg, S., *Atomic and Molecular Spectroscopy: Basic Aspects and Practical Applications*, 2nd Ed., Springer-Verlag, Berlin, 1992.

CHAPTER 3

# How to Modify the Light Output of the Laser

## Chapter Overview

Modifications of laser output can be achieved to vary the polarization of the emission, generate pulses of short duration and high power with variable repetition rates, and create new frequencies of emission. In this chapter we detail various modification schemes to devise new laser emission characteristics, including such protocols as polarization control, cavity dumping, Q-switching, mode-locking, synchronous pumping, pulse compression, second- and third-harmonic frequency generation, and optical parametric pumping.

## Modifying Laser Light

Laser light cannot only be generated but can also be modified, for a laser is a light source whose output can be temporally and spatially controlled. There are many possible modification protocols. For example, the energy of the light can be modified by passage through nonlinear optical elements. (Nonlinear optical effects depend on the response of a material to the square, or to a higher power, of the incident photon's electric field and are discussed in greater detail later in the chapter.) If the laser light were continuous, that is, CW, we could make it pulsed, choosing from a very wide range of pulse rates and widths. If the light were pulsed, we could make it continuous. If the laser light had an energy based on frequency $\nu$, we could produce light of $2\nu$. We could obtain pulses of very high energy by creating pulses of very short duration, say femtoseconds. Our next task will be to discuss a variety of means to modify the output of lasers to obtain new characteristics of laser emission.

Several ways to modify laser light are outlined in Table 3-1. One modification involves polarizing the light emitted by the laser. Another variation involves regulating the direction of the light in the laser cavity, a necessity in the design of ring lasers. Perhaps the most common control of laser light is the production and modification of light pulses. Finally, the energy output of a given laser is determined by the construction material of the laser, but, by passing the output laser light through nonlinear optical elements, various order harmonics of the laser frequency can be produced, yielding light of different energy. We will discuss these various modifications in the order presented in Table 3-1. Let us begin then with polarization of laser light and what can be achieved through polarization effects.

## Polarized Light

### Polarization

Recall our discussion of the polarization of light in Chapter 1, specifically Figure 1-1 which shows light with a specific polarization. Polarization effects are useful in chemical studies for two central reasons. One application is to study how matter absorbs polarized light or how matter fluoresces under the stimulus of polarized light. A second application is to utilize polarization effects to construct shutters capable of passing or blocking the passage of laser light. By opening and closing such shutters, pulsed laser light can be obtained from CW lasers. Electro-optic (E-O) shutters used for the production of pulsed laser light utilize the properties of polarized light to achieve this modification of laser emission.

### Plane Polarization

Figure 3-1 illustrates a plane polarized photon propagating in the $z$ direction with its plane of polarization at an angle $\theta = 45°$ with respect to the $x$ and $y$ axes. Note that the view is down the axis of propagation, $z$. In such a case, the oscillating $E$ field vector $\mathbf{E}$ can be viewed as the resultant of the vector addition of the oscillating $E_x$ and $E_y$ components. If we now discuss the electric field of the photon in terms of its components $E_x$ and $E_y$, we can easily describe several important phenomena associated with polarized light. Focusing on the oscillatory properties of $E_x$ and $E_y$, we can ask what happens to $\mathbf{E}$ if the $E_x$ and $E_y$ components are in or out of phase, that is, have maxima, zeros, or minima at different $z$ positions. Whether the components are physically in or out of phase will depend on the relative velocities of the $E_x$ and $E_y$ components in the medium. Note that in this discussion the maximum values of $E_x$ and $E_y$ are equal. Said another way, the amplitudes of the $E_x$ and $E_y$ waves are equal. If the velocity of $E_x$, $v_x$, equals that of $E_y$, $v_y$, then the two light waves are in phase. If $E_x$ and $E_y$ are in phase, each component travels the same amount as the photon propagates

**Table 3-1**
Controlling Laser Light

| *Effect* | *Technique* | |
|---|---|---|
| Polarization | Absorption<br>Reflection<br>Birefringence | |
| Light direction | Optical diode | |
| Pulse production | *Q*-switching | Active<br>acousto-optic<br>electro-optic<br>Pockels cell<br>Kerr cell<br>rotating mirror<br>Passive<br>saturable absorbers |
| | Cavity dumping | Active<br>flipped mirror<br>acousto-optic<br>electro-optic<br>Pockels cell<br>Kerr cell |
| Pulse modification | Mode-locking | acousto-optic<br>saturable absorbers |
| | Synchronous pumping | mode-locked laser<br>dye laser |
| | Pulse compression | nonlinear fiber optics<br>parallel gratings |
| | Optical parametric pumping | nonlinear optical response |
| Frequency modification | 2nd, 3rd harmonic generation | nonlinear optical response |

down the $z$ axis. Thus, **E** remains at $\theta = 45°$ with respect to $x$ and $y$, and the photon is linearly polarized. This is illustrated in Figure 3-1(a).

## Elliptical and Circular Polarization

If in some medium $v_x \neq v_y$, then the wave describing the $E_y$ component is displaced along the $z$ direction from the $E_x$ component. This is illustrated in Figure 3-1(b). At any point along the $z$ direction, the magnitude of $E_x$ does not equal that of $E_y$ and the resultant **E** makes, in general, some angle other than 45° from either the $E_x$ or the $E_y$ axis. The insert in Figure 3-1(b) illustrates this case. Now if one drew a figure tracing the path followed by the tip of the **E** vector as the $E_x$ and $E_y$ waves moved through the medium, and then projected that figure onto a plane perpendicular to the $z$ direction, the projected figure would be an ellipse. Such a projection may be imagined from the insert in Figure 3-1(b). The light resulting from such a travel is called elliptically polarized. If $v_x \neq v_y$ such that the zero amplitude of the $E_y$ wave occurs just when the $E_x$ wave is at its maximum amplitude, the projected figure of the **E** vector tip would not be an ellipse but rather a circle. The light produced in this case is called circularly polarized. Another way to describe circularly polarized light is as follows. If one thinks of the $E_x$ and $E_y$ waves as sine waves,

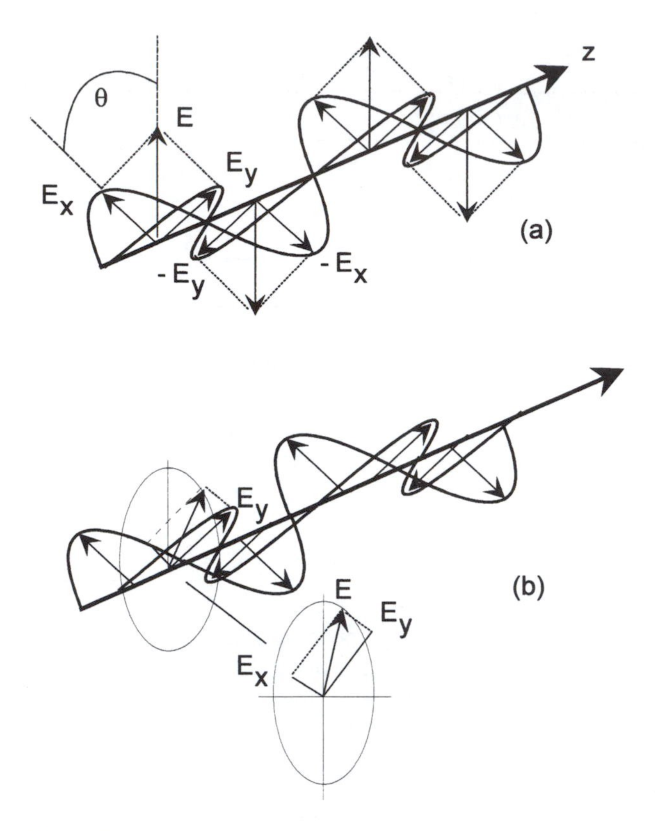


**Figure 3-1**
*(a)* Components of a photon's electric field vector when the photon's plane of polarization makes an angle with respect to some chosen axis system. In this diagram, a plane polarized photon is oriented with its plane of polarization at 45° to the *x*-*y* axes. The direction of propagation is along the *z* axis. The **E** vector can be considered to be the vector addition resultant of the components $E_x$ and $E_y$. For plane polarized light, the $E_x$ "wave" is in phase with the $E_y$ "wave," which is illustrated by the two waves having the same starting point along the *z* direction. The vector addition of $E_x$ and $E_y$ at any position along *z* will result in the **E** vector still lying in the plane located 45° from the *x* and *y* axes. *(b)* Circularly or elliptically polarized light results when the $E_x$ and $E_y$ waves are not in phase but can be thought to have different starting points along the *z* axis. Vector addition then results in the **E** vector and the plane of polarization it defines making some angle $\theta$ with the *x* and *y* axes. The insert details the vector addition. If the phase difference is 90°, which might mean, for example, that the $E_x$ component is at its maximum while the $E_y$ component is at zero, the tip of the **E** vector will trace out a circle as the wave travels along *z*. Such light is circularly polarized. Any other phase difference results in elliptically polarized light.

then the $E_x$ wave has maxima at 90°, 270°, and so on. If the $E_y$ wave has zero amplitude when the $E_x$ wave is at its 90° or 270° points, the light produced on passage through the medium is circularly polarized. Thus, circularly polarized light is produced when there is a 90° or 270° phase difference between the $E_x$ and $E_y$ components.

If the phase difference between $E_x$ and $E_y$ is 90°, the tip of the **E** vector rotates to the right as one looks at the oncoming beam. In this situation we will call the light *right circularly polarized*. For a phase difference of 270°, the light is called *left circularly polarized* (note there are other conventions based on looking down the beam output). Circularly polarized light is a special case of elliptically polarized light with 90° or 270° phase differences. For any other phase differences (>0), the light produced on passage through the medium is elliptically polarized.

A special note should be made here. The combination of left and right circularly polarized light results in linearly polarized light! This result is significant in that linearly polarized light can be thought of as the combination of LCP and RCP light. It might seem odd, but quite often the effects of linearly polarized light are discussed in terms of what happens to the "LCP and RCP beams."

## Production of Polarized Light

### Polarized Light by Absorption

An easy way to obtain polarized light is through the absorption of light by specific molecules. Recall that absorption of light occurs along a preferred direction within a given molecule. Suppose we consider a rod-like molecule and assume that the molecule will only absorb light whose polarization is parallel to the long axis of the molecule. This scheme is illustrated in Figure 3-2. Thus, when the **E** vector of the incident photon makes an angle $\theta = 90°$ with the preferred absorption direction, light is highly absorbed. Conversely, when $\theta = 0°$, no light is absorbed. In essence, only light of the correct polarization passes through the assembly of absorbing molecules. Today we are all familiar with Polaroid sheets, first made by Edwin Land who put appropriate absorbing molecules in a suitable plastic medium and thus fixed the preferred absorbance and transmittance directions.

### Polarized Light by Reflection

We can also produce polarized light by reflection. This is possible when light meets the following two

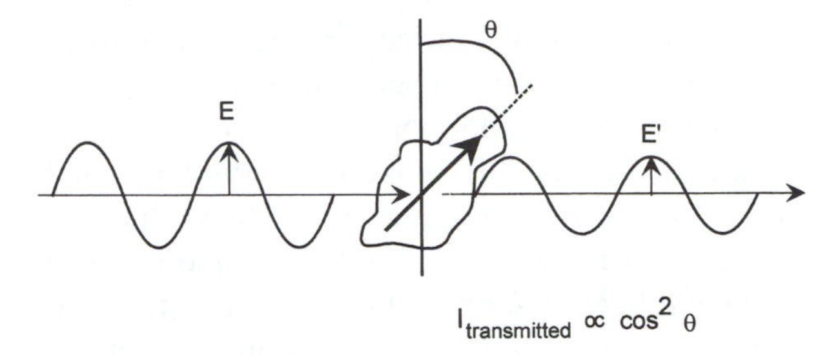


**Figure 3-2**
Production of polarized light by a molecule absorbing light. For the rodlike molecule illustrated, light is absorbed only when the polarization of the light is parallel to the long axis of the molecule. In particular, the intensity of the transmitted light is proportional to the factor $\cos^2 \theta$, where $\theta$ is the angle that the **E** vector of the incident photon makes with the longitudinal axis of the molecule. When $\theta = 90°$, the incident polarized light is perpendicular to the molecular absorption dipole, no light is absorbed, and light of that polarization is maximally transmitted by the molecule.

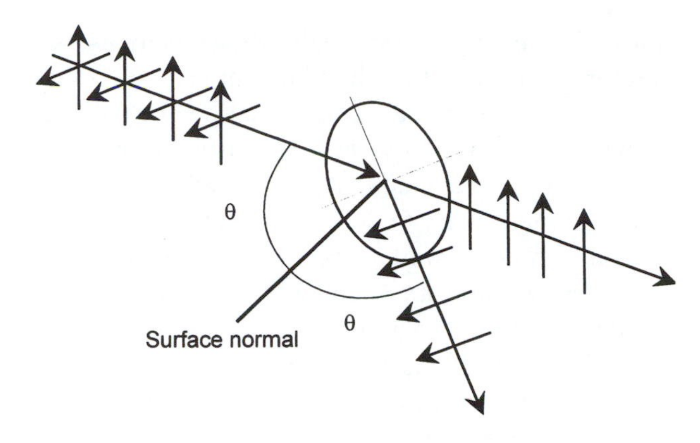


**Figure 3-3**
Production of polarized light by reflection. Let two mutually perpendicular **E** vectors represent unpolarized light incident on a substance. Let one polarization be parallel to the surface of the substance, which may be a crystal or liquid whose refractive index is $n$. The light whose polarization is parallel to the surface and whose angle of incidence with respect to the surface normal is the Brewster's angle will be reflected and travel along the surface. This case defines Brewster's angle. The other polarization passes through the substance in the same direction and with its original polarization, thus producing polarized light. The Brewster's angle varies with the refractive index of the medium.

conditions: (1) the electric vector of the light is parallel to the surface on incidence and (2) the direction of propagation of light makes a special angle called the Brewster's angle with a line perpendicular to the surface (the surface normal). When these conditions are met, the light of the incorrect polarization is reflected, and light of parallel polarization is transmitted. Thus, the transmitted light is now linearly polarized. This is illustrated in Figure 3-3. Brewster's angle depends on the refractive index $n$ of the medium. For glass, $n \approx 1.5$ and $\theta_{\text{Brewster}} \approx 57°$.

How are these principles utilized to produce polarized light in a laser? Recall that unpolarized light consists of light with all possible polarizations. Just as unpolarized photons generate unpolarized stimulated emission, polarized photons generate polarized stimulated emission. Suppose we consider the subset of two mutually perpendicular **E** vectors as a simple model for the unpolarized stimulated light emission initially produced in a laser cavity. Then for light incident on a glass surface at Brewster's angle only one of the vectors will reflect, the other will pass through the medium. Consider such a window in the laser's cavity at Brewster's angle to the cavity axis and placed in front of the highly reflective end mirror. All unpolarized stimulated emission produced in the laser cavity that is incident upon the window is transmitted to the end mirror, except for the light whose polarization satisfies the Brewster's angle condition. For our simple model, one vector is transmitted, the other reflected. The transmitted light reaches the end mirror and is reflected back to the Brewster's window. This light, matching the Brewster's angle, again passes through the window entering the laser cavity. We have now produced polarized light in the laser cavity, which in turn produced polarized stimulated emission and which undergoes amplification multiple times before leaking through the output mirror. This two-vector model can be generalized to an **n**-vector model of all possible polarizations. This is a common way of constructing a laser with polarized output. Thus, if a laser is described as having Brewster's windows in its optical path, you can be confident that the output of that laser is polarized.

## Polarized Light by Birefringence

Nature also provides a way to polarize light. A so-called *birefringent crystal* causes light to travel at different velocities depending on the direction within the crystal. For such an optically anisotropic material, the refractive index of the substance varies with the direction of the incident light with respect to the crystal axes. Calcite, mica, and topaz are examples of birefringent materials. Since velocity is a vector, the effect of the birefringent crystal on light travelling in the crystal can be described by the components of the velocity. This is illustrated in Figure 3-4. If all three components $v_x$,$v_y$, and $v_z$ of the velocity are necessary to describe the motion, then the crystal can be viewed as having three refractive indices, $n_x$, $n_y$, and $n_z$. Light will travel at different velocities along the *x*, *y*, and *z* directions. Such a birefringent crystal is also called *biaxial*. If only two unique velocity components are necessary because two of the components are equal, say $v_x = v_y \neq v_z$, then $n_x = n_y \neq n_z$, and the crystal is *uniaxial*. Moreover, whether the crystal is biaxial or uniaxial, it is also anisotropic. *Anisotropic* is the term used to refer to materials whose physical properties depend on the direction in which they are measured. *Uniaxial* and *biaxial* describe the specific nature of the anisotropy. In contrast, materials with only one refractive index (for example, crystals characterized by a cubic structure) are *isotropic* substances.

Now let's see how birefringent materials aid in producing polarized light. Again let us view randomly polarized light as two mutually perpendicular **E** vectors. When unpolarized (random) light travels through the birefringent crystal, one of the light waves will be retarded more than the other since its polarization is in a different direction (90°) from the other. These different speeds which cause the different refractive indices of the crystal also cause the angle of refraction for each light wave to differ. If the exit surface of the crystal is cut in such a manner as to internally reflect one of the polarizations completely, only light of the other polarization exits the crystal. Thus, the crystal has produced polarized light. The two polarizations are often called either the *p* (parallel) and *s* (slow) *polarizations* or simply the *fast* and *slow polarizations*. The names "fast" and "slow" have their origin from whether the polarization of interest arises from the faster or slower light waves.

Birefringent crystals allow us to construct devices known as *wave plates* for the control of polarization. Consider a linearly polarized light wave as two equal magnitude components in phase. Let this linearly polarized light wave pass through a birefringent crystal with its plane of polarization at 45° to the *p* ($E_x$) and *s* ($E_y$) axes of Figure 3-1. Consider the

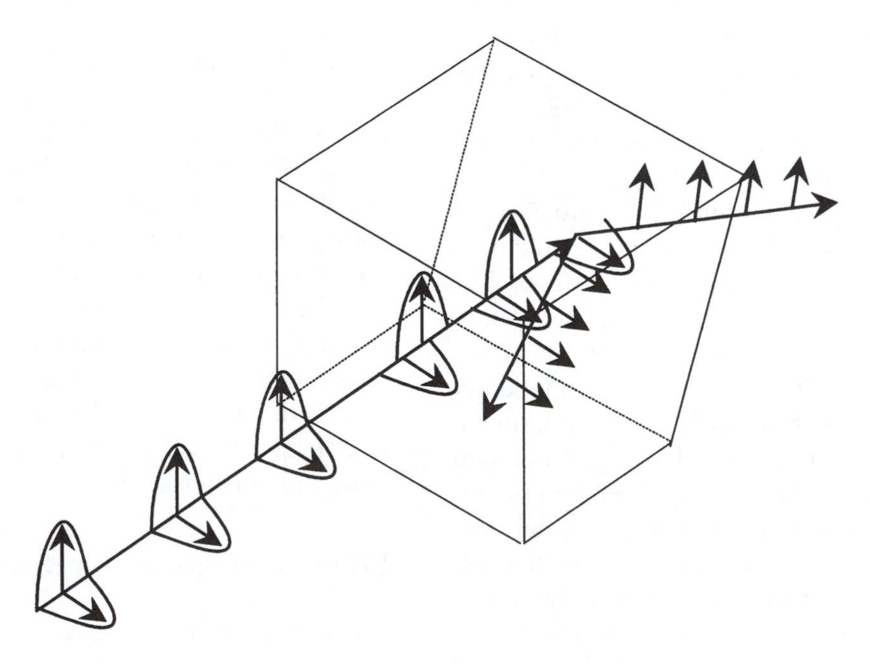

**Figure 3-4**
A birefringent crystal will have two refractive indices if the crystal is described as uniaxial and three refractive indices if the crystal is described as biaxial. When unpolarized light (represented by two mutually perpendicular waves) enters a birefringent crystal, one of the waves slows down with respect to the other because its refractive index is different. When the waves reach the exit surface of the crystal, each will be refracted differently again because of their refractive index difference. If the exit surface of the crystal is cut at an appropriate angle, one of the waves can be made to reflect entirely within the crystal, leaving the other wave to exit and yield polarized light. The type of polarization—plane, circular, or elliptical—can be controlled by the path length in the crystal.

birefringent crystal to have a thickness $T$. Now let us follow what happens to the components along the $p$ and $s$ directions. The $p$ wave transit time is given by:

$$\left(\frac{1}{v_p}\right)T = \left(\frac{n_p}{c}\right)T \tag{38}$$

and the $s$ wave transit time is given by:

$$\left(\frac{1}{v_s}\right)T = \left(\frac{n_s}{c}\right)T \tag{39}$$

When we look at the differences in time, we obtain:

$$\Delta t = \left(\frac{1}{c}\right)(n_s - n_p)T \tag{40}$$

From this time difference we calculate a phase difference by noting how much the light waves are changed travelling for this time difference. Recall Equation 4 which describes a light wave travelling in one direction ($z$) with the possibility of a nonzero phase angle $\phi$. If we consider the phase angle of the $p$ light wave to be zero, then the phase angle difference between the $p$ and $s$ waves can be accounted for by the inclusion of a nonzero $\phi$ in the equation for the $s$ wave. For a wave of frequency $\nu$, the phase angle (= phase angle difference) can be calculated by:

$$\Phi = 2\pi\nu\Delta t = 2\pi\nu\left(\frac{1}{c}\right)(n_s - n_p)T$$

$$= \frac{2\pi}{\lambda_o}(n_s - n_p)T \tag{41}$$

Here $\lambda_o$ is the wavelength of incident light in a vacuum ($c = \nu\lambda_o$) which is essentially that $\lambda$ in air. The phase difference depends on the crystal thickness $T$. If a crystal is cut to a thickness such that:

$$(n_s - n_p)T = \frac{\lambda_o}{4} \tag{42}$$

or an integral multiple thereof, the crystal is called a *quarter-wave plate*. In terms of phase angle differences, a quarter-wave plate causes a 90° phase angle. Such a phase angle produces circularly polarized light which is illustrated in Figure 3-1(b) when the $E_x$ and $E_y$ waves are 90° apart. For double the crystal thickness, the phase angle is 180° and the plane of polarization, the **E** vector, is rotated by 45°. Such a crystal is called a *half-wave plate*. It is worth repeating the fact that a half-wave plate will rotate the plane of incident plane polarized light waves by 45°.

Birefringence filters find many important uses in modifying laser light output. For example, consider a filter constructed of several wave plates arranged in a series. We can choose the thickness of each plate so that:

$$(n_s - n_p)T = m\lambda_o \tag{43}$$

where $m$ is any integer. This equation describes a filter which is effectively a $m(\lambda/2)$ wave plate. With this filter placed inside the laser cavity, any wavelength light $\lambda$ not equal to $\lambda_o$ will create elliptically polarized light of wavelength $\lambda$. Light of such polarization is inefficiently reflected off the laser cavity mirrors with great loss of intensity. We can maximize the transmission of $\lambda_o$ from the cavity by rotating this birefringence filter to replace the elliptically polarized light with linearly polarized light, which is efficiently reflected off the cavity mirrors. Birefringent filters are extremely valuable, often critical to the operation of most dye lasers.

The phenomenon of birefringence plays a very important role in modifying laser output not only in the filter described above, but also in nonlinear optical techniques soon to be discussed.

## Shutters

There are basically three common ways to pass or block laser light: electro-optic shutters, which depend on aspects of polarized light; acousto-optic shutters, which depend on light diffraction; and saturable absorber shutters, which depend on relaxation times of radiationless energy transfer within organic dye molecules. Optical diodes are one-way light shutters allowing light to pass only in one direction. We will examine the principles underlying the basis for each of these categories of shutters.

### Electro-Optic (E-O) Shutters

Electro-optic devices work by rotating the plane of polarization of light passing through them as a function of the frequency of the electric field applied to the device. We will discuss two common types of electro-optic devices used as shutters: the Pockels cell and the Kerr cell.

### *Pockels Cell*

To appreciate how the Pockels cell works, we must keep in mind our earlier discussion of birefringence and biaxiality. The phenomenon that certain uniaxial birefringent crystals can be made biaxial by the application of an electric field is known as the Pockels effect. The origin of the effect lies in a second order nonlinear optical response of the crystal medium. (Nonlinear optical effects are discussed later in this chapter.) All we need to appreciate at this point is that in the biaxial crystal created by the applied electric field, light will have three possible velocities along three different directions in the crystal. Figure 3-5 will aid us in understanding the Pockels cell shutter.

In this figure, incident plane polarized light passes through the Pockels crystal. (Potassium dihydrogen phosphate is an example of a crystal that exhibits the Pockels effect. Table 3-2, to be discussed later, lists additional crystals that exhibit nonlinear optical properties that can be used for construction of Pockels cells.) Again we will think of the incident plane polarized wave as two equal $E_x$ and $E_y$ component light waves in phase. The Pockels cell crystal is inherently birefringent and uniaxial, which means that, without any voltage on the crystal, the initial $E_x$ and $E_y$ components would yield components $E_x'$ and $E_y'$ in the crystal, both of which would travel at the same velocity and emerge from the crystal still in phase with the same initial polarization. If the exit polarization filter were set at 90° to the axis of the incident polarization, no light would pass through the cell.

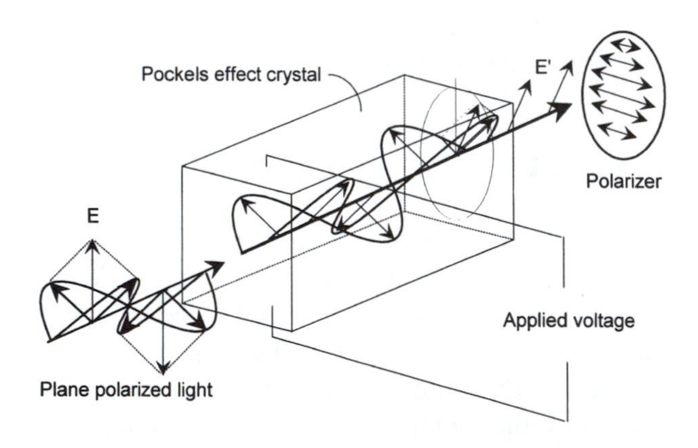


**Figure 3-5**
A Pockels cell polarizer. An applied electric field creates a biaxial crystal from a uniaxial one. Light incident on the electric-field-induced biaxial crystal may exhibit three different velocities along the three different directions of the crystal, and each directional component of light travels at a different velocity through the biaxial medium. As a consequence, plane, circular, or elliptically polarized light can be produced from the incident plane polarized light in such a Pockels effect crystal depending on the applied electric field. With an exit polarizer used in conjunction with the Pockels effect crystal, the intensity of the transmitted light can be varied from incident to zero.

Now suppose a voltage were placed on the crystal to make it biaxial. Since the refractive indices along the $E_x'$ and $E_y'$ directions can be controlled by the applied electric field, the velocity of the $E_x'$ component can differ from that of the $E_y'$ component. The velocity difference can be described by a phase difference between the $E_x'$ and $E_y'$ component light waves, and we note that the phase difference depends on the applied electric field. A voltage that produces a 180° phase difference yields linearly polarized light whose plane of polarization can be made to equal that of the exit polarizer. The Pockels cell operated between this voltage and no voltage will be a simple on/off shutter. A voltage which produces a 90° phase difference yields circularly polarized light whose polarization plane can be considered to match that of the exit polarizer half of the time. This condition provides only half of the initial light intensity on exit. Intermediate voltages produce elliptically polarized light with a variable amount of light intensity upon exit. Thus, a Pockels cell can be constructed as either an on/off shutter or as a variable intensity control, making it quite a versatile device for modifying laser light output.

### *Kerr Cell*

A Kerr cell is a second type of polarizing device. A Kerr cell is very analogous to a Pockels cell modulator except a dipolar liquid, rather than a birefringent crystal, serves as the nonlinear optical medium. Let's examine the origins of the Kerr effect. Normal liquids are unordered—the orientation in space of one molecule bears no relationship to any of its closest neighbor molecules. (As an aside, liquid crystals differ in this respect, for they form liquid phases where molecules *do* orient with respect to each other over many molecular lengths. Polarization control devices are made from liquid crystalline materials, but we will not discuss them here.) Most liquids do not respond to electrical or magnetic fields to create any sense of long-range orientational ordering. However, liquids comprised of molecules with large dipole moments, nitrobenzene for example, will order in the presence of very intense electric fields. In the presence of such oriented molecules, the liquid becomes anisotropic, and, in effect, noncentrosymmetric. As a consequence, the nonlinear optical effects responsible for the Pockels cell effect now become possible. The nonlinear optical effects induced in liquids oriented by electric fields are collectively called the Kerr effect. The parallel phenomenon induced by magnetic fields is called the Cotton effect. As a curiosity, the Pockels effect depends only on a second-order nonlinear effect, while the Kerr effect depends on a third-order nonlinear effect. The fact that the Kerr effect is a third-order effect might be reasonable from the point of view that the Pockels effect starts with a birefringent material, while the Kerr effect has to first create one. Because Kerr cells depend on a third-order effect, they require greater applied voltages to work than do Pockels cells. In practice, Kerr cells are not often used to modify laser outputs.

## Acousto-Optic (A-O) Shutters

What is an acousto-optic shutter? This type of shutter depends on the diffraction of an incident laser beam by a "lattice" of regions of different refractive indices created by sound waves in an otherwise transparent liquid medium. The creation of refractive index "regions" is accomplished by sound waves set up in the liquid by a voltage applied to a piezoelectric crystal. The crystal, in contact with the liquid, expands or contracts in response to the voltage and creates pressure pulses. The pressure pulses are sound waves whose frequency and amplitude are determined by the frequency of the voltage applied to the piezoelectric crystal. A region of higher pressure in the liquid will have a different refractive index than a lower pressure region. If light enters the cell while regions of different refractive indices exist in the liquid, diffraction can occur. If we control the frequency of the piezoelectric crystal, we change $\lambda_{sound}$ and $\theta_{Bragg}$, and hence the on/off period of this shutter. This shutter directs light either along the 1st order direction, $2\theta_{Bragg}$ measured from incident, or along the 0° direction. If the acousto-optic device is arranged as in Figure 3-6, then diffracted light reflects off the mirror in such a way as to pass through the acousto-optic cell. The light is again diffracted back in the direction of the incident light. If the incident light were the light in the laser cavity, then the light in the cavity as well as the laser's output light would be periodic, depending on the frequency of the piezoelectric crystal. Pulsed laser light is created in this process. If the frequency of the modulator is chosen to be one-half of the intracavity mode spacing frequency of the laser ( $= c/2L$), constructive interference occurs. We will further detail this situation when we discuss mode-locking.

## Optical Diode

The electro-optical device known as an optical diode actually operates as an optical isolator since it only permits light to pass through it in one direction. When would such a device be useful? An isolator is needed for the unidirectional operation of a ring dye laser. In such a laser, light must be amplified in only one direction in a cyclic path between mirrors, and not in the other direction. If a reflection due to a mirror in the laser were to cause light to travel backwards, lasing amplification would also occur in the backwards direction. If the forward and the back-

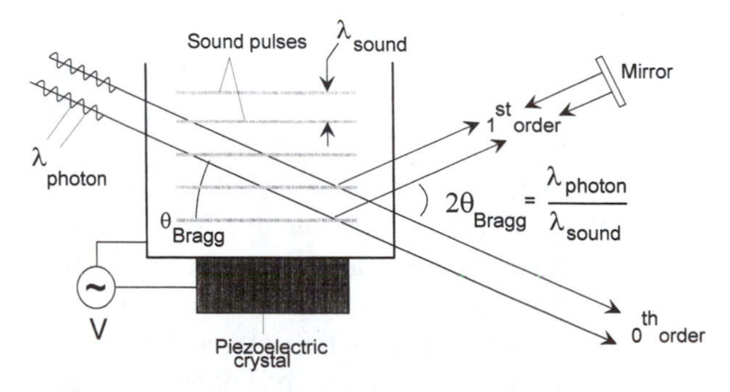


**Figure 3-6**
An acousto-optic cell. Acoustic waves are established in the medium by application of the voltage to the piezoelectric crystal (transducer). The diffraction that occurs at an angle $2\theta_{Bragg}$ from the zeroth order (incident) beam depends on $\lambda_{sound}$ which in turn is controlled by the voltage input to the acoustic transducer. By placing a mirror in the path of the diffracted first order beam, light can be made to return in the direction of its origin by simply being rediffracted again in first order.

ward beam powers were allowed to interfere in the laser, a destruction of the laser medium could occur. This clearly is an undesirable situation! In this example, the optical diode only allows light to travel in one direction in the ring dye laser.

Let's consider how an optical diode functions as a one-way shutter. Optical diodes often operate via the Faraday effect. The Faraday effect is the induction of optical activity in a sample caused by the interaction of the sample with a strong magnetic field. A sample exhibiting optical activity will rotate a plane of polarized light. Suppose laser light were incident on the diode from both the left and right directions. Let vertically polarized light incident from the left pass through the diode without loss by arrangement of a left-hand side polarization filter. Light incident from the right is first selectively polarized, and its plane of polarization is rotated by the Faraday effect of the diode to a horizontal polarization. When this light arrives at the left vertical polarizer, it is not passed. The device thus restricts light passage to only one direction (the left direction here), in effect creating an optical isolator.

### Saturable Absorbers

A saturable absorber is a fascinating device with no moving parts. This type of shutter requires the placement inside the laser cavity of a container of a dye with special optical properties. The dye is capable of absorbing at the $\lambda_{max}$ of the laser emission, but the extent of absorption of the dye decreases with laser intensity. For low powers of stimulated emission within the laser cavity, the dye absorbs the radiation, and thereby, in effect, blocks transmission to the exit mirror. However, as the power of stimulated emission builds up in the cavity, the dye becomes transparent because it cannot absorb any more photons. All future photons pass through unabsorbed.

What is the origin of this effect? The rate at which the dye can lose its absorbed photon energy by radiationless transfer to return to lower absorbing states is slower than the rate at which photons strike the sample. We often speak of the dye as having undergone a process of *bleaching*. As a consequence, the stimulated emission is transmitted through the saturable absorber to the output mirror where it can appear as a pulse of light. Once the power decreases as the population inversion is depleted through the emission of a laser pulse, then the dye can absorb additional radiation. In effect, the rate at which the dye loses its absorbed photon energy by radiationless transfer to return to the lower absorbing states has become competitive with the rate at which photons are now striking the sample. The presence of more dye molecules in lower energy states leads again to absorption of the stimulated emission and blocking of the passage of light. This type of shutter is often used to "*Q*-switch" lasers and will be discussed shortly.

## Pulse Production

### Making Pulses

Turning a CW laser on and off with a shutter produces pulses, and from our discussion above we now know a considerable amount about a wide variety of ways to produce pulses. Throughout our presentation we have assumed knowledge of what constitutes a pulse of light, but now let's make sure we know what is meant by the *pulse duration* and *pulse frequency* of laser energy. These concepts are illustrated in Figure 3-7.

The objective of our next discussion is to show how pulses of different widths and frequencies can be produced using various shutters in conjunction with the techniques called *Q-switching*, *cavity dump-*

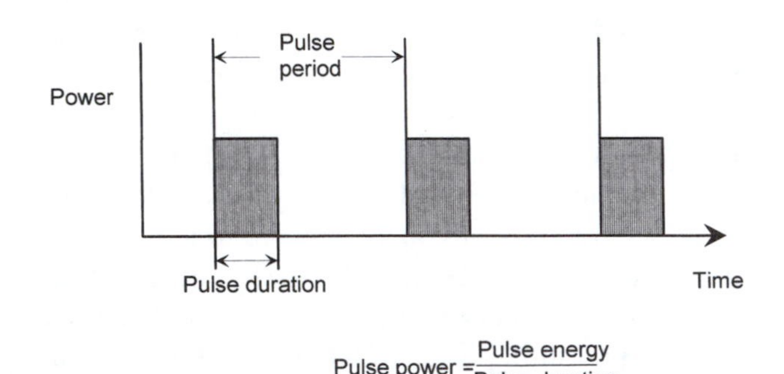


**Figure 3-7**
Characteristics of laser pulses. When a population inversion cannot be continuously sustained, a pulse of laser emission results. The time over which lasing occurs is denoted as the *pulse duration*, the time interval between pulses is defined as the *pulse period*, and the ratio of the pulse energy to the pulse duration is known as the *pulse power*. Lasers whose output is normally continuous can be pulsed by placing an on/off shutter in the laser beam. The characteristics of such pulses are also described by the terms defined here.

**Table 3-2**
Types of Modified Ion Lasers in Terms of Power, Pulse Width, and Pulse Repetition Rate

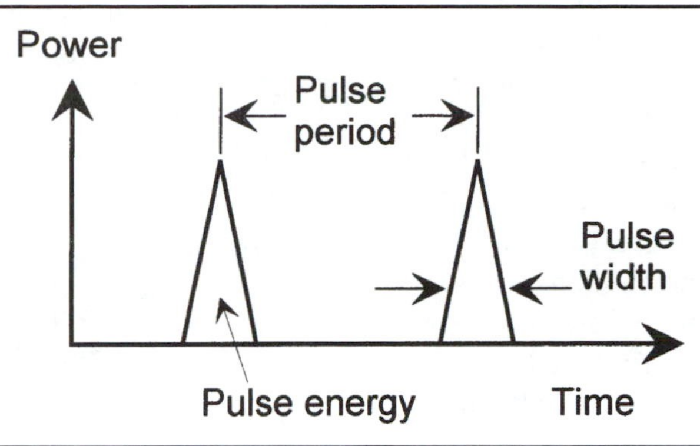


| *Laser System: Modified Ion Laser* | *Power* | *Pulse Width* | *Period of Pulse Repetition* |
|---|---|---|---|
| Mode-locked | 100 W | 100 ps | >10 ns |
| Cavity-dumped | 10–100 W | > 10 ns | Variable |
| Mode-locked and cavity-dumped | > 10 W | > 0.1 ns | Variable |
| Synchronously pumped | > 100 W nJ/pulse | > 1 ps | > 10 ns |
| Synchronously pumped and mode-locked | > 1kW | > 10 ps | Variable |

*ing*, *mode-locking*, *synchronous pumping*, and *pulse compression*. Perhaps before reading further, you might want to look at Table 3-2 to gain some preliminary idea of what the various pulse production techniques can do. At this point, Table 3-2 is best looked at as a preview, however. Generally the purpose of pulse production techniques is to achieve short-duration high-power pulses. Keep in mind that if a modest-power CW laser can be pulsed to achieve even a millijoule of output in the time period of a nanosecond, the power in that pulse is $10^{-3}$joule in $10^{-9}$ seconds, which is a megawatt—an enormous power!

Why produce such pulses? Perhaps we are interested in a high-power pulse to initiate a chemical reaction. Various chemical or physical events occur in very short times and are only accessible with high powers. Or, perhaps we are interested in producing nonlinear optical effects in some chemical medium. To do so requires extremely high power. All of the pulse techniques specified above have greatly expanded the utility of the simple laser. Let's begin to see how by first discussing the technique of Q-switching.

## Q-Switching

From electrical engineering, the $Q$ factor of a resonant circuit measures the energy stored in the circuit relative to the energy lost by the circuit for each passage (cycle) of current through the complete circuit. The $Q$ factor, or quality of the circuit, is denoted by the symbol $Q$. A circuit can be operated to change the condition of energy storage from a high to a low value and vice versa. The process in either direction is $Q$-switching. If the quality $Q$ of the laser cavity, the ratio of energy stored in the cavity to that lost (leaked) by the mirrors, is changed from a low $Q$ to a high $Q$, the laser is referred to as *$Q$-switched*. If the change of the laser's $Q$ is from high to low, this too is $Q$-switching, but such a laser is referred to as *cavity-dumped*. Table 3-3 summarizes the conventions for specifying these quality changes. Either technique produces high-power pulses.

**Table 3-3**
Quality Changes

| | |
|---|---|
| Low to high | $Q$-switching |
| High to low | Cavity dumping |

How can these *Q*-switchings be accomplished? As a model for a *Q*-switched laser, consider a balloon which has two holes in it, one a pinhole, the other sizable. We attempt to fill the balloon by adding helium gas at essentially a constant rate from a compressed gas cylinder. The input helium is equivalent to the pump energy of a laser. We note that the balloon barely fills up because of the existence of the holes. Little helium gas is stored in the balloon, and we regard the quality of the balloon as low. With some searching to locate the leaks, we find the larger hole and patch it with a piece of tape. The balloon now fills rather nicely, and we designate its quality as high. In essence, we have modified the quality of our "balloon laser" from low to high. However, we keep filling and the tape patch eventually fails, allowing much of the helium to escape in a quick rush out of the balloon. The sudden loss of gas is equivalent to a pulse of laser light, and the system changes from high *Q* to low *Q*. Note, however, that the system operated mainly as a low *Q* device, which only momentarily became a high *Q* device before losing its quality (power).

Once again, this model describes *Q*-switching. The laser medium is kept at low-energy storage, low *Q*, because of a large loss of pump energy until the leak is plugged. Our analogy, while not bad, misses one important point: As pump energy is being lost (i.e., as gas is leaking out), the population inversion is still growing. The population inversion can grow without the occurrence of stimulated emission as long as spontaneous emission is slow compared to the growth of the inversion. Once stimulated emission starts in a system of a very large population inversion, a very large pulse of laser light results, equivalent to the sudden helium leak.

The shutters described above can function to provide the tape patch and the sudden removal of the tape. For *Q*-switching, shutters may be divided into two classes: active or passive. Figure 3-8 illustrates the basic construction of both types of Q-switch shutters. A saturable absorber placed in a laser cavity is a passive shutter in that it has no moving parts and cannot be externally controlled. Its shutter time depends on the relaxation time of the material used to absorb the light. Often organic dyes are used, and of course each dye has a different relaxation time. But after the choice of absorber dye has been made, no other control of the "shutter time" is possible.

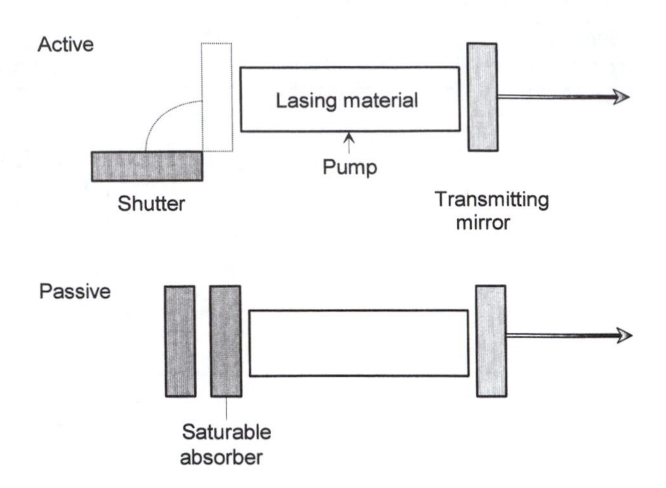


**Figure 3-8**
Active and passive *Q*-switch operation. An *active* shutter can be externally regulated to control the timing of *Q*-switching. A saturable absorber dye placed in the laser cavity accomplishes *passive* *Q*-switching with no external controls.

Active shutters based on the E-O and A-O devices described above can be externally controlled to open and close to create variable pulse durations and frequencies. A rotating mirror placed in the laser cavity can act as another type of active shutter that is strictly mechanical. Whenever the mirror face becomes perpendicular to the laser cavity axis, it reflects light in the cavity and stimulated emission builds up and transmits out the exit mirror. For any other position of the mirror, light is reflected in a nonuseful direction and a condition of low *Q*, no lasing, exists. Rotating mirror switches are simple, but are limited to relatively long pulse durations due to limits on the velocity of rotation of the mirror. The pulse frequency of a rotating mirror shutter is measured in microseconds, while E-O and A-O shutters yield pulse frequencies measured in nanoseconds.

We should note here that there are limits to how fast a *Q*-switch can be, and those limits depend on the rate of spontaneous emission. The rate of spontaneous emission of the laser material must be compatible with the buildup of the population inversion during the low *Q* period of the cycle. Not all lasers can be *Q*-switched because of this rate limitation.

Again, why *Q*-switch a laser? *Q*-switching produces "giant" energy pulses. Typical pulse durations of a *Q*-switched laser are in nanoseconds, and the increase in output energy is easily a hundredfold. Power outputs in the gigawatt (GW) range with durations of 10 ns are not uncommon in *Q*-switched lasers.

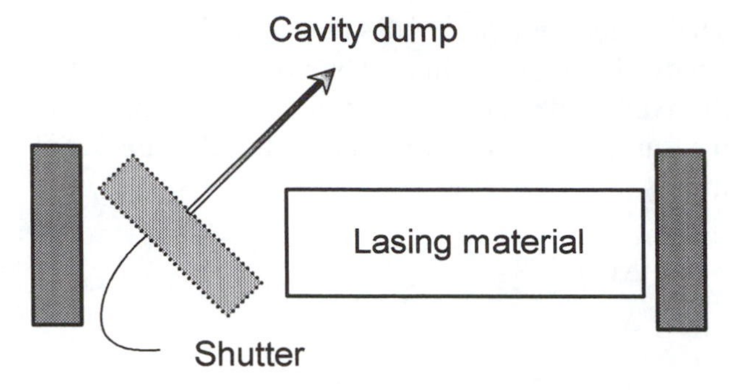


**Figure 3-9**
Configuration for cavity dumping of laser output. Two fully reflective mirrors define the laser cavity and prevent exit of the laser emission from the cavity. A third coupling mirror can be periodically switched into the beam to reflect the light in a different direction and out of the cavity. Laser emission produced in this fashion exits from the laser in a pulsed mode.

### Cavity Dumping

Next let us consider a cavity-dumped laser where the $Q$ is changed from high to low. The balloon laser analogy used to describe $Q$-switching can be extended to the concept of cavity dumping. In this situation we must quickly remove and replace the tape covering the large balloon leak to let out many small gusts of gas. We do not deflate the balloon as we did in $Q$-switching, but rather try to keep it as full as possible. Our analogy suggests that the laser cavity maintains a relatively high $Q$ because it is always lasing, and only periodically are photons allowed to "leak" out. This is illustrated in Figure 3-9. Cavity dumping is accomplished using the shutters we have discussed, particularly the acousto-optic modulator.

Cavity dumping is a convenient way to take a CW argon ion laser and produce short, powerful pulses. We should note that with cavity dumping the average energy of these lasers remains about the same, but the power in each pulse is much larger because of the short pulse duration. The pulse duration of a cavity-dumped laser is approximately the round-trip travel time in the laser cavity. The round-trip time for a 15 cm long laser is about a nanosecond. When we dump all the circulating power, even though the time to dump may be extremely short, we need a finite time for the circulating power to build up again. The time to replenish is also on the order of the nanoseconds required for a round trip of the laser cavity.

## Laser Pulse Modification

Once a laser is modified to produce pulses, there are three common means of further modifying a laser's pulse: mode-locking, synchronous pumping, and pulse compression. These further modifications are used to create even shorter duration pulses or to select just a single pulsed mode. Mode-locking is generally used with ion lasers, synchronous pumping is used with dye lasers. Pulse compression is a means to narrow virtually any pulse duration, but is often employed to create subpicosecond to femtosecond pulses from ring dye lasers.

### Mode-Locking

Let us consider the idea of mode-locking first. This method of pulse production generates high-energy pulses of ultrashort duration. The typical laser has axial modes that we know are spaced in frequency:

$$\Delta\nu_{\text{axial}} = \frac{c}{2L} \tag{44}$$

where $2L$ is the round-trip length for the laser. Through mode-locking we seek to put all of the cavity oscillations in phase and thereby coherently superimpose their electric field amplitudes. Recall that coherence, in particular temporal coherence, is one of the laser's most remarkable features and is the fundamental basis for the success of mode-locking. To see how this works, consider two equally intense light waves in phase. Such waves can add their individual electric field amplitudes, $E$, and, at every point in time, the resultant amplitude would be $E + E$. An important point to remember is that intensity of a light wave is proportional to the square of its electric field amplitude. Thus the intensity of the two superimposed waves is $(E + E)^2 = 4E^2$, or four times the intensity of just one wave. If the two waves were not in phase, the resultant intensity could range from as low as zero to a magnitude greater than that for just one wave alone—but never four times as great. A simple analysis suggests that if $2q + 1$ modes of equal intensity are in phase, then the peak intensity is $(2q + 1)^2$ larger than a single mode intensity. This is $(2q + 1)$ times more power than the sum of the intensities of all the modes oscillating independently.

How does mode-locking generate ultrashort pulses? Without presenting the derivation, the pulse width in time, or the pulse duration of a laser, is approximately the inverse of the frequency distribution of the laser's output. For a laser with $q$ output modes:

$$\text{pulse duration} = \frac{1}{q\Delta\nu} = \frac{1}{q\left(\frac{c}{L}\right)} = \frac{L}{qc} \tag{45}$$

To obtain very short times, we can use either a laser of short length, $L$, or one with a large number of modes, $q$. In practice, the latter technique is used.

Mode-locking is accomplished by opening and closing a shutter, often an acousto-optic modulator. What parameters govern the timing of the shutter operation? If the frequency of the acousto-optic modulator is chosen to be one-half of the axial mode separation frequency (Equation 31), the portion of the first diffracted light wave that returns for a pass through the modulator will only pass through if its frequency is unchanged or equal to initial, $\pm c/2L$. In other words, the frequency of the diffracted light wave must match one of the frequencies of the axial modes existing in the cavity. The result of this process is a train of pulses considerably narrowed in time duration and separated in time by the mode spacing.

For example, suppose we have a laser whose cavity has $L = 60$ cm and which has an output of 100 mJ in $10^{-6}$ s at 600 nm with $\Delta\lambda = 30$ nm. The axial mode spacing is:

$$\Delta\nu_{\text{axial}} = \frac{c}{2L} = \frac{3 \times 10^8 \text{ m/s}}{1.2 \text{ m}}$$

$$= 2.5 \times 10^8 \text{ Hz} \tag{46}$$

The pulse separation time is $1/\Delta\nu_{\text{axial}}$, $2L/c$, or $4 \times 10^{-9}$ s, 4 ns. For a laser whose output bandwidth in terms of wavelength is $\Delta\lambda$, the frequency bandwidth is:

$$\nu = \frac{c}{\lambda}$$

$$\Delta\nu_{\text{band}} = c\left(\frac{-1}{\lambda^2}\right)\Delta\lambda_{\text{band}} \tag{47}$$

$$|\Delta\nu_{\text{band}}| = c\left(\frac{\Delta\lambda_{\text{band}}}{\lambda^2}\right)$$

The number of axial modes that fit in the frequency bandwidth can be estimated by dividing the width of the axial mode into the width of the bandwidth, both measured in terms of frequency. Thus the number of modes is:

$$q = \frac{|\Delta\nu_{\text{band}}|}{\Delta\nu_{\text{axial}}}$$

$$= \frac{3 \times 10^8 \text{ m s}^{-1} 3 \times 10^{-8} \text{ m}}{(6 \times 10^{-7} \text{ m})^2 2.5 \times 10^8 \text{ s}^{-1}} = 10^5 \tag{48}$$

This requires a pulse duration of:

$$\text{pulse duration} = \frac{T}{q} = \frac{4 \times 10^{-9} \text{ s}}{10^5}$$

$$= 4 \times 10^{-14} \text{ s} = 40 \text{ fs} \tag{49}$$

This mode-locked laser has a pulse width of only 40 fs in duration—an enormous narrowing. In general, mode-locked lasers typically yield picosecond pulses, though newer solid-state lasers now achieve 100 fs pulses routinely.

If we look at the power in this pulse, we will find it to be gigantic because of the short duration time:

$$\frac{\text{energy}}{\text{pulse}} = \frac{\text{energy output in } 10^{-6} \text{ s}}{\#\text{ of pulses in } 10^{-6} \text{ s}}$$

$$\#\text{ of pulses} = \frac{10^{-6} \text{ s}}{\text{pulse separation time}}$$

$$= \frac{10^{-6} \text{ s}}{4 \times 10^{-9} \text{ s per pulse}} = 250 \text{ pulses} \tag{50}$$

$$\frac{\text{energy}}{\text{pulse}} = \frac{100 \text{ mJ}}{250 \text{ pulses}} = 4 \times 10^{-4} \text{ J/pulse}$$

$$\frac{\text{power}}{\text{pulse}} = \frac{\text{energy/pulse}}{\text{pulse duration}}$$

$$= \frac{4 \times 10^{-4} \text{ J/pulse}}{40 \times 10^{-15} \text{ s}} = 10 \text{ GW/pulse}$$

This is an immense power, of course, but only for a short duration time. The chemistry of materials subjected to such powers and probed for such short times is beginning to be an incredible story.

## Synchronous Pumping

Another pulse modification technique used in conjunction with dye lasers is synchronous pumping. Here a mode-locked laser is used to pump a dye laser whose output pulses are significantly shorter than the already short mode-locked output pulse. The lasing cavity of the dye laser is precisely adjusted to have an axial mode spacing that is an integral multiple of the mode-locked pulse separation frequency. In other words, the dye laser pulse spacing (or the round-trip time of the dye laser) is a multiple of the mode-locked laser cavity frequency $c/2L$. Mode-locked laser pulses raise the dye molecules in the dye stream to excited states to generate a population inversion for lasing. A laser pulse propagating in the dye laser cavity has just the right timing to traverse the cavity and arrive back at the dye stream just as the mode-locked pump laser initiates lasing in the fresh dye. The travelling dye pulse quickly stimulates the emission of the dye lasing line(s), depleting the population inversion. In other words, the mode-locked laser excites the dye molecules, but before they can lase on their own accord, a second laser pulse of exactly the dye laser line stimulates the excited population to lase faster than "normal." Thus, the output dye laser's pulse is very short, typically a few picoseconds. The synchronously pumped dye laser was one of the first short-pulse lasers and still is a workhorse laser in time-resolved spectroscopy.

With our discussion of pulse production techniques concluded, you might wish to review the summary of these techniques presented in Table 3-2, which gives typical output powers, pulse widths, and pulse repetition rates.

## Pulse Compression

Interest in chemical phenomena on very short time scales, 10 to 100 femtoseconds, has led to ways of generating pulses in the picosecond to femtosecond regimes. One of these methods is called *pulse compression* and is schematically illustrated in Figure 3-10. Typically a synchronously pumped dye laser with its inherent short pulse output is the starting point for pulse compression. Recall that a dye laser has a relatively broad frequency or, as it will be convenient for this discussion, output wavelength distribution. For this modification technique, the dye laser

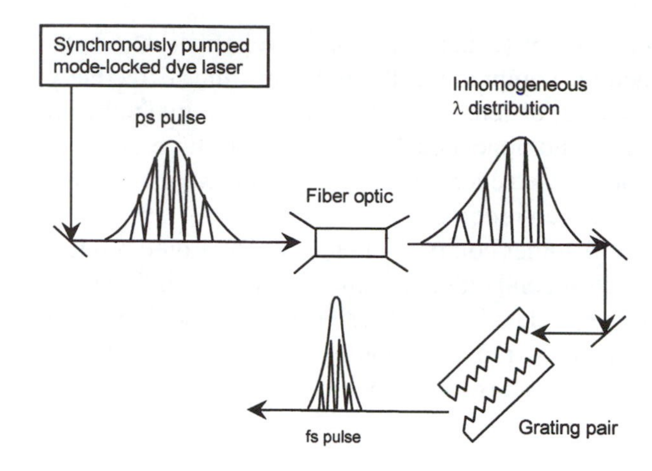


**Figure 3-10**
Pulse compression. The short pulse output of a synchronously pumped mode-locked laser is passed through an optical fiber composed of glass. The pulse, which consists of a small but finite distribution of wavelengths (or frequencies), is broadened inhomogeneously as a consequence of the wavelength dependence of the refractive index of glass. Diffraction gratings are then arranged to increase the optical path length of those wavelengths of light that arrive at the gratings first and to decrease the optical path length for the later-arriving wavelengths of light. This action results in a compression of the pulse in time to generate pulses on even the femtosecond timescale.

output is passed through a short length of optical fibers that broaden the pulse inhomogeneously in wavelength (or frequency). The resultant pulse is often referred to as a *chirped pulse* or *signal*. The optical fiber is typically glass, whose refractive index varies with wavelength. Thus the wavelengths contained in the initial pulse will travel through the fiber at different velocities and take different times to traverse the same fixed distance. The wavelengths whose refractive indices in the medium are the smallest will traverse the fiber first, and the resultant chirped pulse is an inhomogeneous distribution of wavelengths separated in time. Perhaps the really clever aspect of this pulse compression scheme occurs next. When this inhomogeneous distribution is incident on one face of two parallel diffraction gratings, compression of the pulse in time occurs. The compression results from a longer path length for the wavelengths that arrive first at the grating and shorter path lengths for those arriving later. How does the time compression result? The grating pair provides a wavelength-dependent optical path. By adjusting the gratings in orientation, the path lengths can be set to make all the travel times closer together, compressing the pulse in time. Effectively, then, the different wavelengths of the initial pulse are first inhomogeneously spread out in time and then compressed even more closely in time. Pulses of 6 fs have been reported.

Another common but now older way to produce femtosecond pulses is with a CPM, a colliding pulse mode-locked laser. We refer the interested reader to Chapter 5 or to the literature for a more extensive discussion of this technique.

## Nonlinear Optical Elements

The last topic in our discussion of modifying the laser's output is the use of nonlinear optical effects to modify laser light both in time and energy.

### Continuum Generation

One of the most important features of a laser is its high monochromaticity. This feature is both a blessing and a curse. Can highly monochromatic light of any desired wavelength be produced with a single laser? Unfortunately, no single laser easily yields such output with a wide energy range. However, a method known as *continuum generation* is routinely used to obtain pulsed light of high and specific energy from a single laser. This approach is based on the observation that very short pulses of very intense laser light (of any wavelength) incident on water, carbon tetrachloride, and other simple liquids, generate white light, that is, a continuum of energies. Highly energetic and monochromatic light can then be selected from the white light, usually through the use of filters. It is somewhat ironic that after investing in the laser to obtain monochromatic light, it is often necessary to return to the full spectrum of white light to obtain desired emission wavelengths.

### Harmonic Generation

Today you are almost certain to hear in any discussion of lasers the terms *frequency doubling* or *second-harmonic generation*. Frequency-doubling materials such as potassium dihydrogen phosphate (KDP) and lithium niobate are often used with specific wavelength lasers to provide additional laser lines of high energy. In particular, frequency doublers are extremely common in use with YAG lasers to generate visible and ultraviolet light from the YAG's inherent IR lasing lines. Frequency doubling and second-harmonic generation (SHG) are just two of the many techniques now possible under the general heading of *harmonic generation*.

Harmonic generation depends on nonlinear responses of a medium to a stimulus. In the case of lasers, the stimulus is the very intense electric field present in a very high intensity light wave. We have noted that light can be thought of as an oscillating dipole such as that illustrated in Figure 3-1. When the molecules of a substance respond in direct proportion to the magnitude of the electric field of the light wave, the response is said to be *linear*. The response of the medium is the polarization created in the medium by the applied electric field of the light. This response can be quantified by:

$$P = \varepsilon_o \chi^{(1)} E \tag{51}$$

According to Equation 51, $P$ should be viewed as the magnitude of the dipoles induced in the medium per unit volume and $\varepsilon_o \chi$ can be viewed as the polarizability of the medium (or molecules comprising the medium). In fact, the right-hand side of Equation 51 is just the first, or *linear* term in Equation 52, a series expansion in powers of the electric field:

$$\mathrm{P} = \varepsilon_o \, [\chi^{(1)} E + \chi^{(2)} E^2 + \chi^{(3)} E^3 + \ldots] \tag{52}$$

Nonlinear effects come into play when the polarization of a medium is not adequately described by the linear term of Equation 52 (that is, Equation 51) but must involve second-order (quadratic) and often higher-order terms as written in Equation 52. What does this mean physically? At high electric field strengths, the dipoles induced in the medium are not only in the same direction as the plane of incident polarization but develop dipole components in other directions. These additional dipole components generate light whose polarization is different from the incident polarization. Furthermore, since the velocity of light can vary with direction in a solid, the medium can generate within itself light of different polarizations and velocities. These internally-generated light waves can interact constructively and destructively to yield new light waves whose frequencies are sums and differences of the initial and generated frequencies.

A few additional comments on harmonic generation are of particular note to chemists. First, the harmonically generated light described above can only be created in noncentrosymmetric crystals, that is, crystals without a center of symmetry. We could also describe such crystals as isotropic. For those who know some group theory, only crystals described by point groups that do not contain the inversion symmetry element can support harmonic light generation. Second, the simplest assumption is to consider only the linear and quadratic terms in Equation 52. With this assumption, we note without proof that the only frequency that is generated by the quadratic field is twice the initial frequency. This observation explains why the phrases *frequency doubling* or *second-harmonic generation* are often applied to nonlinear materials.

We have noted that only noncentrosymmetric crystals can generate harmonics. A few of the crystals capable of harmonic generation are listed in Table 3-4. Harmonic generation has found ever-increasing applications in conjunction with lasers. Indeed, the development of lasers was vital to the development of nonlinear materials, since, before the advent of lasers, no light sources were sufficiently intense to induce harmonic generation. Now the development of nonlinear optical devices and of lasers goes hand in hand.

**Table 3-4**
Common Nonlinear Crystals

| *Material* | *Common Name or Acronym* | *Wavelength Range/μm* |
|---|---|---|
| $KH_2PO_4$ | KDP | 0.35 – 4.0 |
| $\beta$-$BaB_2O_4$ | BBO | 0.19 – 2.6 |
| $LiB_3O_5$ | LBO | 0.16 – 2.6 |

## Optical Parametric Oscillator

The search for a laser whose output can be tuned to a desired frequency continues. For example, no single laser has achieved a tuning range across the entire visible spectrum. Perhaps it will prove impossible to develop a truly broadband laser, but the goal of developing such a light source has been a powerful motivator for the discovery of new lasers and ways to modify the output of existing lasers. The optical parametric oscillator (OPO) is one such inspired modification.

An OPO is not itself a laser. When driven by a laser, such a device will output tunable coherent light by nonlinear optical processes. We can qualitatively understand how OPOs work by noting that the nonlinear effects in the oscillator must satisfy the energy conservation statement:

$$\nu_{\text{pump}} = \nu_{\text{signal}} + \nu_{\text{idler}} \qquad (53)$$

An actual OPO consists of a crystal capable of nonlinear optical responses that is placed inside a cavity with mirrors forming an oscillator. A high-intensity laser of frequency $\nu_{\text{pump}}$ is focused on the crystal. An idler or second frequency could be provided by a second laser and is also focused on the crystal. More commonly, the idler frequency is just the noise in the crystal itself that arises from the vibrations of the crystal lattice. The idler light wave mixes with the second-order dipole moments created by the intense pump light wave to yield what is called the signal light wave of frequency $\nu_{\text{signal}}$. For this interaction to produce an intense signal light wave, two key conditions have to be met. One condition that we have already noted is the energy-matching expression of Equation 53. A second condition that is necessary for the signal wave to represent constructive interference involves a matching of the phase of the idler wave with the phase of the pump wave. Recall that harmonically generated light can travel in the crystal with different speeds. Depending on the relative speeds, parallel waves can be in or out of phase. Phase matching ensures constructive interference and the production of an intense signal wave. Phase matching also allows the frequency of the signal wave to be tuned. Several factors which will only be mentioned here contribute to the phase matching and the tuning range of the OPO: the crystal used, the temperature of the crystal (as temperature affects the lattice vibrational frequencies), and the angle of the incident pump light wave polarization with respect to the symmetry axes of the noncentrosymmetric crystal. Typical tuning ranges for some of the more common OPO crystals are included in Table 3-4. The future impact of OPOs and devices derived from nonlinear optical properties cannot be fully imagined at this time.

## SUMMARY

A laser alone is a simple light source. When combined with the wide variety of devices designed to modify its output, the laser becomes an incredible light source whose utilization continues to promote new discoveries.

## SUGGESTED EXPERIMENT

A suggested reference to an experiment that illustrates some of the principles described in this chapter:

Wirth, F. H., "Dye Laser Experiments for the Undergraduate Laboratory; Experiment 5—Frequency Doubling," Laser Science, Inc., Cambridge, MA. An experiment to illustrate the use of ammonium dihydrogen phosphate crystals to generate the second harmonic of visible light (630 nm) in the ultraviolet region (315 nm).

## GENERAL REFERENCES

Andrews, D. L., *Lasers in Chemistry*, Second Edition, Springer-Verlag, Berlin, 1990.

Beach, D. P., Shotwell, A., and Essue, P., *Applications of Lasers and Laser Systems*, PTR Prentice Hall, Upper Saddle River, NJ, 1993.

Demtroder, W., *Laser Spectroscopy: Basic Concepts and Instrumentation*, 2nd Ed., Springer-Verlag, Berlin, 1996.

Halliday, D., and Resnick, R., *Physics*, Part II, John Wiley & Sons, New York, 1968.

Hecht, J., *The Laser Guidebook*, Second Edition, McGraw-Hill, New York, 1992.

Higgins, T. V., "Nonlinear crystals: where the colors of the rainbow begin," *Laser Focus World*, 1992, 28(1), 125–133.

Higgins, T. V., "Nonlinear optical effects are revolutionizing electro-optics," *Laser Focus World*, 1994, 30(8), 67–74.

Higgins, T. V., "Optical modulation controls the properties of light," *Laser Focus World*, 1995, 31(1), 83–88.

Jenkins, F. A., and White, H. E., *Fundamentals of Optics*, Third Edition, McGraw-Hill, New York, 1957.

Myers, A. B., and Rizzo, T. R., Eds., *Laser Techniques in Chemistry*, John Wiley & Sons, New York, 1995.

Negus, D., and Seaton, C., "Ease of use and versatility mark modelocked Ti:sapphire design," *Laser Focus World*, 1992, 28(2), 69–72.

O'Shea, D. C., Callen, W. R., and Rhodes, W. T., *Introduction to Lasers and Their Applications*, Addison-Wesley Publishing Company, Inc., Reading, MA, 1977.

Saeed, M., DiMauro, L. F., and Tornegard, S., "Amplifier pumps enhance ultrafast studies," *Laser Focus World*, 1991, 27(2), 57–70.

Svanberg, S., *Atomic and Molecular Spectroscopy: Basic Aspects and Practical Applications*, 2nd Ed., Springer-Verlag, Berlin, 1992.

CHAPTER 4

# Details of Several Common Lasers

## Chapter Overview

Laser sources are commonly classified according to the nature of the lasing material in which a population inversion is established. The active medium ranges from simple monatomic gases to solid-state materials containing transition metal ions embedded in a host lattice (either an ionic crystal or an amorphous glass). In fact, lasing action has been achieved in thousands of substances and in nearly every state of matter. Six major categories of lasers are commonly identified: monatomic gas lasers, molecular gas lasers, metal vapor lasers, solid-state lasers, semiconductor lasers, and dye lasers. Within these classifications, several specific laser systems have gained scientific and industrial importance. In this chapter we explore the characteristics and principal uses of these laser sources.

## Common Lasers

A recent survey [1] of the industrial and scientific laser market reported that gas lasers—both monatomic and molecular gas lasers—represent almost 60% of the total global laser market. The workhorse of the gas laser category is clearly the $CO_2$ laser, representing almost 45% of the market, with 98% of the use representing industrial and medical applications. Helium-neon, helium-cadmium, nitrogen, ion, and excimer-inert gas halide lasers are other gas laser systems with established niches in the laser community. Solid-state lasers (e.g., titanium:sapphire and Nd:YAG lasers) and diode lasers are the next most common lasers, contributing about 25% and 15% of the laser market, respectively. Dye lasers represent one segment of the remaining commercially available laser systems. To understand the origin of the vast array of laser tools and the specific applications for which each system is well-suited, we now describe some details and characteristics of several of the more commonly used lasers. Specifically we discuss the lasers listed in Table 4-1.

**Table 4-1**
Common Lasers

| | |
|---|---|
| Helium–Neon | Nd:YAG |
| Argon–Ion | Ti:Sapphire |
| Excimer | Diode |
| Carbon Dioxide | Dye |
| Nitrogen | |

## Helium-Neon Laser

The helium-neon laser was the first gas laser to be demonstrated and commercially developed. Ali Janvan, Donald Herriott, and William Bennett first constructed the helium-neon laser in 1961 at the AT&T Bell Laboratories in Murray Hill, NJ. As the name suggests, the active medium of this gas laser is composed of a mixture of the monatomic gases helium and neon in a variable and usually proprietary ratio of 5–12 (He) to 1 (Ne) [2]. The total pressure of the gas mixture is generally on the order of a fraction of a torr to several torr. The transition responsible for stimulated emission occurs in neon, with helium serving to promote lasing action. While electrical discharge of pure neon leads to the production of over 130 lasing lines, a more efficient excitation of neon via collisional activation with excited helium achieves substantially enhanced output. It is the combination of an easily excited atom, the two-electron helium atom, with an atom of more extensive energy levels that enables the laser to operate efficiently and simply. The principles governing laser emission are described below.

The energy level diagram in Figure 4-1 illustrates the transitions involved in the helium-neon laser. This is basically a four-level laser. The ground-state $1s^2$ electron configuration of helium is promoted via electric discharge to both metastable $1s^1 2s^1$ states, the singlet $^1S$, and the lower-energy triplet $^3S$. The mole ratio of helium and neon increases the probability of a collision between a metastable helium atom and a ground-state neon atom. Such a collision is desirable because of the coincident matching of the energy of the metastable helium atom and the energy for the $1s^2 2p^6 \rightarrow 1s^2 2p^5 4s^1$ and $1s^2 2p^6 \rightarrow 1s^2 2p^5 5s^1$ transitions in neon. The population of these highly excited levels of neon, with unpopulated intermediate energy states, is the basis for lasing action. The complicated distribution of neon excited states leads to numerous lasing lines. The strongest transitions are the $2p^5 5s^1 \rightarrow 2p^5 3p^1$ decay to yield visible radiation at 632.8 nm (but also at 629, 635, and 640.1 nm) and the $2p^5 5s^1 \rightarrow 2p^5 4p^1$ and $2p^5 4s^1 \rightarrow 2p^5 3p^1$ decays to generate

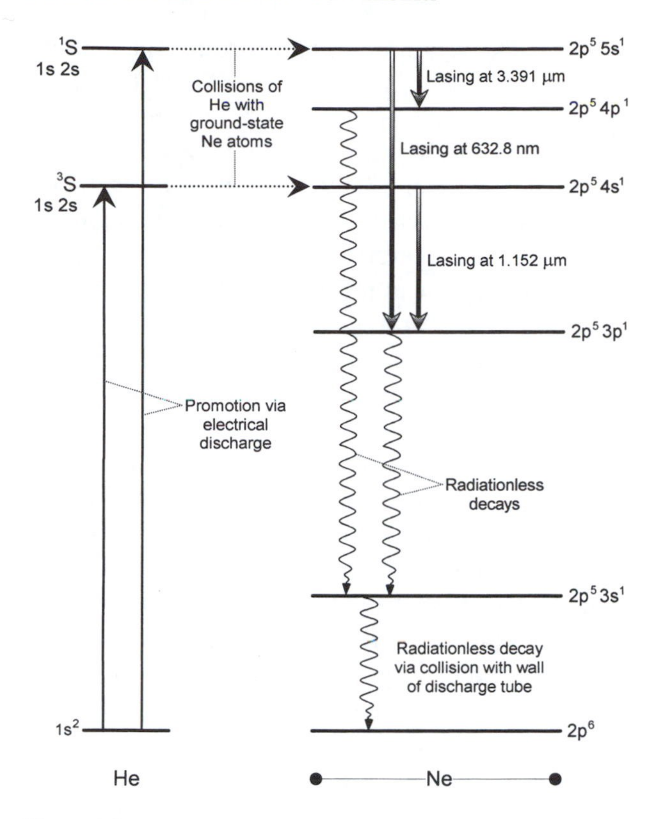


**Figure 4-1**
Energy level diagram for a helium-neon laser, a four-level laser system. Ground-state helium atoms are promoted to singlet and triplet excited states via electrical discharge. Collisional energy transfer of the excited-state He atoms to ground-state neon atoms populates excited states above the lowest excited state for neon, generating a population inversion. Lasing emission occurs at three wavelengths: 632.8 nm, 1.152 $\mu$m, and 3.391 $\mu$m. Two radiationless decay steps, the latter via collisional deactivation with the walls of the discharge tube, return neon to the ground state.

infrared emission at 3.391 $\mu$m and 1.152 $\mu$m, respectively. Weaker transitions produce 543.5 nm (green), 594.1 nm (yellow), and 604 and 611.9 nm (orange) radiation. Optics selection and cavity tuning dictate the lasing line that dominates.

Two-step radiationless decays (either $2p^54p^1 \rightarrow 2p^53s^1 \rightarrow 2p^6$ or $2p^53p^1 \rightarrow 2p^53s^1 \rightarrow 2p^6$) follow lasing to repopulate the ground-state electron configuration. In either pathway, the efficiency of the $2p^53s^1 \rightarrow 2p^6$ transition is enhanced by collisions of the excited neon atoms with the walls of the laser cavity. The population inversion, and hence the laser gain, is enhanced by an efficient radiationless depopulation step. The relatively longer lifetime of the $2p^55s^1$ and $2p^54s^1$ states compared with the lifetimes of the lower-energy $2p^53p^1$ and $2p^54p^1$ states further enhances the population inversion. A resonant collision is said to occur when an efficient energy transfer between excited He and ground-state Ne leads to the production of the $2p^55s^1$ or $2p^54s^1$ excited states of Ne and the ground-state configuration of He.

The simple construction of a helium-neon laser includes a discharge tube containing the gas mixture, a power supply for electrical discharge, and two mirrors cemented to the ends of the discharge tube defining the optical cavity. This prealignment of the mirrors provides the user with easy operation. The overall cross section of the discharge tube is limited (generally around 3 mm in diameter) to provide sufficient contact of the excited neon atoms with the wall's surface to maintain the necessary population inversion for lasing action. Helium-neon lasers operate continuously and deliver power in the 1-5 milliwatt range, depending on the wavelength of emission. The low power of these lasers is balanced by their relatively low cost, simplicity of design, and low degree of hazardousness. Developments using the green 543.5 nm line are especially promising because of the low cost of this emission relative to that of other green-emitting lasers. Similar operating principles govern the helium-cadmium laser, with cadmium transitions at 442 nm in the blue and at 325 nm in the ultraviolet.

Numerous applications have made use of the relatively inexpensive nature, good beam quality, and continuous output of He/Ne lasers. These attributes are useful in optical scanning devices, in printing applications, and in character recognition equipment. The long coherence lengths of He/Ne lasers, typically on the order of 20-30 cm, make them especially desirable for optical alignment. With special techniques to restrict oscillation to a single cavity mode, coherence lengths can even be extended to kilometers.

## Argon Ion Laser

The argon laser is the most common example of a gas laser composed of a single inert monatomic gas as the lasing medium. As the name indicates, the argon ion, and not the neutral atom, is required to sustain a population inversion of stimulated emission. A sealed tube of argon gas with a precisely controlled pressure of approximately 1 torr is critical to the operation of the laser. The design features of this plasma tube are detailed in Figure 4-2. Ionization of neutral argon atoms is achieved by applying a continuous and controlled discharge through the gas medium. Depending on the current, ground-state and/or excited-state $Ar^+$ and $Ar^{2+}$ ions are formed upon electrical discharge. The excited states generally responsible for lasing action are those arising from two successive collisions of a ground-state ion with electrons (i.e., ground-state $Ar^+ + e^- \rightarrow$ metastable $Ar^+$; metastable $Ar^+ + e^- \rightarrow$ upper-excited-state $Ar^+$). To minimize deactivating collisions of argon ions and electrons with the walls of the plasma tube, a magnetic field is applied to the laser tube to keep the laser plasma in the center of the tube.

Laser emission is observed for both singly- and doubly-charged argon ions at a number of discrete wavelengths over the interval between 330 and 530 nm. For the $Ar^+$ ion, the primary lasing lines occur at 488.0 nm and 514.5 nm. The stimulated emissions are produced as the $3s^23p^44p^1$ excited state radiates to reach the $3s^23p^44s^1$ state of $Ar^+$ which then reaches the $3s^23p^5$ state via a spontaneous emission process at 74 nm. Neutral argon atoms are regenerated by electron capture from the $Ar^+$ $3s^23p^5$ state. These lines are illustrated in Figure 4-2. For the $Ar^{2+}$ ion, lasing transitions in the near-ultraviolet are observed at 334.0, 351.1, and 363.8 nm. A high-resolution *etalon* (see the Glossary) is used to select a single frequency for laser operation.

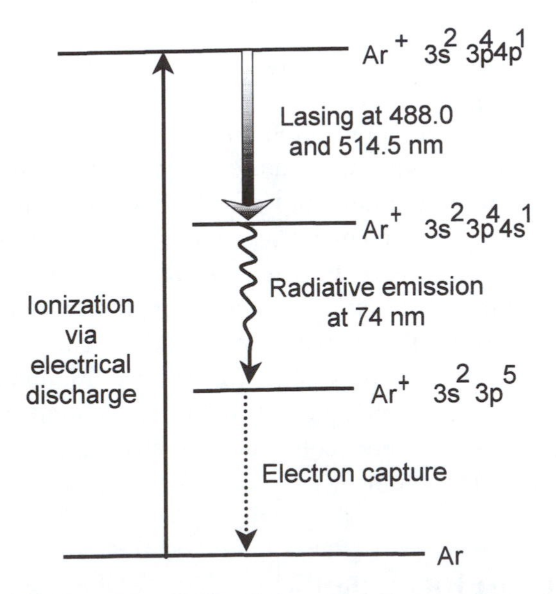


**Figure 4-2**
The origin of argon ion lasing lines. Ground-state neutral argon atoms with filled electron configurations are ionized via electrical discharge. Subsequent electron impact generates excited-state lasing $Ar^+$ ions via two successive collisions of $Ar^+$ with electrons. Stimulated emission to yield 488.0 nm and 514.5 nm lasing emission is followed by spontaneous emission at 74 nm and electron capture to regenerate the ground-state neutral argon atoms.

The design of argon ion lasers must consider several critical features. For example, proper gas pressure must be maintained to sustain lasing power, because gas atoms are trapped by the plasma tube under the hostile conditions of laser operation. A careful choice of plasma tube material is thus required to minimize adsorption of argon on the tube wall. A tube material of low porosity, typically quartz, graphite, or beryllium oxide, will not only reduce interactions of the tube with argon, but also reduce trapping of contaminants that will outgas to contaminate the argon. The bore material must also have a low vapor pressure and be highly pure, so as not to contain extraneous materials of high vapor pressure. At typical operating temperatures within the tube (e.g., 120°C), BeO has a vapor pressure less than $10^{-11}$ torr and thus is a suitable bore material.

Argon atoms, like all noble gases, possess filled electron configurations which require high energies (and therefore high-current discharges) to ionize the atoms. As a consequence, the plasma tube must also be designed to dissipate the large amounts of heat generated during electrical discharge of the neutral gas atoms. High-power argon ion lasers utilize water for cooling. To efficiently transfer heat from the plasma tube to the cooling water, a bore material of high thermal conductivity is essential. Uniform and efficient cooling along the length of an ion laser tube contributes to long-term laser power stability. Recent technological developments have resulted in air-cooled argon ion laser devices which necessarily have a lower output power, but are simpler to install and more convenient to operate. While forced-air cooling eliminates the need for plumbing, vibrations may be induced by the cooling fan, diminishing the quality and stability of the laser output.

Output powers from milliwatt to watt levels for single-line UV operation are obtainable in visible CW argon ion lasers. A number of design features enhance the frequency stability and output power that can be achieved when using an etalon for single-frequency operation. The choice of etalon material (particularly with regard to the magnitude of the thermal expansion coefficient), the desired wavelength of lasing, and the presence of a temperature-controlled etalon housing all affect etalon performance.

The krypton ion laser operates in a similar fashion to the argon ion laser. A lower efficiency of lasing limits maximum output power to about 5 watts, but the energy levels for excited-state krypton ions generate intense emission in the red at 647.1 nm, as well as other significant lines across a broad visible spectrum in the blue, green, and yellow. The similar principles of operation for these two lasers have led to the use of mixtures of argon and krypton in a single laser cavity (a *mixed-gas laser*) to generate strong lasing lines across the entire visible spectrum.

Both argon ion and krypton ion lasers are popular laser choices for numerous spectroscopic applications where high power is needed. Because of their high-power capabilities, ion lasers are also commonly used to pump dye lasers. Modifications of the laser output, such as mode-locking and cavity dumping, are frequently used to produce trains of short (e.g., picosecond) pulses with enhanced peak power for pumping operations. Medical applications of argon ion lasers take advantage of the intense emission lines in the blue-green region of the visible spectrum for studies involving the hemoglobin in red blood cells and for investigations of cell structure using fluorescent dyes that absorb in the visible region. The broad laser emission across the visible spectrum also makes argon ion and krypton ion lasers ideal sources for the display and entertainment industry when a balanced output of primary colors is desired. Disk mastering for CD-ROM production currently uses the 458 nm output of argon ion lasers and the 413 nm emission of krypton ion lasers. In this application, the shorter the wavelength of laser emission, the smaller the spot size on the CD, and hence the larger the amount of data that can be written on the disk.

## Excimer Lasers

Excimer lasers form diatomic species in excited states as the active medium. The diatomic species which form by chemical reaction only exist in the excited state through weak attractive forces between two atoms which do not form a chemical bond in the ground state. The term *excimer* technically applies to homonuclear *exc*ited-state d*imers* only, with the term *exciplex* reserved for heteronuclear *exc*ited-state com*plexes*. However, in practice, the term excimer is used for both types of excited-state species. What kinds of substances do not dimerize or form compounds in the ground state, yet may form short-lived dimers or complexes with sufficient energy? The noble gases, which are relatively inert in their ground state as a result of their filled electron configurations, are ideal candidates for excimers. Typically, the active medium in an excimer laser is an excited-state rare gas dimer, such as $Ar_2^*$, $Kr_2^*$, or $Xe_2^*$, or a rare gas oxide, fluoride, or chloride (e.g., $ArO^*$, $ArF^*$, $KrF^*$, $KrCl^*$, $XeF^*$, or $XeCl^*$) generated by an electrical discharge through a mixture of the rare gas and either oxygen, fluorine, or chlorine gas. These excimer species are ideal lasing media, as a population inversion is guaranteed by the absence of a populated ground state. Figure 4-3 emphasizes this point by presenting the molecular potential energy curves for an excimer and its ground-state species.

Commercial excimer lasers use separate gas supplies for large systems, but often premix gases for smaller lasers. An inert buffer gas, such as helium or neon, is present at very high levels (90% to 99% of the total pressure) to mediate energy transfer. The rare gas is present at levels of 1% to 9% of the total pressure, with oxygen, fluorine, or chlorine gas comprising the rest. An electrical discharge is used to initiate a chain of reactions that produce the metastable excimer. Since the excimer is the upper level of the laser transition, the excimer laser is therefore an example of a two-level laser with exceptionally high efficiency.

A variety of ultraviolet laser lines are available from excimer lasers, commonly 193 nm for ArF, 222 nm for KrCl, 248 nm for KrF, 308 nm for XeCl, and 351 nm and 353 nm for XeF. Appendix II provides a more complete listing of available excimer laser wavelengths. Average powers range from 1 to 100 W. The mechanism for lasing ensures that excimer lasers are pulsed devices with pulse durations on the nanosecond timescale. Megawatts of peak power can be attained from the high individual pulse energies (often several joules), making excimer lasers the most efficient and powerful UV laser sources available.[1] The major limitation to the operation of excimer lasers is the cost of replenishing the

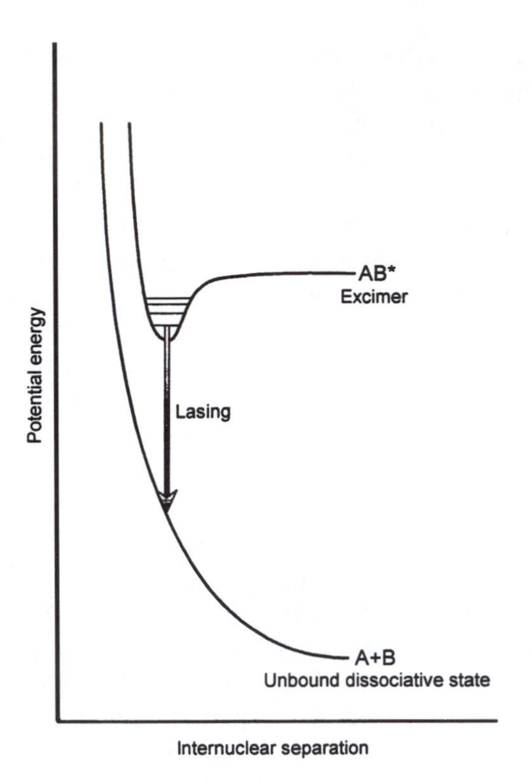


**Figure 4-3**
The potential energy curves for an excited-state dimer or excimer $AB^*$ and its associated ground-state species, $A + B$, to illustrate the origin of the excimer lasing lines. A population inversion always exists between the unpopulated ground-state dimer and the higher-energy bound excimer state. An excimer laser, thus, constitutes an efficient two-level laser system.

gases in the laser cavity and the general noxiousness of the gases. However, excimer laser cavities are designed to be repeatedly refilled and are sealed to accommodate both gas consumption and degradation during operation. Optical components must be designed to resist corrosion by chlorine or fluorine, typically requiring Teflon coatings and/or solid nickel, brass, or ceramic construction.

Excimer lasers have found numerous uses in industry, in chemical research, and in medical applications because of the characteristic high power and ultraviolet nature of the emission. Applications requiring vaporization (photoablation) or the breaking of chemical bonds often employ excimer lasers for their power and UV energy. *L*aser *i*nduced *d*etection *a*nd *r*anging (LIDAR) systems also incorporate excimer lasers for remote sensing applications needing high power to drive low-probability events. Applications of LIDAR systems will be discussed as a case study in Chapter 7.

## Carbon Dioxide Laser

The carbon dioxide laser is a molecular gas laser with infrared emission that arises from lasing transitions between vibrational-rotational levels in the $CO_2$ molecule. As a linear triatomic molecule, $CO_2$ exhibits four fundamental vibrational modes or motions. These fundamental modes have three different frequencies, each of which gives rise to a series of allowed quantum vibrational-rotational states. Figure 4-4 summarizes the key features of the energetics of the $CO_2$ laser. The vibrational mode known as the *symmetric stretch* of the C=O bonds occurs through a concerted in-plane movement of the oxygen atoms in opposite directions. An *asymmetric stretch* of the C=O bonds is an in-plane motion of the oxygen atoms in the same direction to similar or different extents. The *bending mode* moves the oxygen atoms to compress the O–C–O bond angle. This bending mode is doubly degenerate as it involves two equivalent motions. One motion involves the oxygen atoms moving in the O–C–O plane to accomplish the compression; the other motion involves the oxygen atoms performing this same movement, but in a plane perpendicular to the initial molecular plane. The overall vibrational motion of a $CO_2$ molecule is described by a linear combination of the four fundamental vibrational modes. The associated energy state of the $CO_2$ molecule is written as *(ijk)*, where $i$, $j$, and $k$ are positive integers or zero, reflecting the number of energy quanta associated with each fundamental mode. The symmetric stretch is designated by $i$, the bending mode by $j$, and the asymmetric stretch by $k$. The three pure first vibrational excited states, i.e, (100), (010), and (001), are higher in energy than the ground-state level by 1340 $cm^{-1}$, 667 $cm^{-1}$, and 2349 $cm^{-1}$, respectively. Note that these energy differences are considerably less than between electronic levels. In each vibrational state, the $CO_2$ molecule is also associated with a rotational state arising from rotation about the carbon center of mass. As a consequence of the smaller energies for rotation than for vibration, each vibrational energy level is split into a number of rotational levels.

Within the laser cavity, $CO_2$ molecules are not promoted directly to higher vibrational levels to initiate lasing action. Instead, the active medium of a $CO_2$ laser also contains $N_2$ gas and an inert gas, generally He, at typical levels of 4 and 5 times, respec-

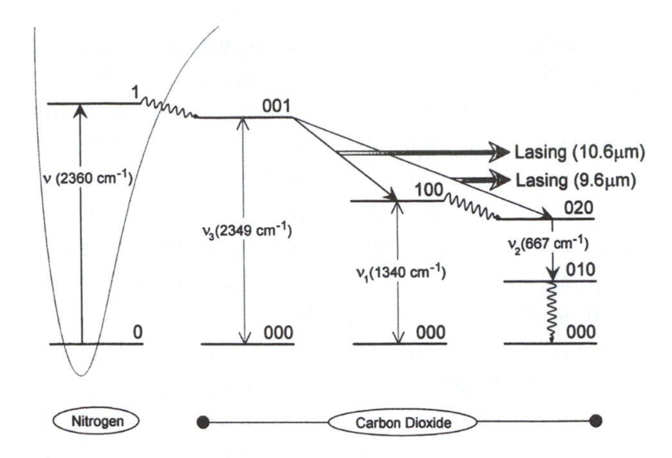


**Figure 4-4**
Energetics of the carbon dioxide laser. Vibrational levels of nitrogen and $CO_2$ are presented with the notation *ijk* denoting the number of energy quanta associated with the $CO_2$ symmetric stretching, bending, and asymmetric stretching modes, respectively. The rotational structure of the vibrational bands is not presented. Nitrogen molecules within the laser cavity are promoted to the first vibrational excited state via electron impact. The energy of this state matches the energy of the 001 vibrationally excited state of $CO_2$, permitting resonant energy transfer to occur when $N_2$ and $CO_2$ molecules in these states collide. Two lasing transitions, principally at 9.6 $\mu$m and 10.6 $\mu$m, populate rotational levels of the 100 and 020 vibrational states. Radiative and collisional decay repopulate the ground vibrational states of $CO_2$.

tively, that of the $CO_2$. Here nitrogen plays an analogous role to helium in the He/Ne laser. Upon electrical discharge, $N_2$ is raised to its first vibrationally excited state. Subsequent relaxation to the vibrational ground state is forbidden by normal spectroscopic selection rules. However, resonant energy transfer to a $CO_2$ molecule can occur through collision because of the similarity in energy levels between the (001) vibrational mode of $CO_2$ and a vibrationally excited $N_2$ molecule. Efficient energy transfer populates numerous rotational levels of the (001) vibrational state. Lasing transitions occur to rotational levels of two excited vibrational states, the (100) level associated with symmetric stretching and the (020) level arising from a bending mode. These levels are not populated via energy transfer from $N_2$ (as a consequence of inconsistent matching of energy levels), and thus a population inversion exists. Laser emission, principally at 10.6 $\mu$m and 9.6 $\mu$m, respectively, is observed for these pathways. However, the closely spaced rotational levels of the (001), (100), and (020) vibrational levels leads to as many as 50 or more emission lines to comprise emission bands between 900 and 1000 $\mu$m and between 1000 and 1100 $\mu$m. Helium is present in the system to act as a buffer gas, principally aiding in heat transfer and in energy transfer from the (100) state to maintain a population inversion. Further radiative and collisional decay deactivates the remaining excited $CO_2$ molecules to the ground state.

Several structural designs are possible for the $CO_2$ laser ,[3] as illustrated in Figure 4-5. The classic construction is a sealed glass tube containing $CO_2$, $N_2$, and He and with longitudinal electrical discharge, as shown in Figure 4-5(a). Mirrors define the resonant cavity. The disadvantage of this design is the likelihood of $CO_2$ dissociation upon electrical discharge to yield CO and $O_2$. However, $CO_2$ can be regenerated by including $H_2$ or $H_2O$ vapor in the sealed tube to react with CO. The recombination reaction is also catalyzed by a nickel electrode. Alternatively, a longitudinal flow of gas, generally in the same direction as the electrical discharge, continually replenishes $CO_2$, as shown in Figure 4-5(b). The speed of gas flow can be varied to generate more power; generally, cooler temperatures are maintained at higher flow velocities, thereby enhancing the efficiency of laser operation. A waveguide configuration for a $CO_2$ laser (with metal electrodes

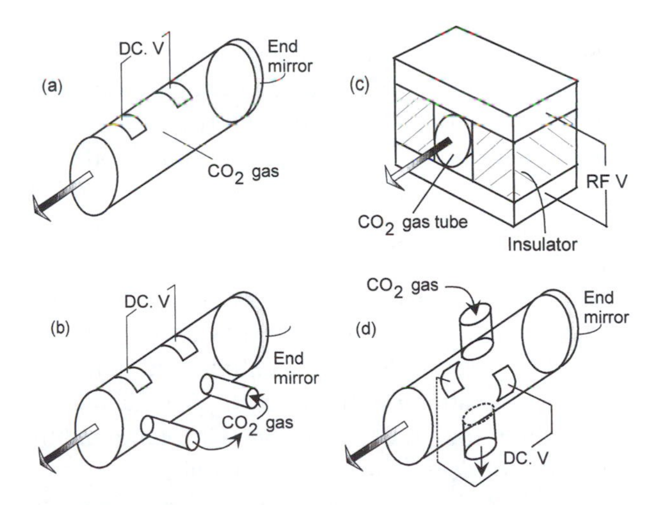


**Figure 4-5**
Major types of carbon dioxide lasers, including: *(a)* a sealed tube with longitudinal electrical discharge, *(b)* a longitudinal flow laser with $CO_2$ flow parallel to the electrical field, *(c)* a waveguide laser with electrodes parallel to the laser cavity to generate the electrical discharge, and *(d)* a transverse flow laser with $CO_2$ flow and electrical discharge perpendicular to the axis of the laser cavity and to each other.

arranged parallel to the laser cavity to generate the electrical discharge) provides a compact size with reduced losses from diffraction; this is shown in Figure 4-5(c). A transverse flow laser (both gas flow and electrical discharge perpendicular to the laser cavity and perpendicular to each other) achieves even higher power, as illustrated in Figure 4-5(d).

The infrared operation of the $CO_2$ laser demands optical materials and coatings that are transparent (or reflective) in this region. Alkali halides such as NaCl or KBr are excellent choices for their transparent qualities in the infrared, but their hygroscopic nature demands protective care. Germanium and gallium arsenide are common choices for laser cavity windows, and copper and molybdenum metal mirrors are often used for their low absorption and high thermal conductivity characteristics. These attributes are important at high power levels. While copper has a higher thermal conductivity than molybdenum, copper is a softer metal and harder to polish (as well as being more expensive).

The tremendous potential power of $CO_2$ lasers, arising from the high efficiency of the lasing transition, is a key reason for their use in industry to weld, cut, and drill materials. One application of particular importance for $CO_2$ lasers is the cutting of titanium metal, an especially hard material for mechanical tools to cut. Heat-treating of metals is another industrial use, as heat from the $CO_2$ laser beam alters the crystalline structure on a metal surface. Similarly, the $CO_2$ laser has attained popularity in surgical applications to cut and cauterize tissue, as a consequence of the absorption of infrared radiation by water in the tissues. Multiline operation of $CO_2$ lasers requires only the inclusion of an etalon or other tuning element in the laser cavity. Tuning is possible in the range of 9.1–11.0 $\mu$m. The versatility of these lasers is further enhanced by the choice of either CW operation or pulsed output, with pulses of 0.1–1 ms duration or shorter with *Q*-switching and cavity dumping procedures. (See Chapter 3 and the Glossary for details of these techniques.)

While $CO_2$ lasers are the most efficient and powerful of the molecular gas lasers emitting in the infrared, other infrared gas lasers are available. CO and $N_2O$ lasers represent short- and mid-IR sources, respectively, with output in the 5.0–6.5 $\mu$m range for CO lasers and in the 10.3–11.1 $\mu$m range for $N_2O$ lasers. The emission of these lasers arises from vibrational-rotational transitions, just as in the $CO_2$ laser. Rotational splitting of the main vibrational transition generates the multitude of emission lines observed. The gas mixture in CW CO lasers typically contains CO, nitrogen, helium, and oxygen or xenon, with a total pressure of 35 torr. $N_2O$ gas lasers have similar gas mixtures, with $N_2O$, nitrogen, helium, and CO as the typical components, and with total pressure around 10 torr .[4]

Far-infrared molecular gas lasers with output in the 40–2000 $\mu$m region contain simple organic molecules with permanent dipoles as the active media. Commercial sources include $CH_3OH$, various deuterium-substituted $CH_3OH$ gases, $C_2H_2F_2$, $CH_3F$, and $CH_3NH_2$. These lasers operate on the basis of transitions between rotational energy levels within the excited vibrational state. Since the energy of these rotational energy levels is strongly influenced by molecular mass, the practice of substituting the deuterium isotope for hydrogen can dramatically alter the emission output. In fact, commercial lasers are designed so that users may efficiently vary the gas contained within the laser cavity to extend the flexibility of these lasers. Precise excitation of the excited vibrational state is achieved by pumping with a single laser line in the 10 $\mu$m range from a CO or $N_2O$ laser.

## Nitrogen Laser

The nitrogen laser is a molecular gas laser with pulsed emission in the ultraviolet region at 337.1 nm. The emission arises from transitions between different electronic energy levels, unlike the vibrational-rotational transitions of the $CO_2$ molecular gas laser. Figure 4-6 presents the energy level diagram for $N_2$, featuring the $X^1\Sigma^+_g$ ground electronic state and the first three excited electronic states denoted $A^3\Sigma^+_u$, $B^3\Pi_g$, and $C^3\Pi_u$, respectively. While details of the names of the electronic states are not important here, the relative energies and lifetimes of the electronic states are significant. It is worth noting, however, that *X*() denotes the ground electronic state, while *A*(), *B*(), and *C*() denote the first, second, and third excited electronic states, respectively.

The active medium of a nitrogen laser is usually pure nitrogen gas at a pressure between 20 torr and 1 atm .[5] How is the upper lasing level of the nitrogen

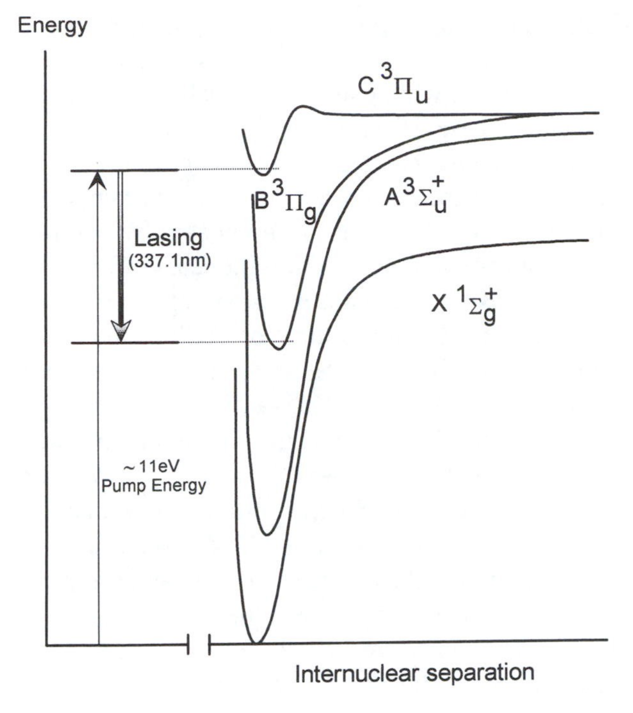


**Figure 4-6**
The source of the lasing line of the nitrogen laser. The ground state and first three excited electronic energy levels (singlet plus two triplets) of $N_2$ are presented in the energy diagram. The upper triplet excited state, characterized by a short lifetime, is populated via electrical discharge of ground-state molecules. Lasing at 337.1 nm occurs to the lower and longer-lived triplet energy state. Thus, a population inversion cannot be sustained, and only pulsed emission results.

laser populated? Electrical discharge is used to pump a population of $N_2$ molecules to the highest $C^3\Pi_u$ state. The lifetime of this excited electronic state is on the order of 40 ns. While a lasing transition occurs to the lower unpopulated $B^3\Pi_g$ electronic level, the population inversion is very quickly depleted, since the lower level has a substantially longer lifetime of 100 $\mu$s. The inability to sustain $N_2$ molecules in the upper state of the lasing transition leads to pulsed emission only. Nevertheless, nitrogen lasers exhibit an extremely high gain, producing a highly intense pulse as essentially all of the excited $N_2$ molecules undergo the lasing transition at the same time. As feedback—that is, amplification of laser light within the cavity via multiple passages of radiation using mirrors—is unnecessary, these lasers are often described as *superradiant*. A single end-mirror in the laser cavity serves to direct the output. Commercial $N_2$ lasers offer variable repetition rates up to 100 Hz with 10 ns pulses of about 250 kW peak power. Lasers with both self-contained and external gas supplies are available.

As sources of intense pulsed ultraviolet radiation, nitrogen lasers have found numerous uses in photochemical studies and photolysis experiments to initiate chemical reaction and generate excited-state species. Spectroscopic uses of nitrogen lasers include coherent anti-Stokes Raman spectroscopy CARS, fluorescence-lifetime measurements, photoacoustic investigations, and LIDAR applications. The high-energy output of nitrogen lasers is also well-suited for pumping dye lasers.

## Solid-State Transition Metal Ion Lasers

The lasing transitions observed in these solid-state lasers arise from transition metal ions embedded in a transparent host ionic crystal or glass. While the energy levels of the transition metal ions are primarily responsible for the lasing emission, the host material influences the electrical environment (the "crystal field") experienced by the ions. Thus, the energy levels and emission wavelength vary with the nature of both the ion and host species. The three most common lasing transition metal ions are chromium, neodymium, and titanium, although the lanthanides holmium, thulium, and erbium have also been doped in various crystalline and glass matrices.[6] Two particularly versatile lasers in this category include the Nd:YAG laser with neodymium ions embedded in a host lattice of yttrium aluminum garnet ($Y_3Al_5O_{12}$) and the Ti:sapphire laser with corundum ($Al_2O_3$) as a host for titanium ions.

### Nd:YAG Laser

The Nd:YAG laser design is based on the inclusion of a small concentration of $Nd^{3+}$ ions in a yttrium aluminum garnet (YAG) crystal. The role of the neodymium ions is to act as a substitute for yttrium ions which have a similar ionic size. In effect, neodymium is present as an impurity in the YAG crystal, typically at a level of 1–2% by weight. The YAG crystalline host provides a medium with high thermal conductivity to quickly remove the heat generated during high-power continuous wave (CW) operation. The lasing medium is generally rod-shaped with a diameter of about a centimeter and a length of several centimeters.

Trivalent neodymium ions exhibit a multitude of energy levels, yet the Nd:YAG laser is a classic four-level laser, as illustrated in Figure 4-7. Neodymium is a rare earth element with an incomplete $4f$ electronic subshell. The multitude of energy levels arises from the coupling of the spin and orbital angular momenta to yield total angular momentum states, all of which are closely spaced in energy. The number of these states is given by $2J+1$, where $J$ is the total angular momentum quantum number. For example, the state $^4I_{9/2}$ has J = 9/2 and represents 10 states closely spaced in energy. Formally these 10 states should have the same energy, that is, be degenerate, but the presence of the host lattice (the aluminum garnet which surrounds each $Nd^{3+}$) slightly separates these 10 energies. The influence of the host on the energy levels of the neodymium ions is referred to as the *crystal field splitting*. The upper energy levels of $Nd^{3+}$ ions are populated by pumping the ground-state atoms with light near 0.73 $\mu$m and 0.8 $\mu$m using flashlamps, broad-spectrum light sources, tungsten arc lamps, or gallium aluminum arsenide (GaAlAs) diode lasers. The 0.73 $\mu$m pump band corresponds to the $^4I_{9/2} \rightarrow {}^4S_{3/2}$ and $^4I_{9/2} \rightarrow {}^4F_{7/2}$ excitation transitions, while the 0.8 $\mu$m pump band is associated with the $^4I_{9/2} \rightarrow {}^4F_{5/2}$ and $^4I_{9/2} \rightarrow {}^3H_{9/2}$ absorption processes. Fast, nonradiative transitions from the upper energy levels of $Nd^{3+}$ to the $^4F_{3/2}$ energy level populate a metastable state and produce a population inversion between the $^4F_{3/2}$ level and the lower unpopulated $^4I_{11/2}$ level. The lasing emission is represented by the $^4F_{3/2} \rightarrow {}^4I_{11/2}$ transition, yielding primarily 1.06 $\mu$m radiation. Crystal field splitting of the $Nd^{3+}$ energy levels by the host produces secondary lasing lines; suppressing the fundamental output mode with special optics allows emission lines at 1.317 $\mu$m and 0.946 $\mu$m to be observed. Fast, nonradiative decay from the $^4I_{11/2}$ level to the ground state ($^4I_{9/2}$) subsequently occurs as the final relaxation step.[7]

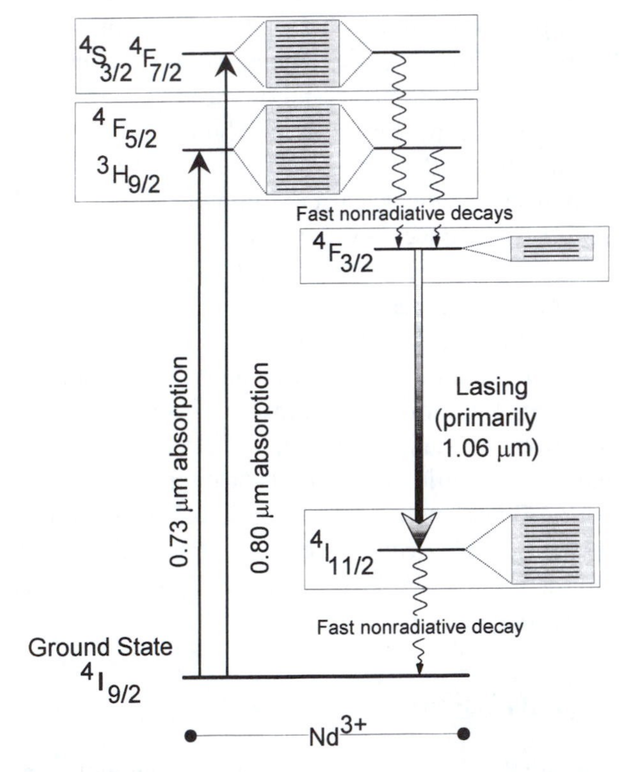


**Figure 4-7**
The energetics of the YAG laser based on the doping of the YAG crystal with $Nd^{3+}$. Several of the excited energy levels of $Nd^{3+}$ are presented. The upper levels are populated via 0.73 $\mu$m and 0.80 $\mu$m excitation pumping from a flashlamp, arc lamp, or diode laser. Fast, nonradiative decays occur to a metastable electronic state, $^4F_{3/2}$, followed by lasing emission primarily at 1.06 $\mu$m to a second lower-energy metastable state. Additional relaxation occurs to the ground state. The Nd:YAG laser constitutes a four-level laser with ground state $^4I_{9/2}$, two metastable states $^4I_{1/2}$ (not shown) and $^4F_{3/2}$, and the upper level states $^4S_{3/2}$, $^4F_{7/2}$, $^4F_{5/2}$, and $^3H_{9/2}$.

Nd:YAG lasers may be operated in either a pulsed or CW mode. Typical output powers for CW operation are generally several watts, but such lasers are capable of 100 W outputs in multiline operation. Subsequent mode-locking or *Q*-switching can then produce subnanosecond pulses with high average power. Nonlinear frequency conversion crystals are often employed for harmonic generation, yielding the second harmonic at 532 nm, the third harmonic at 355 nm, and the fourth harmonic at 266 nm. (See Chapter 3 and the Glossary for further details of these techniques.)

Clearly, Nd:YAG lasers are versatile, powerful light sources with a kaleidoscope of wavelengths, intensities, powers, and modes of operation. What are some of the practical concerns and drawbacks of these lasers? Most notably, the inefficiency of pumping via a flashlamp or arc lamp leads to requirements for larger power supplies to attain higher laser powers, thus necessitating the removal of generated heat using forced-air cooling or circulating water. Hence, powerful Nd:YAG lasers can be inherently bulky with massive cooling systems. Thermal lensing of the YAG crystal often occurs with increased heating. Here the temperature-dependent refractive index of the host crystal causes the material to act as a lens whose focus changes with temperature. Both of these complications are overcome with the use of diode lasers as pump sources, contributing to the active development of diode-laser pumping technology. Finally, the YAG crystal itself grows slowly, posing time constraints and crystal size limitations in laser construction.

Despite these limitations, Nd:YAG laser systems are incorporated in a wide spectrum of applications. The high power and short wavelengths of Nd:YAG lasers lead to numerous applications in materials processing, particularly for drilling, spot welding, marking, and soldering. The combination of short pulses and short wavelengths for Nd:YAG emission also make these lasers suitable for surgical application. Optical fiber delivery of the 1.06 $\mu$m radiation enhances the use of Nd:YAG lasers in the medical field. Contributing to the versatility of research applications is the capability of generating the second-, third-, and fourth-harmonic frequencies of the fundamental mode, enabling Nd:YAG lasers to pump tunable visible light lasers. Research applications using Nd:YAG lasers are well-documented and include such areas as remote sensing, Raman spectroscopy, and mass spectrometry. The performance capabilities of these lasers have led to extensive uses as military range-finders (measuring target distances by round-trip times of laser pulses) and target designators (marking targets with coded laser spots). The diversity of Nd:YAG laser applications will continue to grow as advances in pumping techniques and crystal host development occur.

## Ti:Sapphire Laser

The titanium-doped sapphire laser is an example of a tunable solid-state laser in which the lasing emission arises from both electronic and vibrational transitions of the $Ti^{3+}$ ion embedded in a corundum ($Al_2O_3$) host lattice. The Nd:YAG laser depends on the energy multiplicities associated with incompletely filled 4*f* electronic subshells. The Ti:sapphire laser is an example of what is now being called a *vibronic laser*, in which the energy levels associated with an incomplete 3*d* electronic subshell combine with the vibrational frequencies of the host lattice to yield bands of closely spaced energy levels capable of lasing.

At concentrations of 0.1% by weight, $Ti^{3+}$ ions replace $Al^{3+}$ ions in the crystal lattice, thereby acting as an impurity. The principles of operation are characteristic of four-level systems and are similar to the Nd:YAG laser. However, instead of discrete energy levels giving rise to narrow lasing lines, as in $Nd^{3+}$, very closely spaced, vibrationally broadened electronic energy states in $Ti^{3+}$ permit tunable emission over a range of wavelengths.

Figure 4-8 illustrates the absorption and emission bands of the $3d^1$ $Ti^{3+}$ ion. The $^2T_2$ ground state

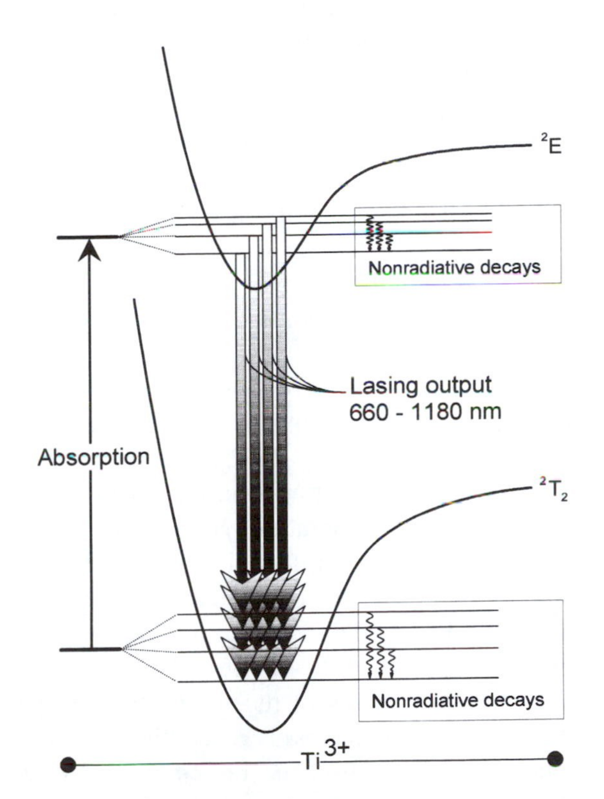


**Figure 4-8**
The energetics of the Ti:sapphire laser based on the energy levels of $Ti^{3+}$. The $^2T_2$ ground state and the $^2E$ excited state represent vibronic bands—bands arising from the coupling of the electronic energy levels of the incompletely filled 3*d* subshell of $Ti^{3+}$ and the vibrational frequencies of the corundum host lattice. The electronic energy levels are thus broadened by closely spaced vibrational sublevels. Lasers or flashlamps provide optical pumping from the ground-state $^2T_2$ level to the excited-state $^2E$ level. A broad lasing output from 660 to 1180 nm is observed.

and $^2E$ excited state are vibronic bands—electronic energy levels broadened by vibrational sublevels. Optical pumping excites the $Ti^{3+}$ ion to the upper levels of the $^2E$ vibronic band. Nonradiative loss of vibrational energy relaxes the system to the bottom of the vibronic band which constitutes the upper lasing level. The lasing transition lowers the ion to the vibrationally excited sublevel of the $^2T_2$ ground electronic state. Lasing emission over the range of 660–1180 nm is possible. Further nonradiative vibrational relaxation returns the $Ti^{3+}$ ion to the bottom of the ground vibronic band.

The Ti:sapphire laser is like a dye laser in that it must be pumped by some intense light source in order to establish the population inversion necessary for its operation. Numerous pumping sources are available. While flashlamp pumping is possible, current commercially available systems use lasers for optical pumping, including argon ion lasers, dye lasers, frequency-doubled Nd:YAG lasers, and copper vapor lasers. The short lifetime (3.2 $\mu$s) of the $Ti^{3+}$ excited state necessitates short-pulse, $Q$-switched, or CW laser sources, or short-pulse flashlamps. The Ti:sapphire laser has the broadest tuning range of any single solid-state laser (in fact, the broadest tuning range for any single solid-state, gas, or liquid laser medium),[8] with fundamental emission from 660-1180 nm and maximum power for emission between 700 and 900 nm. Tunable visible light is also generated from near-IR output through a variety of techniques, including harmonic generation, Raman shifting, and frequency mixing. Second-harmonic generation yields lasing wavelengths of 350–470 nm, third-harmonic generation yields from 235–300 nm, fourth-harmonic generation yields near 210 nm, and frequency mixing yields from 240–250 nm. Both CW and pulsed operations are possible, with a variety of repetition rates and peak powers. Newer commercially available lasers allow choices between picosecond or femtosecond pulse widths to be made with a few simple adjustments.

One of the few limiting features of Ti:sapphire lasers is the quality of the crystal. Recent research reveals that impurities from the presence of the $Ti^{4+}$ ion lead to loss of gain for laser action because of interfering absorption of the laser output by these ions. Annealing at 2000°C eliminates the presence of $Ti^{4+}$ ions in the crystal. A figure-of-merit parameter used to quantify the quality of the crystal is defined as the absorption of the pump light divided by the absorption of the laser light. Figures of merit above 100 are currently possible for commercially available sapphire crystals.

Although lasing in Ti:sapphire systems was first observed in 1982 and commercial Ti:sapphire systems were only introduced in 1988, these lasers have received wide acceptance in a multitude of research applications. The IR output and tunability naturally suggest numerous infrared spectroscopic uses, potentially competing with dye laser systems. In addition, LIDAR applications for remote sensing of water vapor in space are feasible with near-infrared laser lines at 727, 760, and 940 nm. Ti:sapphire lasers could also be useful in photodynamic therapy (PDT) research as new photosensitizing dyes are developed, because of the effective penetration of near-infrared radiation by human tissues. Experts in the optical data storage industry believe that Ti:sapphire lasers will play a significant role in the data-storage market as a consequence of the broad tunability of laser emission. Simultaneous operation of a single Ti:sapphire laser in the ultraviolet (at 400 nm), visible (at 527 nm), and infrared (800 nm) regions could replace the use of three different lasers that are now required for writing, reading, and erasing optical disks. Modifications of Ti:sapphire lasers have provided femtosecond laser pulses which are increasingly finding new uses to probe molecular events in real time. Application of Ti:sapphire lasers is an active field of research and promises innovative and exciting developments.

## Semiconductor Diode Lasers

Semiconductor diode lasers are solid-state devices that use the junction between $p$- and $n$-doped crystals as a lasing medium. The energy levels of the entire solid-state lattice influence the observed lasing transitions, in contrast to the role of isolated transition metal ions in the lasing emission of other solid-state lasers such as Nd:YAG and Ti:sapphire lasers. To understand the principles of operation of semiconductor lasers, we will review the essential concepts underlying the electronic properties of solids, particularly those of semiconducting materials.

## Origin of Semiconducting Behavior

Crystalline solids are characterized by a regular, periodic array of the atoms comprising the solid. From quantum mechanical considerations, the allowed states for electrons in this crystal lattice occur in well-defined bands of electronic orbitals, rather than in single atomic orbitals at discrete energies. At a given temperature, those energy levels fully occupied by electrons are known as *filled bands*; the highest occupied band (whether partially or fully occupied) is denoted as the *valence band*. The unoccupied quantum states comprise the *conduction band*. Solids are characterized as metals, insulators, and semiconductors based on the extent of occupation of the valence band and on the energy difference between the valence and conduction bands. Those solid materials classified as metals possess partially filled valence bands in which electrons move freely upon the application of an electric field. The free movement of such electrons throughout the conduction band in the crystal lattice gives rise to the observed conductivity of the metal. In contrast, insulators have filled valence bands and characteristically large band gaps, such that thermal excitation energies are insufficient to promote electrons in the filled valence band to the empty conduction band. No movement of electrons occurs under the influence of an electric field, leading to no measurable electrical conductivity. For semiconductors, however, thermal excitation can promote electrons from the filled valence level to the conduction band. Conductivities of intermediate magnitude may then be observed. This small activation energy to excite electrons to the conduction band leads to temperature-dependent electrical conductivities for semiconductors. Figure 4-9 summarizes the band structures of metals, insulators, and semiconductors.

The Group IV elements silicon and germanium are common semiconductors. Simple binary compounds between Group III and Group V elements are isoelectronic with Group IV elements and are also semiconducting materials. These binary semiconducting structures include gallium arsenide (GaAs), gallium phosphide (GaP), aluminum arsenide (AlAs), aluminum phosphide (AlP), and indium phosphide (InP). Ternary and quaternary alloy systems also exhibit semiconducting behavior. These materials contain variable but controlled amounts of one or two pairs of elements with the same number of valence electrons. Some examples of such alloys include the ternary compounds $Ga_{(1-x)}Al_xAs$, $In_{(1-y)}Ga_yAs$, $GaAs_xP_{(1-x)}$, $Pb_{(1-x)}CdS$, and $Pb_{(1-x)}SnSe$, and the quaternary compound, $In_{1-x}Ga_xAs_{1-y}P_y$.

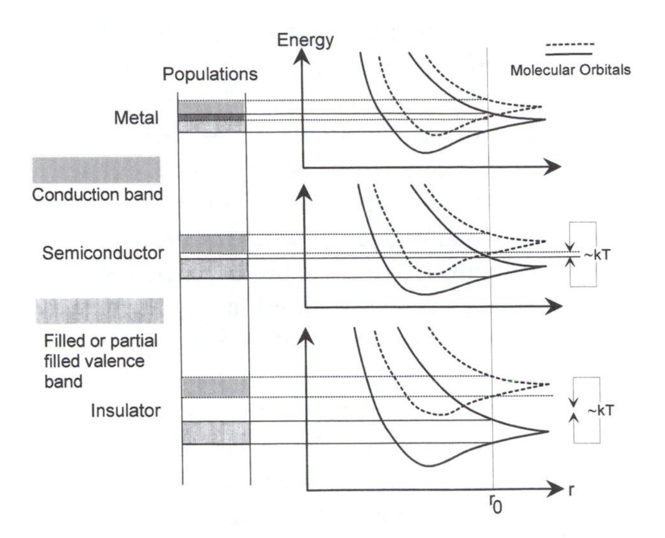


**Figure 4-9**
The origin of the band structures of metals, insulators, and semiconductors. The solid lines depict the range of energies of the delocalized molecular orbitals that result from the overlap of filled or partially filled atomic orbitals. Such quantum states comprise the filled or partially filled *valence band*. The dashed lines indicate the corresponding range of energies of the delocalized molecular orbitals that result from the overlap of empty atomic orbitals. These quantum states constitute the *conduction band*. The spread of energies in the valence and conduction bands varies with the internuclear distance. The energy spacing equivalent to $kT$ is presented for comparison. The bands and the energy spacings yield three types of substances: *metal*, *semiconductor*, and *insulator*. In the case of metals, electrons pass freely from the partially filled band to the conduction band. For semiconductors, the small energy spacing between the valence and conduction bands ensures that thermal excitation can promote electrons between these bands to induce electrical conductivity. The large energy gap between the valence and conduction bands of the insulator is responsible for its lack of electrical conductivity.

## Altering the Electronic Structure of Semiconductors

The promotion of electrons to the conduction band of semiconductors produces a partially filled valence band. The loss of an electron from the valence level is said to generate a "hole" or a positively charged carrier of electricity. Thus, the conductivity of a semiconductor arises not only from the mobile electrons promoted to the conduction band, but also from the increased mobility of electrons in the valence band. The concentration of negative mobile electrons in the conduction band is denoted as *n*, while the concentration of positive mobile holes in the valence band is given as *p*. For pure semiconductors, the magnitudes of *n* and *p* are equal and generally small. By introducing a small amount of impurities (< 1 part per million) in the semiconductor, however, an excess of mobile electrons or holes can be produced. The impurity, or dopant, must substitute for an atom in the host lattice and must contain a different number of valence electrons than the semiconductor atom. To increase the conduction-band electron concentration, *n*, an element with a larger number of valence electrons replaces the semiconductor atom to produce an "n-type" semiconductor. Silicon (4 valence electrons, $3p^4$) doped with phosphorus (5 valence electrons, $3p^5$) generates an extra electron which is not needed in the covalent network of tetrahedral Si-Si bonds and which can be easily excited into the conduction band. Alternatively, the valence-band hole concentration, *p*, can be increased with the substitution of an atom with fewer valence electrons than the semiconductor atom it replaces. The "*p*-type" semiconductor created by the doping of silicon with aluminum or gallium (3 valence electrons each, $3p^3$, $4p^3$, respectively) conducts electricity by the free movement of holes in the valence band.

Impurity atoms may be introduced into semiconducting binary compounds to produce both *p*- and *n*-type semiconductors. For example, in the compound GaAs, a *p*-type semiconducting material results with the substitution of electron-poor dopants for either Ga (with an element like Zn) or As (with an element like C). Likewise, extra valence electrons are introduced to form *n*-type semiconductors by substituting electron-rich impurities for either Ga (e.g., with Si) or As (e.g., with Se). Alternatively, the balance between the concentrations of valence-band holes and conduction-band electrons can be disrupted by altering the 1:1 stoichiometry of the binary structure.

## Semiconductor Junctions

The basis for lasing action in semiconductor diode lasers centers on a complex multilayer structure composed of semiconductors of variable composition. Figure 4-10 is a schematic drawing of a semiconductor laser construction. The key component is the *p–n* junction created by mounting parallel layers of *p*-type and *n*-type semiconductors at a separation distance of about 1 $\mu$m. A current is induced by applying an external voltage via electrical leads attached to the junction. This process is called *forward biasing* when a negative voltage is applied to the *n*-type semiconductor and a positive voltage is applied to the *p*-type material. Electrons are transferred from the conduction band of the *n*-type material and trapped by the holes in the valence band of the *p*-type semiconductor. Such recombination releases energy equivalent to the band gap in the form of a photon. What distinguishes a diode laser from a simple light-emitting diode, however, is the presence of a cavity to provide light amplification and the presence of a population inversion. The multilayer structure achieves both objectives. By placing layers with lower refractive indices adjacent to the *p–n* junction, light is confined to the active region. The polished ends of the semiconductor chip serve as the end mirrors for a cavity with a length that is a half-integral number of the wavelength of light emitted, $n/2 \times \lambda$. The emitted light is not well-collimated, however. To ensure a population inversion, a high forward bias is applied to the junction. Furthermore, the electrons and holes are confined within the active region by depositing adjacent to the *p–n* junction only those layers of semiconductor materials with higher band gaps than the *p–n* junction itself. Thus, electrons and holes cannot diffuse into the adjacent layers. Light amplification which can result as stimulated emission effectively competes with absorption for the photons generated by recombination. Thus, diode lasers rely on both the optical and electron confinement dictated by the multilayer structure of the semiconductor laser chip.

The variable stoichiometry of many semiconducting materials generates an extensive range of lasing wavelengths.[9,10] The binary semiconductors GaAs and ZnSe emit at the fixed wavelengths of 904 nm and 525 nm, respectively. The ternary semiconductor lasers $Ga_{0.5}In_{0.5}P$ and $In_{0.2}Ga_{0.8}As$ also have characteristic laser lines at 670 nm and 980 nm, respectively. Emission over the range of 620-895 nm is possible by varying the stoichiometry in $Ga_{1-x}Al_xAs$ lasers, while infrared operation between 1100 and

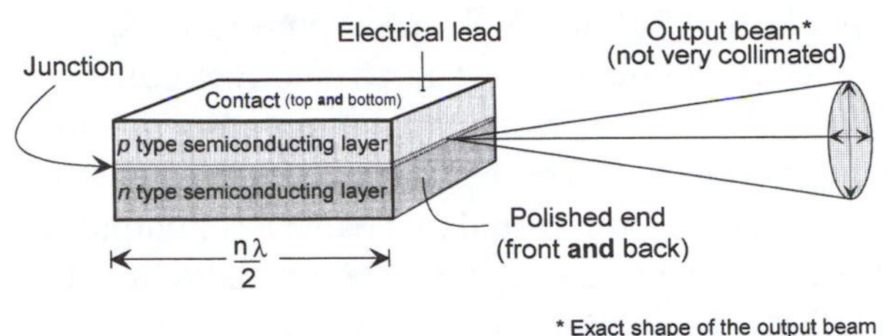


**Figure 4-10**
Schematic drawing of a semiconductor laser construction illustrating p-n junction lasing. A *p*-type semiconductor contains a dopant with fewer electrons than its host, creating mobile holes in the valence band. An *n*-type semiconductor contains mobile electrons in its conduction band that arise from impurities with more electrons than the host solid. An external voltage applied via electrical leads induces a current that causes electrons from the conduction band of the *n*-type semiconductor to be transferred and trapped by the holes in the valence band of the *p*-type semiconductor. The recombination of electrons within the holes releases energy in the form of the photons of the output beam of the laser.

1650 nm is obtained with $In_{1-x}Ga_xAs_yP_{1-y}$ lasers. While the gallium arsenide diode laser family operates at a fixed wavelength, the lead salt diode lasers are tunable since the lasing wavelength is temperature dependent. These lasers require cryogenic temperatures (15–90 K) in order to emit from approximately 3–30 $\mu$m. Representative examples include the quaternary PbEuSeTe (output from 3.3–5.8 $\mu$m) and the ternary PbSSe (4.2–8.0 $\mu$m), PbSnTe (6.3–29.0 $\mu$m), and PbSnSe (8.0–29.0 $\mu$m) lasers.

Semiconductor diode lasers offer compact size, low cost, ease of fabrication, a range of emitting wavelengths, and even the possibility of battery operation. While output powers are generally low (milliwatt levels), these lasers have rapidly revolutionized the optical information and communications industry. Compact disc players, laser printers, laser pointers, and fiber optic communications all employ diode lasers. The recent achievements of emission in the red region of the visible spectrum suggest that diode lasers may soon encroach upon many of the applications currently using helium-neon lasers.

## Dye Lasers

In contrast to the examples of gas and solid-state lasers presented thus far, dye lasers use liquid solutions as active media. The basis for lasing action is a population inversion created in organic dye molecules during a fluorescence emission process. Suitable dyes are generally large polyatomic organic compounds with structural characteristics enhancing fluorescence emission, particularly multiple ring systems and conjugated double bonds for extensive electron delocalization. The combination of dye and solvent dictate the laser wavelengths that may be attained. Dye lasers that operate with emission wavelengths ranging from the ultraviolet to the near-infrared region are readily available, making these lasers versatile devices in both research and industrial applications.

The lasing action of organic dyes can be explained through a consideration of an energy level diagram for a typical dye, as depicted in Figure 4-11. Each of the electronic energy levels is defined by an energy continuum—a range of possible vibrational and rotational sublevels closely spaced in energy and arising from the polyatomic nature of the dye. These levels are further broadened by interactions with a solvent. As a consequence, the transition from the ground electronic state $S_0$ to the first excited singlet electronic level $S_1$ is not characterized by a single discrete energy difference, but by a range of energy values. Thus, the $S_0 \rightarrow S_1$ transition results in a broad absorption band. Upon promotion to the $S_1$ state, the excited dye molecule undergoes a rapid (picosecond timescale) radiationless decay to the lowest energy levels of the $S_1$ state. Further relaxation occurs to one of the intermediate sublevels in the $S_0$ ground electronic state through the emission of a photon. This emitted radiation, or fluorescence, is of longer wavelength than the absorption process as a consequence of both the fast nonradiative decay in the $S_1$ state and the intermediate energy level within the $S_0$ continuum. Furthermore, the energy continuum of the $S_0$ state gives rise to a broad fluorescence spectrum. The energy level diagram of Figure 4-11 suggests that the dye laser is a four-level laser. Two nonlasing transitions exist: from the upper to the lower $S_1$ levels, and from the intermediate to the ground $S_0$ states. The laser emission results when a population inversion is present for the transition between the lower $S_1$ level to the intermediate $S_0$ state. Note, however, the population inversion of this lasing transition can be reduced by the competing process of intersystem crossing. Singlet-to-triplet transitions ($S_1 \rightarrow T_1$), although low in probability, de-

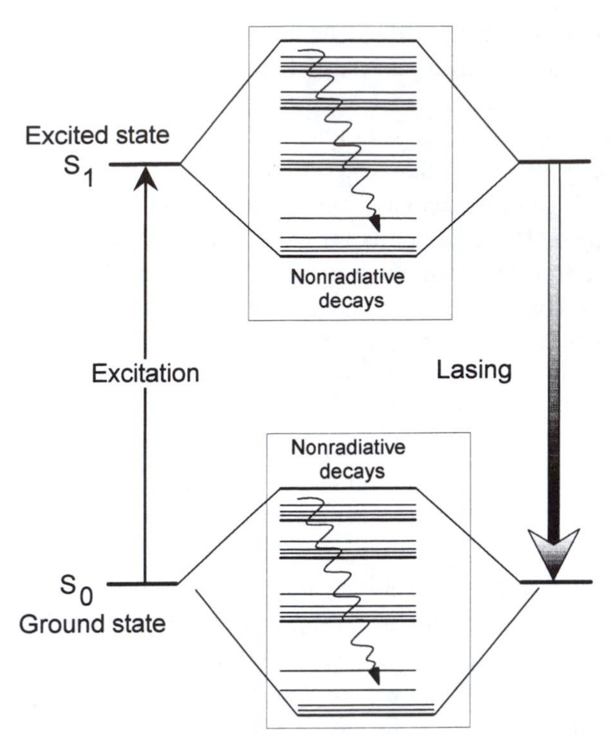


**Figure 4-11**
Energy level diagram for a typical organic dye laser, a four-level laser system. The ground-state and excited-state electronic energy levels of the dye are broadened by the rotational and vibrational substructure of the polyatomic dye. Interaction of the dye with the solvent medium further broadens the energy levels, leading to a broad range of absorption and fluorescence wavelengths, i.e., a broad absorption and fluorescence band. Nonradiative decays are observed within both the $S_1$ and $S_0$ states. The lasing transition corresponds to a population inversion created between the lower levels of the $S_1$ state to the upper or intermediate levels of the $S_0$ state.

plete the $S_1$ state and ultimately repopulate the ground $S_0$ state through phosphorescence, which is emission from a triplet state to the singlet ground state. Laser efficiency can be significantly reduced by this action.

Excitation of organic dyes to the $S_1$ state is achieved by optical pumping. The primary consideration in selecting an optical pump is a match of the pump's output wavelength with the dye's absorption maximum. Typically, the pump source is a second laser—generally a nitrogen laser or a monatomic gas laser—although xenon flashlamps may also serve as optical pumps. High average powers are possible with excimer (e.g., XeCl at 308 nm) and copper vapor lasers (511 and 578 nm) as pump sources, and frequency-doubled and -tripled neodymium lasers (532 and 355 nm, respectively) are also potential pumping lasers.[11] The CW or pulsed nature of the pump source is transferred to the dye laser output. To obtain monochromatic dye laser emission, a wavelength-dispersing device such as an etalon, prism, or diffraction grating is inserted into the laser cavity. Thus, dye lasers offer tunable emission over the range of fluorescence wavelengths characteristic of the dye. Substituted coumarins, for example, exhibit wavelength maxima over the range of 350–550 nm; various rhodamine derivatives are ideal for lasing in the 570–650 nm region. Infrared output from 740–970 nm is optimally obtained with numerous carbocyanine dyes.

Figure 4-12 illustrates several tuning curves for a selection of dyes optically pumped by different sources. Continuous tuning of a dye laser throughout the visible region may be achieved by a change of dye and generally a change of cavity mirrors. For most dye laser systems, the optical pump beam is directed onto a stream or jet of dye. Thus, an interchange of dyes requires a complete replacement of the solution in the flow system. Alternatively, a number of dye laser systems operate with dye solutions contained in standard fluorescence cuvettes, simplifying the procedures for dye changes. All dyes, however, have limited lifetimes. This photochemical stability is typically measured in terms of the number of watt-hours of pumping that are possible without substantial (>25%) degradation of laser output. Lifetimes may range from a few watt-hours to several thousand watt-hours.

On the horizon are dyes incorporated into a polymer matrix which will provide great stability, ease of handling, and convenient usage.

The output characteristics of dye lasers are a complex interplay of such factors as optical pump source, dye choice, dye concentration, solvent characteristics, and laser operating conditions. Thus, the observed wavelength range as well as the temporal nature of the output (CW vs. pulsed), the intensity, the spectral bandwidth, and the peak power of the lasing emission are all variable with each dye laser configuration. Nevertheless, such complexity contributes to the versatility and flexibility of dye lasers. For example, the wavelength tunability and narrow

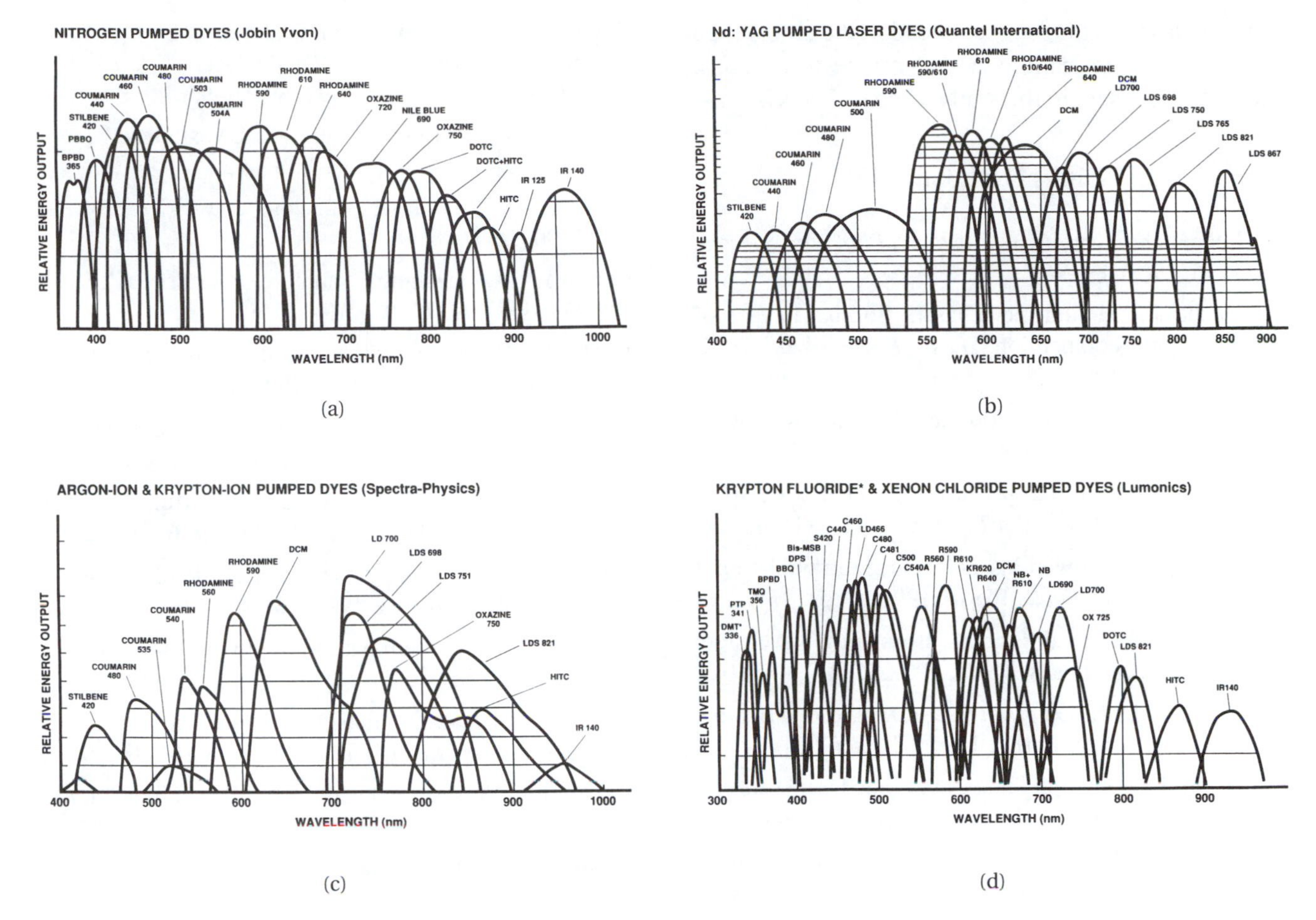


**Figure 4-12**
Tuning curves for laser emission from dye lasers with various optical pumps. The wavelength of the lasing emission varies with the selection of the dye optically pumped by the laser source. The optical pumps correspond to: *(a)* a nitrogen laser, *(b)* a Nd:YAG laser, *(c)* argon ion and krypton ion lasers, and *(d)* krypton fluoride and xenon chloride excimer lasers. Reprinted with permission from Exciton, Inc. Copyright 1989 Exciton, Inc.

spectral bandwidth of these lasers have contributed to their extensive role in spectroscopic applications. Also, the ability to modify the temporal characteristics of dye laser output with such methods as cavity dumping, mode-locking, synchronous pumping, and pulse compression has advanced the utilization of dye lasers in kinetic studies. (See Chapter 3 or the Glossary for details of laser modification techniques.) Applications in the medical industry continue to grow, particularly since dye laser light of high intensity and narrow spectral bandwidth can be directed through optical fibers to target tissues and thus treat cancer, eye diseases, kidney and gallbladder stones, and skin disorders.

## SUMMARY

Clearly the chemist has an extensive array of laser sources from which to select for specific industrial, research, and medical applications. The choice of laser will depend on the characteristic features of the laser emission (emission wavelength, tunability, CW or pulse mode, peak power, etc.), parameters of the laser operation (cooling costs, safety considerations, facility of operation, space limitation, etc.), and particular features of the experimental design (requirements for power, monochromaticity, stability, remote or harsh environment operation, etc.). Further advances in the performance of existing laser systems and laser technology will lead to the emergence of new laser applications and continue to ensure a vital role for lasers in the study of chemical systems.

## LITERATURE CITED


1. Leggett, K., "Gas lasers grow up", *Photonics Spectra*, 1995, 29(9), 84–96.
2. Hecht, J., "Helium neon lasers flourish in face of diode-laser competition," *Laser Focus World*, 1992, 28(11), 99–108.
3. Hecht, J., "Carbon dioxide lasers span power spectrum," *Laser Focus World*, 1992, 28(9), 87–96.
4. Hecht, J., "Infrared gas lasers cover a wide spectrum," *Laser Focus World*, 1993, 29(3), 115–124.
5. Hecht, J., "Nitrogen lasers produce ultraviolet light simply," *Laser Focus World*, 1993, 29(5), 87–91.
6. Hecht, J., "Rare earths create useful long-wavelength lasers," *Laser Focus World*, 1993, 29(11), 135–142.
7. Andrews, D. L., *Lasers in Chemistry*, Second Edition, Springer-Verlag, Berlin, 1990, Chapter 2.
8. Hecht, J., "Tunability makes vibronic lasers versatile tools," *Laser Focus World*, 1992, 28(10), 93–103.
9. Hecht, J., "Diode-laser performance rises as structures shrink," *Laser Focus World*, 1992, 28(5), 127–143.
10. Hecht, J., "Long-wavelength diode lasers are tailored for fiber optics," *Laser Focus World*, 1992, 28(8), 78–89.
11. Hecht, J., "Versatility keeps dye lasers alive," *Laser Focus World*, 1992, 28(7), 59–74.


## GENERAL REFERENCES


Andrews, D. L., *Lasers in Chemistry*, Second Edition, Springer-Verlag, Berlin, 1990.

Baumann, M. G. D., Wright, J. C., Ellis, A. B., Kuech, T., and Lisensky, G. C., "Diode lasers," *Journal of Chemical Education*, 1992, 69, 89–95.

Carts, Y. A., "Nitrogen lasers aid UV studies," *Laser Focus World*, 1989, 25(10), 83–92.

Carts, Y. A., "Ruby lasers shine on," *Laser Focus World*, 1990, 26(7), 83–91.

Carts, Y. A., "Titanium sapphire's star rises," *Laser Focus World*, 1989, 25(9), 73–88.

Hecht, J., "Carbon dioxide lasers span power spectrum," *Laser Focus World*, 1992, 28(9), 87–96.

Hecht, J., "Diode-laser performance rises as structures shrink," *Laser Focus World*, 1992, 28(5), 127–143.

Hecht, J., "Excimer lasers produce powerful ultraviolet pulses," *Laser Focus World*, 1992, 28(6), 63–72.

Hecht, J., "GaAs Diode Lasers: A Decade of Progress," *Lasers & Optronics*, 1991, 10(1), 54–61.

Hecht, J., "Neodymium lasers prove versatile over three decades," *Laser Focus World*, 1992, 28(4), 77–94.

Hecht, J., "Tunability makes vibronic lasers versatile tools," *Laser Focus World*, 1992, 28(10), 93–103.

Hecht, J., "Versatility keeps dye lasers alive," *Laser Focus World*, 1992, 28(7), 59–74.

Messenger, H. W., "Metal-vapor lasers display versatility," *Laser Focus World*, 1990, 26(4), 87–92.

Messenger, H. W., "Solid-state lasers develop new capabilities," *Laser Focus World*, 1990, 26(8), 81–97.

Miyake, C., "Pulsed Ti:Sapphire Lasers—Can They Beat Dye?" *Lasers & Optronics*, 1990, 9(10), 45–50.

O'Shea, D. C., Callen, W. R., and Rhodes, W. T., *Introduction to Lasers and Their Applications*, Addison-Wesley Publishing Company, Inc., Reading, MA, 1977.

Thyagarajan, K., and Ghatak, A. K., *Lasers: Theory and Applications*, Plenum Press, New York, 1981.

# PART II

# The Laser as a Probe

Lasers are well-established research tools frequently used to elucidate both reaction pathways and molecular structures. To gain an appreciation of the significant impact of laser photons as probes of chemical reactions and structures, the next seven chapters encompass a broad perspective of the major areas of such laser applications. In each chapter we present specific case studies to illustrate the exciting potential of the laser as a chemical probe. Two chapters focus on particular processes that are accessible for study with the use of lasers—reactions in the fast/ultrafast time regime and reactions induced by the absorption of multiple photons. Five additional chapters focus on specific techniques that have advanced in their scope because of the advent of lasers—fluorescence, capillary electrophoresis, light scattering, mass spectrometry, and photoacoustic applications. Through these discussions we hope to demonstrate the diversity of investigations employing laser photons as chemical probes and challenge you to imagine new avenues for laser interrogation.

CHAPTER 5

# Ultrafast Chemical Processes

## Chapter Overview

The phenomenal range of timescales for chemical reactions demands an extensive repertoire of experimental techniques to monitor reaction rates. Perhaps most challenging to the kineticist are those reactions occurring at extremely fast rates, generally considered to be in the range of $10^{-9}$ to $10^{-15}$ s. What kinds of processes are characterized by such rapid reaction times? The transfer of protons or electrons, the internal motion of proteins, the intra- and intermolecular transfer of energy, the making and breaking of chemical bonds, and the rotation of molecules are a few important examples of phenomena often characterized by ultrafast timescales. Thus, the ability to generate laser pulses of short duration is particularly useful in studying these reactions and processes. In this chapter we examine several approaches to monitoring the kinetics of reactions in the picosecond to femtosecond ultrafast regime.

## Ultrafast Techniques

Ultrafast laser techniques have become significant tools for fundamental chemistry research. Once again, using the coherence properties of laser light, in this case more temporal coherence than spatial coherence, chemists are able to visualize chemical transformations directly rather than inferring chemical transformations through subsequent product analysis. As improvements occur in methods of generating ultrashort pulses, especially with advances in the wavelength ranges and power levels that can be attained, the application of ultrafast laser techniques to even the most complex of chemical systems can be imagined as routine practice. Let us begin with a brief review of several key aspects of the study of chemical kinetics.

### Kinetics

Reactants form products via an overall process known as a chemical reaction. We all know that we can measure the extent of a reaction, that is, the degree of conversion of reactants to products. We can also measure how fast a reaction occurs, i.e., its reaction rate. What other characteristics of a reaction are we interested in specifying? A fundamental goal of the study of reaction rates is not to merely determine the velocity of a specific reaction, but to propose and test a reaction mechanism, a set of elementary chemical reactions that in total add up to the observed reaction.

One step along the journey from reactants to products will typically be slow. We call this portion of the reaction mechanism the *rate-determining step* and on a macroscopic thermodynamic basis ascribe to that step an energy of activation. Key to the interpretation of the rate-determining step is the concept of an "activated complex"—an intermediate "molecule," formed from reactants, that decays into products. What aspect of this concept is often most intriguing to chemists? In particular, chemists ponder how and on what timescale the "activated complex" or intermediate "molecule" decays to products. The traditional view of these ideas is embodied in the simple unimolecular and bimolecular reactions written below as Equations 54 and 55:

$$ABC \rightarrow [ABC]^{\#*} \rightarrow A + BC \qquad (54)$$

$$A + BC \rightarrow [ABC]^{\#} \rightarrow AB + C \qquad (55)$$

We note that $A$, $B$, or $C$ might be atoms, molecular fragments, or complex molecules. Moreover, the symbol $[\ ]^{\#}$ denotes the activated complex and the symbol $[\ ]^{\#*}$ denotes an activated complex that is in an excited electronic state.

What is the origin of the energy, inferred by the initial reaction arrow, that is used to generate an excited electronic state? Several sources are possible. One source, collisional activation, results from collisions between particles which in turn draw their energy from a thermal heat bath of some sort—a Bunsen burner, hot plate, etc. Alternatively, the source of activation energy that we want to discuss here is light, where photons of specific energies are absorbed to initiate the reaction. This light-driven process is usually referred to as *photoinitiation* and is particularly pertinent to studies of unimolecular decays as illustrated in Equation 54. Understanding the details and timescale of the "decay" of the activated complex to form products is the subject of molecular reaction dynamics. What makes this subject so fascinating at this point in time is that tools are now available to study the decay process as it happens in real time.

## Timescales

Let us review the timescales of physical events as presented in Table 5-1. As fleeting as some of these times seem, they do have finite dimensions. Consider the time it takes a photon to travel 3 meters—10 ns. This timescale becomes important in electrical circuits. For example, consider the dilemma of designing a circuit with a nanosecond switching time, such as required in fast computers, when electrons take on the order of nanoseconds to traverse conducting leads. These timescales pose real limits and challenges. Alternatively, suppose you want to follow in real time the motion of entity *A* (a molecule) as it leaves the activated complex $[ABC]^{\#*}$. The distance 3 nm is about 2 carbon-carbon bond lengths and is certainly a reasonable distance over which to make measurements to follow the process of *A*. How long does it take *A* to traverse 3 nm? From the discussion above, we see that the answer is on the order of picoseconds. If you want to track *A* as it moves away from *BC*, you have to make measurements within picoseconds or faster. Historically, for example, many chemists have been interested in tracking the movement of the iron atom in and out of the plane of the porphyrin ring system in hemoglobin upon ligation. The movement might only involve a distance of 0.1 nm, and thus the time necessary is clearly subpicosecond. Can we make measurements in these time frames? With lasers, yes we can!

## Chronology of Fast Light Sources

Before we see exactly how ultrafast reactions are monitored, let us present a brief history of fast light sources in Table 5-2. The development of flashlamps paved the way for serious approaches to tracking fast molecular events, but obviously left much room for improvement to attain the goal of taking "snapshots" in picoseconds. The ruby laser made an immense jump of six orders of magnitude toward the picosecond goal; ring dye lasers made another jump of six orders of magnitude, reaching below the picosecond realm into femtoseconds. With femtosecond timescale measurements possible, fast molecular events could now be tracked. Indeed, the femtosecond world is virtually routine with mode-locked Ti:sapphire and OPO laser systems available.

The types of events that occur on picosecond and femtosecond timescales are summarized in Table 5-3.

**Table 5-1**
Timescales for Some Physical Processes

| *Process* | | *Photon* | *Molecule* |
|---|---|---|---|
| Speed (m s$^{-1}$) | | $3 \times 10^8$ | $3–5 \times 10^2$ |
| Time to travel | 3 m | $10^{-8}$ s (10 ns) | $10^{-2}$ s (10 ms) |
| | 3 $\mu$m | $10^{-14}$ s (10 fs) | $10^{-8}$ s (10 ns) |
| | 3 nm | $10^{-17}$ s (10 as) | $10^{-11}$ s (10 ps) |
| Time for one vibration of CO | | | 20 fs |
| Franck-Condon electronic transition time | | < 1 fs | |

**Table 5-2**
Chronology of Fast Light Sources

| *Decade* | *Pulsed Light Source* | *Pulse Width* |
|---|---|---|
| 1950s | Flashlamps | ≈ ms |
| 1960s | Ruby laser | ≈ ns |
| 1970s | Nd:YAG laser, flashlamp pumped | ≈ ps |
| | Synchronously pumped dye laser | |
| 1980s | Colliding pulse mode-locked ring laser, CPM with | 50 fs |
| | pulse compression | 5 fs |
| 1990s | Mode-locked Ti:sapphire laser | 50–5 fs |
| | Optical parametric oscillator | |

**Table 5-3**
Physical Processes Occurring on Ultrafast Timescales

| *Order of Picoseconds* | *Order of Femtoseconds* |
|---|---|
| Relaxation processes in liquids | Transition state processes |
| Energy transfer, fluorescence | Optically induced phase transitions in solids |
| Reactions in solutions | Diffraction from optically formed diffraction gratings |
| Induced chemical transformations | Induced chemical transformations |

## The Pump-Probe Technique

If we want to study any of the fast molecular events noted in Table 5-3, we recognize that a fast means of probing a system is necessary. A very fast pulsed laser is a necessary tool, but a "fast" analytical technique is also an indispensable component. Numerous analytical techniques are available to indirectly measure the concentrations of reacting species and emerging products. Those physical properties of substances that are linearly related to the amount or concentration present are the most convenient probes for quantitative and kinetic analyses. For example, many kinetic studies rely on the detection of changes in the absorbance of reactants or products to monitor reaction rates. Conventional absorption spectrophotometers permit reactions with half-times greater than 10 s to be studied readily. For reactions in solution, relaxation methods enable the monitoring of the kinetics of reactions with half-lives on the order of microseconds. A specialized absorption technique—the pump-probe method—has been developed for those ultrafast processes not suited to conventional or relaxation monitoring methods. Central to the pump-probe technique is the use of two pulses originating from the same laser source, which is used both to initiate the molecular event of interest with the pump pulse and to examine the subsequent changes in the sample with the probe pulse. If the absorption of the incident photons of the pump pulse is monitored, the technique is often called *F*emtosecond *T*ransition-State *S*pectroscopy, or FTS.[1] If some emitted photon is studied after the target molecule (or atom) has been excited by an incident photon from the pump pulse, the approach is known as *L*aser-*I*nduced *F*luorescence, or LIF.[2]

Note that the probe pulse might be the output of a tunable dye laser driven by the initiating pump laser. A tunable probe laser is quite important, particularly when the molecular species created by the pump laser is no longer sensitive to the pump laser's energy but needs a different energy provided by the tunable probe to undergo the next step in the process being studied. Figure 5-1 illustrates a typical setup, whereby a beam splitter creates two ultrashort pulses from the initial laser output. A stepping motor enables a variable time delay to be established between pump and probe pulses. Kinetic processes at various time intervals after the initial excitation can then be explored. Discrimination of both simple and complex reaction mechanisms is possible with the pump-probe technique.

## Ultrafast Laser Systems

The challenge of these kinetic studies comes in supplying the pump and probe photons on picosecond to femtosecond timescales. In Chapter 3, we describe numerous ways to generate ultrafast pulses, particularly on the picosecond timescale. The ultrafast fs laser pulses are usually produced using a colliding pulse mode-locked dye ring laser, or CPM, perhaps in conjunction with pulse compression. A simple schematic of an apparatus to carry out ultrafast pump-probe experiments is shown in Figure 5-2.

In a CPM, an initial pulse from a mode-locked laser is split into two beams, each travelling in opposite directions. By controlling the path lengths of the two beams, the two pulses impinge on a saturable absorber in such a way that they create a pulsed standing wave in the absorber which acts as the output laser pulse. The pulse action depends on the time required to repeatedly saturate the absorber. Although the colliding pulses primarily cancel each other out, the remaining pulses have a very short duration. In fact, the CPM is capable of yielding pulses of 50-100

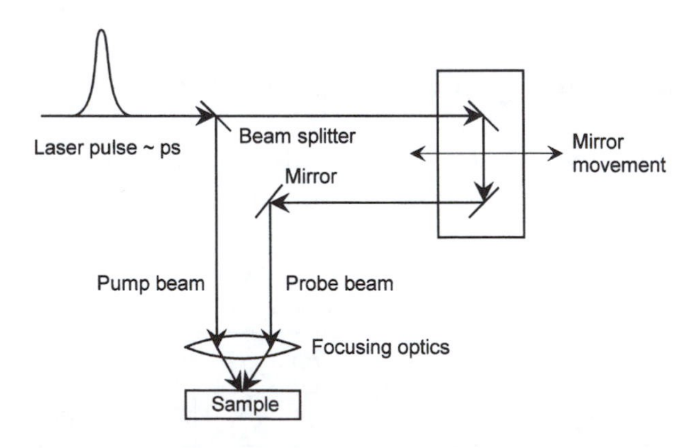


**Figure 5-1**
Schematic diagram of a typical arrangement of instrumentation for kinetic investigations using the pump-probe technique. Ultrashort pulses on the picosecond timescale from the same laser source are focused on a beam splitter to generate the pump and probe beams. The pump beam initiates the molecular event of interest in the sample. The probe beam, arriving at the sample after a time delay established by varying the optical path length, is used to examine changes in the sample induced by the pump beam excitation.

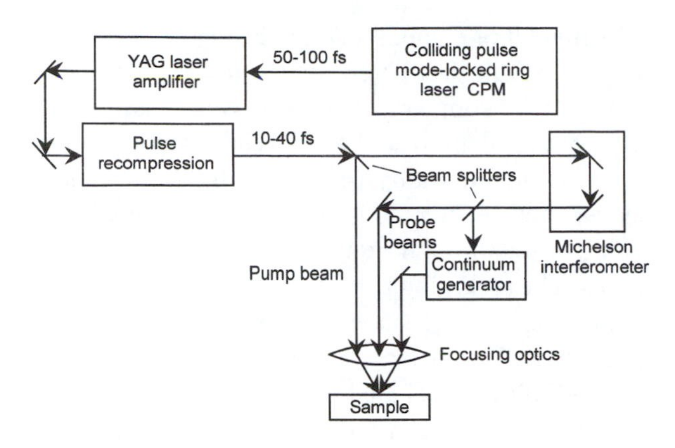


**Figure 5-2**
The femtochemistry apparatus for pump-probe experiments. The ultrafast pulses are generated in the CPM laser and are amplified in the YAG pulsed dye amplifier. After passing through the amplifier, the pulses, which are broadened temporally, are recompressed with a prisms arrangement. The initial pulse of the mode-locked laser is split into the pump and probe beams with controlled optical path lengths. The probe beam can be used alone or passed through a continuum generator to produce white light from which the desired probe wavelength is selected by filters.

fs duration. In conjunction with pulse compression, even shorter laser pulses of 6–10 fs can be produced. The necessary time separation of the pump and probe pulses is accomplished by beam splitting the ultrafast pulse beam and then allowing one beam to pass through a beam delay based on a Michelson interferometer. Recall that, on these timescales, if a beam travels even 0.3 microns longer than the other beam, it arrives 1 fs later when focused at the same place. Examples of investigations using the colliding pulse mode-locked ring laser systems are found in the case studies in this chapter.

## Preview of Case Studies

The extensive range of kinetic applications for the pump-probe technique is dramatically illustrated by the three case studies presented here. The first case study describes how the fast conformational changes accompanying the gain and loss of carbon monoxide by hemoglobin are measured and characterized. The second case study discusses how the unimolecular decomposition of ICN can be followed in real time and how the potential energy surfaces for the ground- and excited-state ICN molecule can subsequently be estimated. The third case study monitors the ultrafast photoinduced transfer of an electron between donor-acceptor species contained within the same molecule. With these examples we hope that you will sense the powerful potential of the pump-probe technique to characterize ultrafast chemical processes.

## Case Study I: Ultrafast Spectroscopy of Ligand Binding Reactions

### Overview of the Case Study

**Objective.** To study the dynamics of ligand-hemoglobin associations and ligand-induced conformational changes using the pump-probe technique.

**Laser Systems Employed.** For an investigation on the picosecond timescale, an $Nd^{3+}$-glass laser generates both the pump and probe beams. Frequency-doubled output at 530 nm (6 ps pulse width) constitutes the pump beam; for the probe beam for the dissociation reaction, fundamental 1060 nm light and frequency-doubled 530 nm emission are combined and focused through a transparent medium to generate a continuum, from which 440 nm light is isolated with a series of filters. An additional CW He/Cd laser with 447.1 nm output is used as a probe beam to monitor the recombination reaction. For an investigation in the femtosecond regime, pulses from a colliding pulse mode-locked ring laser system are subjected to pulse compression to produce output with a 50 fs duration. The 600 nm rhodamine *G* output is further spectrally broadened and then filtered to generate pump pulses at 580 nm (to initiate photodissociation) and probe pulses at 438 nm (to monitor the formation of the deoxy species).

**Role of the Laser Systems.** To provide ultrafast excitation to induce protein-ligand dissociation and to monitor the subsequent absorption changes of the sample due to dissociation reactions, conformational changes, and ligand rebinding.

**Useful Characteristics of the Laser Light for this Application.** Short-duration pulses on the picosecond or femtosecond timescale, monochromaticity, wavelength tunability.

**Principles Reviewed.** Photodissociation, ligand binding, pump-probe techniques, colliding pulse mode-locked techniques.

**Conclusions.** With ultrashort laser pulses and the pump-probe technique, resolution of the complex mechanism of ultrafast dissociations of protein-ligand complexes is possible.

## Ligand Binding Reactions

In recent years, time-resolved spectroscopy with picosecond and even femtosecond optical pulses has been the preferred method for the study of ligand binding to proteins. In particular, ultrafast spectroscopic techniques have been applied in kinetic investigations of the formation and dissociation of liganded heme proteins. Small ligands such as $O_2$, CO, and NO bind reversibly to pentacoordinated hemoglobin and undergo a reversible photodissociation upon laser excitation of the complex. The kinetics of both the dissociation and rebinding processes can be followed spectroscopically by monitoring the absorbance at the characteristic $\lambda_{max}$ of both the liganded species and the uncomplexed photoproduct. The very fast timescale of these reactions ensures that only localized conformational changes will be detected, with no interferences from large-scale movement of the protein or irreversible chemical transformations. An examination of the information revealed from the selected picosecond and femtosecond absorption studies discussed below will serve to illustrate the enormous potential of ultrafast spectroscopic techniques.

## A Study of Protein-Ligand Dynamics

We initially focus on a classic investigation of liganded hemoglobins employing an elaborate experimental setup to monitor several events in distinct time regions, specifically, the picosecond dissociation of carbon monoxide hemoglobin (carboxyhemoglobin) and the longer processes of protein rearrangement and ligand rebinding.[3] In this study, photodissociation is accomplished with a single 6 ps excitation pulse at 530 nm obtained by frequency doubling the output of an $Nd^{3+}$-glass laser. To generate a probe beam, unconverted fundamental 1060 nm light is combined with a second frequency-doubled beam and focused through a 20 cm cell of ethanol. (The focusing of picosecond pulses through transparent media to create a continuum pulse of comparable duration was a discovery of Alfano and Shapiro in 1970.[4]) The resulting broadband continuum beam passes through a variable delay and a series of filters to isolate 440 nm light which is then focused through a quartz etalon having 6 ps segments. The emerging picosecond probe beam arrives at the sample at the same time as the 530 nm excitation pulse. With a vidicon detector interfaced to an optical multichannel analyzer, the picosecond dissociation kinetics are monitored in the characteristic Soret band of porphyrin complexes near 440 nm. Changes in the hemoglobin structure associated with ligand dissociation are reflected in a red shift of this absorbance band. To monitor the subsequent recombination of CO and hemoglobin, as well as any accompanying protein structural transformations, a continuous wave He/Cd laser beam at 441.7 nm is used that is spatially coincident with both the excitation and probe pulses at the sample. A photomultiplier detects absorbance changes at times greater than 50 $\mu$s.

The ultrafast timescale of events is not the sole reason for the elaborate array of lasers and optical devices to study the dynamics of ligand dissociation from hemoglobin. In particular, the three-dimensional and quaternary structure of this protein contribute to the complexity of the investigation. Hemoglobin is an intricate protein consisting of four polypeptide chains packed in a tetrahedral array via noncovalent interactions to form a rather spherical molecule.[5] Two of the chains, the $\alpha$-subunits, contain 141 amino acid residues, while the remaining two chains, the $\beta$-subunits, are 146 residues in length. Each chain is associated with one heme group, a nonpolypeptide unit of Fe-protoporphyrin IX, enabling four $O_2$, CO, or $CO_2$ molecules to bind per hemoglobin molecule. While interactions between $\alpha$ chains or between $\beta$ chains are limited, the $\alpha$–$\beta$ subunit interactions are prominent and change upon ligation to markedly alter the quaternary structure of hemoglobin.

In deoxyhemoglobin, electrostatic interactions between eight pairs of oppositely charged groups ("salt links") constrain the subunits to form a taut conformer, the *T* (tense) form. To minimize steric repulsions between the proximal histidine (occupying the fifth coordination site of $Fe^{2+}$) and nitrogen atoms of the porphyrin, the iron atom is forced about

0.06 nm out of the heme plane in the *T* state. In contrast, in hemoglobin complexes the iron atom of the heme must move back into the porphyrin plane to form a strong bond with the ligand as shown in Figure 5-3. The movement of the iron atom necessarily induces a change in the orientation of the proximal histidine residue, disrupts the salt links, and creates a thermodynamically less favorable interaction of the histidine residue and porphyrin ring. However, the energy stabilization gained by the iron-ligand linkage more than compensates for these energy costs. The fully coordinated complex is designated as the *R* (relaxed) form to indicate this less compact conformation. The overall $T \rightarrow R$ transition occurs on a microsecond timescale.[6]

Why should these slower ligand-induced conformational changes affect the kinetics of the ligand binding and dissociation reactions? Recall that hemoglobin contains four binding sites and that deoxyhemoglobin molecules generally exist in the *T* state. As successive ligand molecules bind to form complexes (denoted HbL, $HbL_2$, etc.), salt links are broken and the *R* form becomes increasingly probable. Likewise, photodissociation of $HbL_4$ at low pulse energies is likely to generate predominantly $HbL_3$ complexes which are essentially in the *R* state. The subsequent recombination reaction observed will thus involve CO and a hemoglobin complex in the R form. As higher degrees of photodissociation occur with higher pulse energies, increased extents of $R \rightarrow T$ conversion are likely. The recombination rate of CO and Hb (now in the *T* form) would thus appear to occur on a slower timescale. Thus, the dependence of the hemoglobin conformation on the degree of ligation has a profound impact on the dynamics of carboxyhemoglobin formation and dissociation.

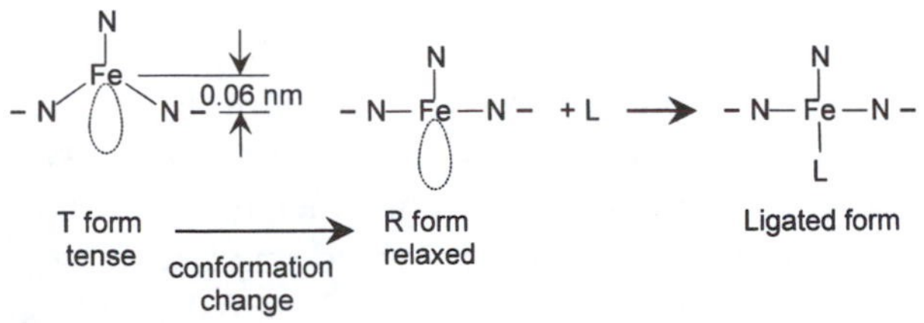

**Figure 5-3**
Some of the conformational changes occurring upon the ligation of hemoglobin. Deoxyhemoglobin molecules exist in the taut conformer, denoted the *T* (tense) form. In deoxyhemoglobin, steric repulsion between the carbon atom of the histidine imidazole ring and nitrogen atoms of the porphyrin ring force the iron atom about 0.6 Å out of the heme plane. Upon ligation, the iron atom moves into the heme plane to form a strong bond with the ligand. The less-compact conformation of the fully coordinated complex is designated as the R (relaxed) state.

The results of the investigation of Noe et al. [3] should be analyzed with the preceding discussion in mind. The photodissociation was investigated at 4°C from 0 to 48 ps. Formation of the dissociated species was observed to follow first-order kinetics with a lifetime of 11 ps. Several events are speculated to occur during this time regime, including vibrational relaxation and intersystem crossing to the dissociative continuum. The 6 ps pulse duration limits further exploration of these processes. The particular laser systems employed also restrict the time period between 48 ns and 50 $\mu$s from being studied. After 50 $\mu$s, however, several events are observed to occur in three distinct time regions depending on the pulse energy of the laser excitation. A first-order process with a lifetime of approximately 140 $\mu$s was observed with pulse train energies of only 5 mJ and is attributed to a recombination of CO with deoxy Hb in the *R* state. At 7.5 mJ, a second slower recombination process with an 11 ms lifetime also appears, consistent with a similar rebinding of CO to *T*-deoxy Hb. Recall that higher pulse energies induce greater photodissociation and that the preferred Hb conformation is dependent on the degree of ligation. With an increase in excitation pulse energy above 12 mJ, this second recombination process is delayed ($\tau_{1/2}$ = 15 ms). What factors could contribute to the delay in the appearance of the *T*-deoxy Hb and CO recombination? The higher degree of complete photodissociation achieved with these higher pulse energies likely induces more substantial quaternary $R \rightarrow T$ conformational changes and more extensive Hb subunit tertiary structural changes, delaying recombination.

## Femtosecond Photolysis of CO-Ligated Hemoglobin

To more precisely resolve the kinetics of formation of the nonliganded Hb species after photodissociation of CO, a spectroscopic investigation using laser pulses of 50 fs duration has recently been conducted.[7] The laser system used to generate these tunable ultrashort pulses was a ring cavity laser using the colliding pulse mode-locked (CPM) technique. In this system, a continuous wave argon ion laser optically pumps a dye laser that is passively mode-locked by means of a saturable absorber. A rhodamine-590 (R–6G) dye in ethylene glycol serves as the gain medium, while a 3,3′-diethyloxadicarbocyanine iodide dye (DODCI) in ethylene glycol serves as the saturable absorber. In the bidirectional ring cavity configuration, counter-propagating pulses collide yielding output pulses which are at 620 nm with a 50 ps FWHM and an energy of ≈ 100 pJ. The fs pulses are subjected to energy amplification and optical pulse compression. Recall our discussion of pulse compression in Chapter 3. The optical pulse compression reduces pulse modulation and generates significantly shorter pulses. After compression, the rhodamine-*G* dye laser output (at 600 nm) is split, and each beam is focused into an 8 mm water cell to generate two white light continua. The emerging spectrally broadened light is further amplified. Pulses of 20–500 fs duration can be obtained by spectral filtering of the amplified beam. A 580 nm beam initiates photodissociation of the carboxyhemoglobin complex. A 438 nm probe beam is selected from the second generated white continuum beam with optical interference filters and is further split to generate both a test beam and a reference beam.

Transient differential absorption spectra were recorded at 438 nm corresponding to the isosbestic point of carboxyhemoglobin and the deoxy form. The formation of the ground-state deoxy species (assayed at 438 nm) occurs with a 250 ± 50 fs time constant. What additional mechanistic information do these results reveal? Martin et al. [8] hypothesize that the timescale for the formation of the deoxy species is indicative of an instantaneous formation of a transient high-spin liganded species that is highly photodissociative. A likely state is one characterized by an electron in the antibonding $d_{z^2}$ orbital which leads to an increase in the Fe–CO bond distance. The kinetic results also support the postulate that displacement of the iron out of the heme plane occurs within a 250 fs time domain. This notion is consistent with the absence of any change in the electronic spectra of the deligated species between 1 and 100 ps. Molecular dynamics calculations of 0.1 nm nuclear displacements further suggest a subpicosecond process.[9] Perhaps an even more fundamental outcome of this study is the revelation of the speed with which chemical bonds can be broken.

### Summary of the Case Study

The availability of ultrafast techniques has led to a better understanding of both the process and the time involved in ligand binding to the available coordination sites of hemoglobin. In particular, we note that:

- Photodissociation to yield a ground-state deoxy species occurs within 250 fs.
- Higher degrees of photodissociation lead to increased extents of $R \rightarrow T$ conversion.
- Recombination of CO with *R*-deoxy Hb occurs with a lifetime of 140 $\mu$s.
- Recombination of CO with *T*-deoxy Hb occurs with a lifetime of 11–15 ms.

## Case Study II: Reaction Dynamics in the Ultrafast Regime

### Overview of the Case Study

**Objective.** To follow the rate of the unimolecular dissociation of a model triatomic compound ICN in real time and gain in the process a mapping of the potential energy surfaces that describe ground and excited states of ICN.

**Laser Systems Employed.** A colliding pulse mode-locked ring dye laser system with pulse compression is used in a pump-probe configuration to generate femtosecond pulses. The 614 nm output of an argon ion pumped dye laser (rhodamine-6G dye) is the basis of the colliding pulse mode-locked ring laser system. Frequency-doubled (614 nm → 307 nm) fs pulses are produced for the pump beam. The 614 nm emission is combined with the 1.06 $\mu$m output of an $Nd^{3+}$:YAG laser to generate the 388.5 nm probe beam.

**Role of the Laser Systems.** The pump beam is used to generate excited ICN molecules that the tunable probe laser beam interrogates as a function of time and wavelength.

**Useful Characteristics of the Laser Light for this Application.** The femtosecond pulse is essential to follow the time course of the dissociation which is complete in picoseconds. Fluorescence induced by the probe laser provides the analytical signal to monitor the disappearance of ICN.

**Principles Reviewed.** Unimolecular dissociations, potential energy surfaces, pump-probe technique, colliding pulse ring laser, laser-induced fluorescence.

**Conclusions.** The study of the unimolecular dissociative kinetics of ICN increases our understanding of the energetics of processes occurring in picosecond regimes.

## Unimolecular Reactions

A unimolecular dissociation is virtually the simplest type of chemical reaction. In such a dissociation, a single molecule separates into metastable molecular fragments or new stable molecules. Equation 54 (presented earlier) gives a generic example. The distinct time = zero starting point for unimolecular reactions simplifies the study of their reaction kinetics. In other words, the timing of the reaction can be uniquely defined when a pump laser pulse creates the excited state or activated complex. For these reasons, our discussion will focus on ultrafast unimolecular reactions.

## Photolytic Decay of ICN

A soon-to-be-considered classic reaction is the unimolecular decay of the triatomic molecule iodocyanogen ICN in the gas phase.[10–14] The actual process is modelled as:

$$\text{ICN} + h\nu\,(\text{pump}) \rightarrow [\text{ICN}]^{\#*} \rightarrow \text{I} + \text{CN} \quad (56)$$

To understand the energetics of this process we turn to the potential energy surfaces presented in Figure 5-4. These potential energy surfaces are of great utility to understand the details of the reaction mechanism and are of great intrinsic theoretical interest in their own right. In fact, a major goal of quantum chemistry is to calculate such potential energy surfaces (PESs). Furthermore, an understanding of molecular electronic spectroscopy depends critically on the PESs. How should we interpret these PESs? The lowest curve, PES = $V_0$, describes the ground-state stable molecule ICN with an equilibrium I-CN bond length of $R_0$. The PES denoted $V_1$ describes the repulsive (i.e., antibonding) state of ICN equivalent to the electronically excited activated complex $[\text{ICN}]^{\#*}$. The PES denoted $V_2$ describes another repulsive state, only this time the repulsive entities are I and an excited $\text{CN}^*$ molecular fragment. Recall that the term *repulsive* implies that no attractive potential is present to hold the atoms or molecular fragments together. If a molecule is excited to a repulsive state, it must dissociate, since there is no "glue" to hold the entities together. In

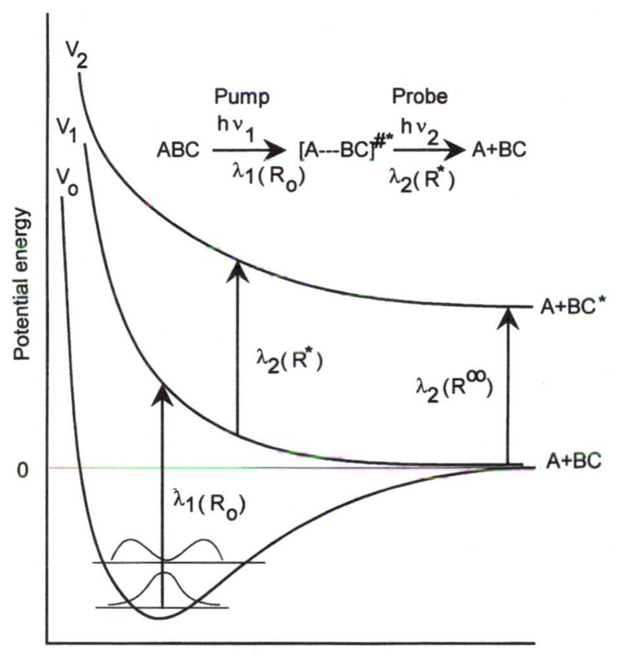


**Figure 5-4**
Potential energy curves for the analysis of the photolytic decay of ICN. The potential energy curve for the ground-state bound molecule ABC is labelled $V_0$, and the potential energy curves for the first and second dissociative energy states are marked as $V_1$ and $V_2$. $V_1$ describes the antibonding or repulsive state of ABC equivalent to the electronically excited activated complex $[\text{ABC}]^{\#*}$, and $V_2$ describes the repulsive state of the fragments A and $\text{BC}^*$. A molecule excited to either $V_1$ or $V_2$, as repulsive states, would "fall apart" from the absence of attractive potential. The pump pulse ($\lambda_1$) takes the molecule from $V_0$ to $V_1$ and the probe pulse ($\lambda_2$) from $V_1$ to $V_2$.

fact, the only question pertinent to a molecule raised to a repulsive state is the length of time that will pass before the molecule dissociates, quantified by the unimolecular reaction rate. Perhaps we should note here that the Franck-Condon principle asserts that the time for the electronic transition to occur (i.e., the time for the photon to be absorbed or emitted) is on the subfemtosecond timescale. Thus, photon absorption or emission is not a rate-influencing process.

For this investigation the wavelength of the pump pulse for ICN was 307 nm, with a laser pulse duration of 40 to 60 fs, obtained by frequency doubling the 614 nm femtosecond output of an argon ion pumped dye laser with rhodamine-6G dye as the lasing medium. This excitation raises the energy of ground-state ICN molecules at their equilibrium internuclear distance of $R_0$(I–CN) = 0.27 nm (270 pm) to the repulsive state $V_1$ at the same internuclear $R$(I–CN) distance (a consequence of the Franck-Condon principle) at clock time $t = 0$. Since $V_1$ is a repulsive state, I should separate from CN. How quickly does this separation occur? What kind of clever experiment can be designed to answer the question? In this study, two types of probe experiments were performed separately. One experiment set the probe pulse to 388.5 nm, the vibrational absorption wavelength for pure CN fragments, by mixing the 614 nm femtosecond output with 1.06 $\mu$m radiation from an $Nd^{3+}$:YAG laser using a second nonlinear crystal. Probe pulses were sent into the sample at various times after the initial pump pulse (that generated $[ICN]^{\#*}$). The time at which the 388.5 nm probe pulse is absorbed defines the complete reaction time, that is, the overall time for the reaction below to occur.

$$ICN \rightarrow I + CN \quad (57)$$

Since molecules travel 300 m $s^{-1}$ (a typical number estimated from the kinetic molecular theory of gases) and since I and CN seem to be "completely" separate in the $V_1$ PES at 0.6 nm, we would anticipate this total reaction time to be:

$$\text{reaction time} \approx 0.6 \times 10^{-9} \text{ m}/300 \text{ ms}^{-1} \approx 200 \text{ fs} \quad (58)$$

We will discuss shortly how the signals are detected in this time regime, but for now we must believe that it is possible to measure such signals. Note in this case that the fate of the probe pulse is followed by absorption spectroscopy.

A second pump-probe experiment used a different sequence of measurements to follow the rate of the decomposition of $[ICN]^{\#*}$ and characterize the steepness or curvature of $V_1$. Recall that an ICN molecule in the repulsive $V_1$ state must dissociate—we want to know the time that it takes to separate the I and CN fragments a requisite distance $R$. These measurements interrogated the excited ICN (generated with the pump pulse) with probe photons of lower energy, specifically from 388.5 to ≈ 400 nm.[2] The fate of these probe photons was followed using laser-induced fluorescence (LIF), for the probe pulses created the electronically excited $CN^*$ fragment that returns to the ground state via fluorescence.

To understand how these fluorescence photons reflect the $[ICN]^{\#*}$ species, let us look in some detail at the representative shapes of $V_1$ and $V_2$ in the range $R_0$ to $R \approx 0.6$ nm. The distinctive shapes of these potential energy surfaces indicate that the energy separation of the PES changes with distance. Thus, a photon of different energy is required to excite $[ICN]^{\#*}$ at $R = 0.3$ nm than at $R = 0.35$ nm. In fact, note that for $R_0 < R < 0.6$ nm, less energy (i.e., photons of longer wavelength) is required for the transition from $V_1$ to $V_2$ as $R$ decreases. We further recall that ICN described by $V_2$ will fluoresce. How can we clock the time $t'$ it takes for $[ICN]^{\#*}$ to slide down $V_1$ to the I and CN products with a separation distance $R(t')$? One method would be to promote $[ICN]^{\#*}$ to $V_2$ and monitor the subsequent fluorescence emission without generating any fluorescence due to direct absorption by CN fragments. Thus, we would need to choose a photon energy away from the 388.5 nm CN absorption but suitable to cause a transition from $V_1$ to $V_2$ at some $R_0 < R < 0.6$ nm, and then measure the fluorescence due to the probe photon as a function of time. When the energy of the probe photon matches the energy of the $V_1$ $V_2$ separation at a certain I–CN distance $R$, the probe photon will induce the $V_1$ to $V_2$ transition and produce a strong fluorescent signal. The appearance of the fluorescent signal will mark the time $t'$ required to separate the I and CN fragments a distance $R$ at $t'$. Remember that the pump photon generates $[ICN]^{\#*}$ at $R_0$ and that the I and CN fragments separate to a distance $R$ over a time interval $t$. The distance $R$ determines the wavelength of the probe photon necessary to induce fluorescence. By repeating this experiment for a range of detuned (that is, $\lambda \neq 388.5$ nm) probe pulse photons, the time-distance information of the travel along $V_1$ will be mapped out. We should expect to see a fluorescent signal for a detuned probe photon as shown in the lower diagram of Figure 5-5 with an increase in signal intensity to a maximum at the confluence of the correct energies and time. Moreover, there should be a shifting of the maximum to shorter times (corresponding to separations $R$ closer to $R_0$) as the probe pulse is detuned closer to the energy of separation between $V_1$ and $V_2$ at $R_0$. We should note that the upper diagram of Figure 5-5 shows the expected signal from the probe pulse tuned to the CN absorption at 388.5 nm. Here the signal should reach a constant level at the time and distance corresponding to "complete" separation of the I and CN entities.

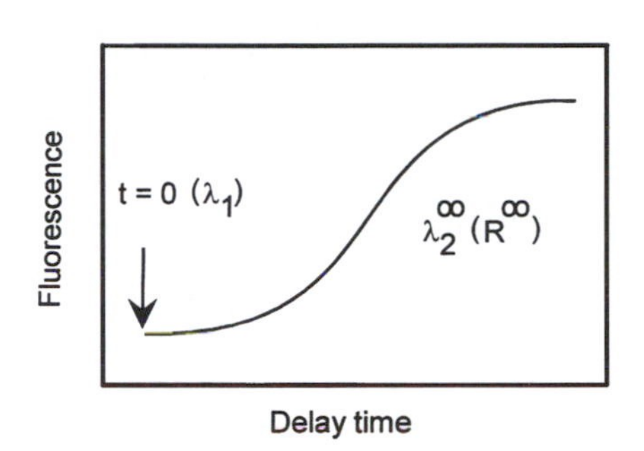


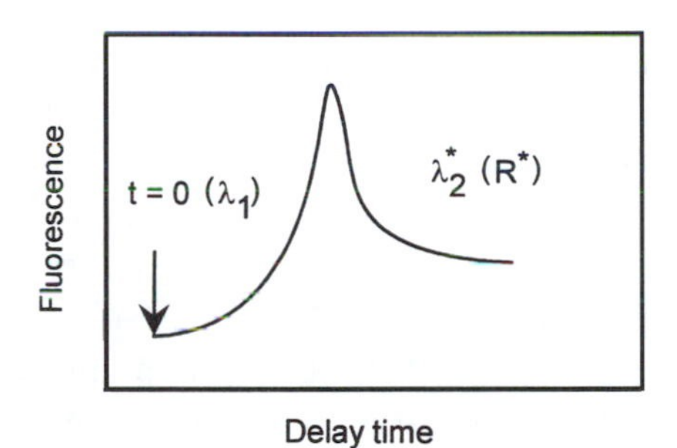


**Figure 5-5**
The transient fluorescence signals from $CN^*$ expected as a consequence of probe pulses generated at various wavelengths. The pump pulse produces the electronically excited activated complex $[ICN]^{\#*}$, and the probe pulse promotes $[ICN]^{\#*}$ to the repulsive state $V_2$ (see Figure 5-3) that fragments into I and the fluorescent $CN^*$. The lower diagram is the time-dependent fluorescence signal expected when the energy of the probe photon matches the energy separation of the $V_1$ and $V_2$ potential energy curves at a certain I-CN distance $R^*$. The magnitude of the delay time (the time between probe pulse and the appearance of the fluorescence signal) reflects the distance of separation of the I and CN products. The upper diagram results when the probe pulse is tuned to the CN absorption at 388.5 nm. The constant level of the fluorescence signal occurs at the time and distance that reflects the complete separation of the I and CN fragments.

# Case Study III: Photoinduced Electron Transfer

## Overview of the Case Study

**Objective.** To monitor the kinetics of photoinduced electron transfer to generate a charge-separated species consisting of a carotenoid radical cation covalently linked to a fullerene radical anion.

**Laser System Employed.** A pump-probe apparatus is utilized to measure the absorption spectrum of the transient charge-separated species generated by electron transfer. A frequency-doubled mode-locked $Nd^{3+}$:YAG laser is employed to synchronously pump a dye laser. Subsequent pulse compression and amplification with a second frequency-doubled $Nd^{3+}$:YAG laser yield 590 nm, 80 fs, and 0.5 nJ pulsed output. The pump and probe pulses are then separated, with the probe pulse focused through a water cell for continuum generation and subsequent frequency selection.

**Role of the Laser System.** To provide ultrafast excitation to: (1) generate the lowest excited state of the fullerene in the carotenoid-fullerene dyad and (2) monitor the absorption of the transient radical pair formed upon subsequent electron transfer from the carotenoid to the excited-state fullerene.

**Useful Characteristics of the Laser Light for this Application.** Short-duration pulses of 150 fs duration, monochromaticity, wavelength tunability.

**Principles Reviewed.** Photoinduced electron transfer, charge separation, pump-probe technique.

**Conclusions.** The pump-probe technique accurately monitors two ultrafast processes: intramolecular electron-transfer chemistry arising from the generation of the fullerene excited state and the subsequent decay of the charge-separated state consisting of carotenoid radical cation and fullerene radical anion. The successful demonstration of photoinduced electron transfer will influence the design of other synthetic electron donor-acceptor systems for a variety of possible applications, including the modeling of photosynthetic solar energy conversion.

## Photoinduced Unimolecular Electron Transfer

As a final example of ultrafast processes, we now focus on an example of light-induced electron transfer from a donor species to an acceptor to create a transient radical pair. This process of photochemical charge separation can be described by a two-step mechanism involving either excitation of the donor (*D*) or acceptor (*A*) to an excited singlet state, followed by electron transfer to create the donor radical cation and the acceptor radical anion:

$$D \xrightarrow{h\nu} D^* \text{ or } A \xrightarrow{h\nu} A^*$$
$$D^* + A \rightarrow D^+ + A^- \text{ or } D + A^* \rightarrow D^+ + A^-$$

One representative example of photochemical charge separation is the early event in the photosynthetic process in plants whereby a special pair of chlorophyll molecules (symbolized by $P_{680}$ for pigments with maximal absorption at 680 nm), activated by excitation transfer from an antenna of chlorophyll molecules, serves as the electron donor to a nearby pheophytin molecule (symbolized by I) as the initial electron acceptor:

$$P_{680}I \xrightarrow{h\nu} P_{680}{}^{*}I \rightarrow P_{680}{}^{+}I^{-}$$

Synthetic systems designed to model or mimic photosynthetic energy conversion typically employ porphyrin or other chlorophyll derivatives as electron donors and efficient electron acceptors such as benzoquinones, naphthoquinones, and anthraquinones.[15,16]

In the following case study, the photoinduced electron transfer is a unimolecular process. The covalently linked electron donor-acceptor system, referred to as a dyad, consists of a carotenoid pigment as donor and a $C_{60}$ fullerene as acceptor. The linkage of two chromophores represents the simplest system to explore the rate and efficiency of photochemical electron transfer.

## Transient Absorption Measurements

The investigators used the pump-probe technique to photoinduce electron transfer and monitor both the formation and decay of the charge-separated state.[17] A dye laser synchronously pumped by a frequency-doubled, mode-locked $Nd^{3+}$:YAG laser generated ultrashort pulses of 80 fs duration at 590 nm. With pulse amplification using a second frequency-doubled $Nd^{3+}$:YAG laser, output of 200 $\mu$J pulse energy and 150 fs duration was produced and split for the pump and probe beams. The pump beam was maintained at 590 nm and reduced in intensity with neutral density filters. Excitation of the dyad at this wavelength generates the lowest excited singlet state of the dyad. The excited singlet state is assigned to the fullerene moiety, Car-$^1C_{60}$, as the carotenoid polyene has negligible absorption at wavelengths above 550 nm. The charge-separated state $Car^{\bullet+}C_{60}{}^{\bullet-}$ is energetically accessible from this short-lived excited state. In order for the probe beam to monitor absorption by the charge-separated state, excitation near the 1068 nm absorption maximum of $Car^{\bullet+}$ must be generated. Excitation of the requisite wavelength is achieved by focusing the probe beam through a water cell for continuum generation and then filtering to select 1080 nm excitation. For maximum sensitivity, the probe beam is split into two equal parts to conduct transient absorption measurements, a form of difference spectroscopy whereby absorption by the sample is detected by comparison to a reference beam.

This investigation yielded measurements of the ultrafast kinetics of both the formation and decay of the transient charge-separated state. The rise of the 1080 nm absorbance was characterized as a first-order process with a lifetime of 800 fs, and the subsequent first-order decay of the 1080 nm signal fit a single exponential with a lifetime of 534 ps. The quantum yield of the charge separation process is estimated to be greater than 99+% due to the lack of observable fluorescence emission and the low quantum yield of either carotenoid or fullerene triplet formation (as determined via absorption measurements on the nanosecond timescale). The rapid and efficient electron transfer in this system suggests that this dyad may be useful in the design of further models for photosynthetic solar energy conversion. Ultrafast laser techniques will be vital to the detection and quantification of the unimolecular photochemistry in such systems.

## SUMMARY

Each advance in reducing pulse time duration allows us to write a more descriptive journal of the path from reactants to products. Major advances in our understanding of unimolecular and bimolecular reactions and their potential energy surfaces are reported every day. We can only guess where knowledge of the femtosecond world will lead.

## FOR FURTHER EXPLORATION

Are you intrigued by the prospect of "real-time" observations of chemical reactions? A reaction dynamicist shares this interest with efforts to characterize the range of intermediates that evolve during the transition from reagents to products. Given the 1–10 ps timescale of a simple exchange reaction, $A + BC \rightarrow AB + C$, femtosecond resolution is necessary to monitor the entire process of bond breaking and bond formation. Ultrashort laser pulses facilitate the clocking of transition states in the femtosecond time regime with the ability to establish a well-defined zero of time and to achieve a time resolution adequate to probe the system for evolving states. Current time resolutions of 6 fs are state-of-the-art.[18] A variety of spectroscopic detection methods on the femtosecond timescale (e.g., absorption, fluorescence, and multiphoton ionization mass spectrometry) permit the nuclear motion from reactants to final products to be followed. The dynamical mapping of different classes of reactions provides further insights into means of controlling reactive pathways.[14]

## DISCUSSION QUESTIONS

1. How could the instrumentation in Figure 5-1 be modified to use a single laser to probe the sample at multiple time intervals?
2. How could the rate of an ultrafast process be modified so the experimenter could measure the reaction rate using either conventional spectroscopy or the pump-probe technique with a slower-pulsed laser?
3. Suppose we had equipment capable of spectroscopic measurements on the timescale of 1 fs. Describe what processes we could study to follow the origin of a fluorescent spectral line.

## SUGGESTED EXPERIMENTS

Suggested references to experiments that illustrate some of the principles described in this chapter:

1. Goodall, D. M., and Roberts, D. R., "Energy Transfer between Dyes," *Journal of Chemical Education,* 1965, 62, 711–714. An experiment to demonstrate the use of fluorescence measurements to monitor the process of energy transfer and to determine the average distance between donor and acceptor molecules. Although designed for a conventional fluorescent spectrophotometer, this experiment could be modified to use laser excitation sources and photomultipliers for detection of fluorescent emission.
2. Grieneisen, H. P., "The Nitrogen Laser-Pumped Dye Laser: An Ideal Light Source for College Experiments; Experiment 4—Laser-Induced Fluorescence in $I_2$ Vapor," Laser Science, Inc., Cambridge, MA. An experiment that illustrates the excitation of $I_2$ fluorescence using a nitrogen-pumped dye laser.
3. Tellinghuisen, J., "Laser-Induced Molecular Fluorescence," *Journal of Chemical Education*, 1981, 58, 438–441. An experiment that illustrates the excitation of $I_2$ fluorescence using a helium-neon laser.

## LITERATURE CITED

1. Zewail, A. H., "Laser femtochemistry," *Science*, 1988, 242, 1645–1653.
2. Zare, R. N., and Dagdigian, P. J., "Tunable laser fluorescence method for product state analysis," *Science*, 1974, 185, 739.

3. Noe, L. J., Eisert, W. G., and Rentzepis, P. M., "Picosecond photodissociation and subsequent recombination processes in carbon monoxide hemoglobin," *Proc. Natl. Acad. Sci.*, USA, 1978, 75, 573–577.
4. Alfano, R. R., and Shapiro, S. L., "Observations of self-phase modulation and small-scale filaments in crystals and glasses," *Phys. Rev. Lett.*, 1970, 24, 592–594.
5. Stryer, L., *Biochemistry*, Second Edition, W. H. Freeman, San Francisco, 1981, Chapter 4.
6. Hofrichter, J., Sommer, J. H., Henry, E. R., and Eaton, W. A., "Nanosecond absorption spectroscopy of hemoglobin: Elementary processes in kinetic cooperativity," *Proc. Natl. Acad. Sci.*, USA, 1983, 80, 2235–2239.
7. Vos, M. H., Martin, J. L., "Femtosecond measurements of geminate recombination in heme proteins," *Methods Enzymol.*, 1994, 232, 416–431.
8. Martin, J. L., Migus, A., Poyart, C., Lecarpentier, Y., Astier, R., and Antonetti, A., "Femtosecond photolysis of CO-ligated protoheme and hemoproteins: Appearance of deoxy species with a 350 fs time constant," *Proc. Natl. Acad. Sci.*, USA, 1983, 80, 173–177.
9. McCammon, J. A., and Karplus, M., "The dynamic picture of protein structure," *Acc. Chem. Res.*, 1983, 16, 187–193.
10. Dantus, M., Rosker, M. J., and Zewail, A. H., "Real-time femtosecond probing of 'transition states' in chemical reactions," *J. Chem. Phys.*, 1987, 87, 2395–2397.
11. Rose, T. S., Rosker, M. J., and Zewail, A. H., "Femtosecond real-time observation of wave packet oscillations (resonance) in dissociation reactions," *J. Chem. Phys.*, 1988, 88, 6672–6673.
12. Rosker, M. J., Rose, T. S., and Zewail, A. H., "Femtosecond real-time dynamics of photofragment-trapping resonances on dissociative potential energy surfaces," *Chem. Phys. Lett.*, 1988, 146, 175–179.
13. Rosker, M. J., Dantus, M., and Zewail, A. H., "Femtosecond clocking of the chemical bond," *Science*, 1988, 241, 1200–1202.
14. Bernstein, R. B., and Zewail, A. H., "Real-time laser femtochemistry: Viewing the transition from reagents to products," *Chemical & Engineering News*, 1988, 66(45), 24–44.
15. Wasielewski, M. R., "Photoinduced electron transfer in supramolecular systems for artificial photosynthesis," *Chem. Rev.*, 1992, 92, 435–461.
16. Gust, D. A., Moore, T. A., and Moore, A. L., "Molecular mimicry of photosynthetic energy and electron transfer," *Acc. Chem. Res.*, 1993, 26, 198–205.
17. Imahori, H., Cardoso, S., Tatman, D., Lin, S., Noss, L., Seely, G. R., Sereno, L., Chessa de Silber, J., Moore, T. A., Moore, A. L., and Gust, D., "Photoinduced electron transfer in a carotenobuckminsterfullerene dyad," *Photochem. Photobiol.*, 1995, 62, 1009–1014.
18. Polanyi, J. C., and Zewail, A. H., "Direct observation of the transition state," *Acc. Chem. Res.*, 1995, 28, 119–132.

## GENERAL REFERENCES

Bernstein, Richard B., and Levine, Raphael D., *Molecular Reaction Dynamics and Chemical Reactivity*, Oxford University Press, 1987.

Rentzepis, P.M., "Advances in Picosecond Spectroscopy," *Science*, 1982, 218, 1183–1188.

Shank, Charles V., "Investigation of Ultrafast Phenomena in the Femtosecond Time Domain," *Science*, 1986, 233, 1276–1280.

Shank, Charles V., "Measurement of Ultrafast Phenomena," *Science*, 1983, 219, 1027–1032.

Williamson, J.C., and Zewail, A.H., "Structural Femtochemistry: Experimental Methodology," *Proc. Natl. Acad. Sci.*, USA, 1991, 88, 5021–5025.

Wirth, Mary J., "Ultrafast Spectroscopy," *Anal. Chem.*, 1990, 62, 270A–277A.

CHAPTER 6

# Multiphoton Spectroscopy: Single Photon Spectroscopy Is Not the Only Way

## Chapter Overview

Normal spectroscopy involves the absorption or emission by a molecule of one photon at a time. With intense laser light, it is possible to observe the absorption of more than one photon at a time by a given molecule. Studying multiphoton events has led to some very elegant and precise spectroscopy, two examples of which are highlighted in this chapter.

## Multiphoton Processes

Most everyone is familiar with spectroscopy, the study of the interaction of electromagnetic radiation with matter. Many spectroscopic studies focus on the absorbance by a sample according to the Law of Photochemical Equivalence, a principle detailing the one-to-one correspondence between a photon and an activated species. What are the consequences of multiphoton absorption—the process whereby a sample absorbs more than one photon per molecule?

As we consider the absorption of multiple photons by a target molecule, we immediately distinguish two cases: (1) sequential absorption of $n$ single photons and (2) concerted absorption of $n$ photons simultaneously. Since the timescale of photon absorption is in the realm of femto- to subfemtoseconds, a time resolution of these two cases is not practical. The schematic shown in Figure 6-1 based on relative energy levels available to the molecule will help to differentiate between these two cases.[1,2] The initial state is denoted $E_n$, the final state $E_m$, and the intermediate state $E_k$. While we have drawn only a single horizontal line to illustrate a single energy state ($E_n$, $E_m$ , or $E_k$), each of these "states" is more accurately represented by either a number of closely-spaced energy levels (for example, vibrational-rotational levels associated with a given molecular electronic energy state) or a spread in energy due to the uncertainty principle. Figure 6-1(a) illustrates a sequential absorption of two photons of equal energy $h\nu_1$, initially exciting the molecule to state $E_k$, then to state $E_m$. Figure 6-1(a′) features the same process, but the photon energies for excitation are not equal. This latter case is common in multiphoton spectroscopy and can easily be accomplished using two different light sources. Concerted multiphoton spectroscopy is illustrated by Figures 6-1(b) and 6-1(b′). In these two cases, the incident photons individually do not have the correct energy to reach a stationary intermediate state such as $E_k$ but still manage to raise the molecule to the energy state $E_m$. Figures 6-1(a) and 6-1(a′) can also be viewed as illustrations of a type of absorption described as *resonant* absorption spectroscopy.

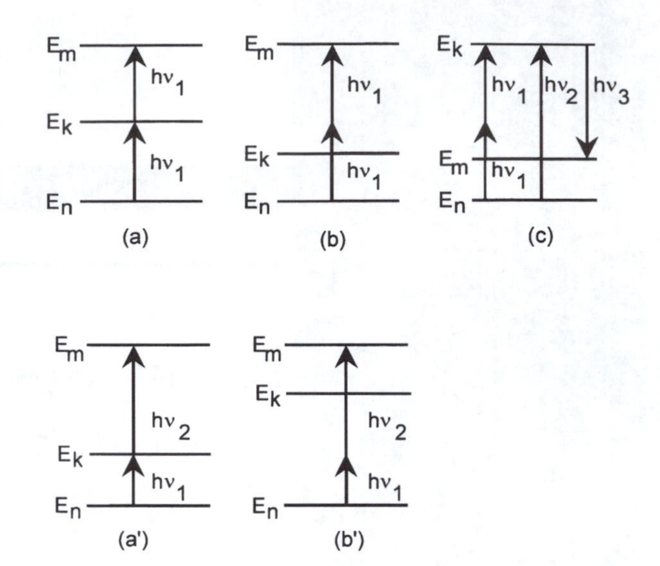


**Figure 6-1**
Schematic energy levels showing possible multiphoton spectral transitions between the energy states $E_n$, $E_k$, and $E_m$. In *(a)*, sequential absorption of two photons of equal energy $h\nu_1$ first excites a molecule in initial state $E_n$ to a stationary intermediate state $E_k$, then to a final state $E_m$. The same process is diagrammed in *(a′)*for two excitation photons of different energy. Concerted multiphoton absorption is illustrated by *(b)* and *(b′)*, where the energy of neither absorbed photon matches the energy difference between $E_k$ and $E_n$, but the total energies of the incident photons match the energy difference between $E_m$ and $E_n$. In diagram *(c)*, a molecule absorbs either two photons each of energy $h\nu_1$ or one photon of energy $h\nu_2$ to reach an intermediate state $E_k$. A photon is then emitted as the molecule reaches the final state $E_m$. The two-photon absorption process is properly called a multiphoton absorption; the one-photon absorption is a classic spontaneous Raman effect (see Chapter 9).

Figure 6-1(c) describes the origin of a typical vibrational Raman Stokes line of Raman spectroscopy (see Chapter 9). A case could be made for Raman spectroscopy to properly be classified as a subfield of general multiphoton spectroscopy. The argument for calling Raman spectroscopy a multiphoton process follows from the idea that a photon is first "absorbed" by the molecule reaching a "virtual" nonstationary state $E_k$ with the result of a second photon being emitted. If $E_k$ is reached by a single photon of energy $h\nu_2$, the spectroscopy is classic spontaneous Raman. If, however, $E_k$ were reached via a multiphoton process, such as the concerted absorption of two $h\nu_1$ energy photons, the process should be called multiphoton spectroscopy. Moreover, if $E_k$ were a stationary state reached by emission of a photon of energy $h\nu_3$, the process could be called multiphoton fluorescence spectroscopy.

The energy levels in Figure 6-1 allow us to speak of other processes as well:[3]

- Stimulated emission from $E_m$ to state $E_n$, in effect a two-photon laser line.
- Stimulated emission from state $E_m$ to resonant or virtual $E_k$ state, again a lasing process, followed by a spontaneous emission to return to $E_n$.
- Spontaneous emission from $E_m$ to $E_k$, followed by another spontaneous emission returning to $E_n$.

## Lasers and Multiphoton Spectroscopy

Multiphoton spectroscopy, even from our brief discussion so far, appears to offer many interesting avenues for detailed study of excited molecules. Why is such spectroscopy so relatively new? As you might surmise, the developments in this area of spectroscopy are linked to the recent advances in laser technology. What does the laser bring to a multiphoton experiment that was not previously available from other light sources? The answer lies most simply in intensity; the immense irradiance available with certain lasers now makes multiphoton spectroscopy a viable experimental technique. Now let us explore in some detail why and how lasers are an integral aspect of multiphoton spectroscopy.

## Multiphoton Electronic Spectroscopy

Our discussion will focus on concerted multiphoton spectroscopy, for it is the concerted aspect that is unique to this type of spectroscopy. Our discussion will also be restricted to electronic spectroscopy, the study of photon absorption in the visible and ultraviolet ranges of electromagnetic radiation. Of course, we must keep in mind that every electronic energy state has a manifold (set) of vibrational and rotational energy levels associated with it, and every change between electronic states also involves vibrational and rotational energy changes as well. Since these latter types of energies can be orders of magnitude smaller than electronic energies, their presence makes each of the $E_{n,m,k}$ states in Figure 6-1 a "fuzzy" line, an energy state with finite energy width.

What is the origin of electronic spectroscopy? The origin is the interaction of the oscillating electric dipole of a photon with the electron charge cloud of a molecule. Since we are interested in the interaction of multiple photons with a single molecule, an interesting question to address would be: How many photons in a typical laser beam can be found in the volume equivalent to that occupied by a representative target molecule?[4] The number of photons in a given volume is calculated by:

$$\text{Irradiance} = \frac{\text{energy}}{\text{area} \cdot \text{time}}$$

$$\text{I} = \frac{(\#\ \text{photons}) \cdot (\text{photon energy})}{\left(\dfrac{\text{volume}}{\text{length}} \cdot \text{time}\right)}$$

$$\text{I} = \frac{(\#\ \text{photons}) \cdot (\text{photon energy})}{\left(\dfrac{\text{volume}}{\text{speed of light}}\right)} \tag{59}$$

$$I = \frac{N h\nu}{\dfrac{V}{c}}$$

$$N = \frac{I}{ch\nu} V$$

where $I$ is the irradiance of the source, $ch\nu$ is the speed of light times the photon energy, and $V$ is the volume of interest. Consider how many photons would be in a volume $V$ equivalent to the size of a toluene molecule $C_6H_5CH_3$:

$$V(C_6H_5CH_3) = \frac{MW}{\text{density} \cdot N_A}$$

$$V(C_6H_5CH_3) =$$

$$\frac{92 \text{ g mol}^{-1}}{0.85\text{g cm}^{-3}10^6\text{cm}^3\text{m}^{-3}6.02 \times 10^{23}\text{molec mol}^{-1}}$$

$$= 1.8 \times 10^{-28} \text{ m}^3\text{molec}^{-1} \quad (60)$$

The number of photons in this volume for an argon ion laser 488 nm ($6.15 \times 10^{14}$ s$^{-1}$) line using Equation 59 is:

$$N = \frac{1.8 \times 10^{-28} \text{ m}^3}{(3 \times 10^8 \text{ m s}^{-1})(6.626 \times 10^{-34} \text{ J s})(6.15 \times 10^{14} \text{ s}^{-1})} I$$

$$N = (1.4 \times 10^{-18} \text{ m}^2\text{s J}^{-1})I \quad (61)$$

For a 1 W argon ion laser, a power level quite useful for many spectroscopy studies:

$$N = (1.4 \times 10^{-18} \text{ m}^2\text{s J}^{-1})\ (1 \text{ J s}^{-1}\text{m}^{-2})$$
$$N = 1.4 \times 10^{-18} \quad (62)$$

Virtually no photons would be found in the molecular volume of toluene at this power level. The conclusion here is that even a 1 W argon ion laser passing through toluene really provides very few photons relative to the number of toluene molecules present. Even in a volume equivalent to a mole of toluene molecules, there would be only $8.8 \times 10^5$ photons compared to $6 \times 10^{23}$ toluene molecules. The interaction of a photon at this power level with a toluene molecule is at best a one-to-one interaction with a 1 in $10^{18}$ chance of occurring. We might note, however, that the interaction is effective when it occurs because of the large "size" of the photon compared to the molecule. Toluene is perhaps 0.5 nm long. The argon ion 488 nm photon is 488 nm "long" considering its wave nature. Thus there is little chance for the toluene to escape from the "massive" photon once the photon is close.

What can we do to improve the chances of multiple photon interaction with a target molecule? Substantially increasing the intensity or power of the laser beam should enhance the probability of multiphoton events. It is not difficult to generate very intense laser beams with mode-locked *Q*-switched lasers. Let us repeat our calculation for an argon ion 488 nm line laser whose irradiance is $10^{16}$ W. Obviously everything scales up by $10^{16}$, and now there would be $9 \times 10^{21}$ photons per $6 \times 10^{23}$ toluene molecules (in a mole). Under these massive power conditions the possibility of two photons simultaneously interacting with a toluene molecule is quite feasible. Figure 6-2 illustrates the argument made here.

While high laser intensity is a requirement, often multiphoton spectroscopy is not conducted with the highest possible intensities available. Often it is more desirable to use a lower power coupled with a more sensitive detection system. The rationale for such an approach is to minimize potential side reactions that can occur with such high irradiances, including processes such as ionization, fluorescence, and the Raman effect.

## Some Multiphoton Spectroscopic Theory

To quantitatively describe multiphoton effects, and hence what we should expect to derive from such spectroscopic studies, we can take three approaches. Two of these are macroscopic and phenomenological, one is microscopic and in effect theoretically derivable. The microscopic approach rests on quantum mechanics and will not be mentioned here except to

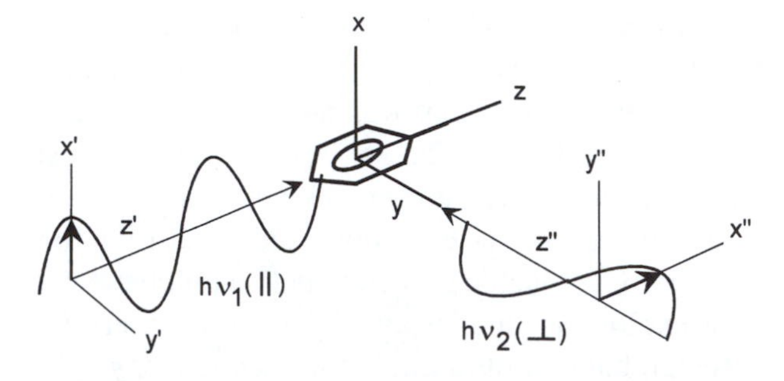


**Figure 6-2**
Schematic drawing showing how one or two photons could interact with a target molecule. For normal intensities a target molecule interacts with only one photon at a time. For the immense intensities possible with lasers, a target molecule could interact with two or more photons simultaneously. Using the energy levels given in Figure 6-1, one-photon absorption could be described as the process: $E_n + h\nu_1 \rightarrow E_k$, where the dipole of the molecule in the energy state $E_k$ is in the $x'$ direction. Similarly, two-photon absorption could only be described as: $E_n + h\nu_1(||) + h\nu_2(\perp) \rightarrow E_m$.

quote a few pertinent formulas whose appreciation adds insight to the magnitude of the multiphoton absorption process. We will spend more time on the macroscopic views, the first derived from rate phenomena, the second derived from electric susceptibilities.

Consider the rates for the processes:[3]

$$M + h\nu = M^*$$
$$M + h\nu_1 + h\nu_2 = M^{**} \qquad (63)$$

The first process would correspond to ordinary spectroscopy. The second process corresponds to concerted two-photon absorption ($h\nu_1$ might equal $h\nu_2$, but that condition is not necessary). If each of the processes can be viewed as an elementary chemical reaction, the rates of formation of the one-photon excited species $M^*$ and the two-photon excited species $M^{**}$ can be written as:

$$\frac{d\,[M^*]}{dt} = k_1\,[h\nu]\,[M] = \sigma I\,(\nu)\,[M]\;;\;\sigma\,\frac{k_1}{c}\;;$$
$$[h\nu] = \frac{I(\nu)}{c}$$
$$\frac{d\,[M^{**}]}{dt} = k_2\,[h\nu_1]\,[h\nu_2]\,[M] = \delta I\,(\nu_1)\,I\,(\nu_2)\,[M]\;;$$
$$\delta\,\frac{k_2}{c^2} \qquad (64)$$

where $c$ is the speed of light. We note here that rather than using the concentration of photons in traditional # per volume units, each photon flux has been included in the rate expression as # per $cm^2$ per s. Moreover, the traditional $k$ rate constants have been replaced by the constants $\sigma$ and $\delta$, called the cross section and absorptivity, respectively.[5] The units of $\sigma$ are $cm^2$ $molecule^{-1}$ $photon^{-1}$ and those of $\delta$ are $cm^4$ s $molecule^{-1}$ $photon^{-1}$. We might note that these constants are proportional to the rate constants and that, for the two-photon process, the rate constant describes how fast the second photon is absorbed after the first photon appears. If the rate of absorption of the second photon is not fast enough, the two-photon experiment fails. For multiphoton experiments where $n$ is the number of photons, Equation 64 would be:

$$\frac{d\,[M^{**}]}{dt} = \delta_n\,(I\,(\nu_1)\,I\,(\nu_2)\ldots I\,(\nu_n))\,[M] \qquad (65)$$

One simple, yet important, general consequence of Equations 64 and 65 is that the rate of production of molecules in the excited state, $E_m$, depends on the product of the intensities of each photon involved, or $I(\nu)^n$ if $n$ photons of energy $h\nu$ are involved. This also suggests that there are multiple ways to successfully accomplish a multiphoton experiment. For example, to achieve a high rate of production of excited state molecules, the intensity of one photon beam could be made large in the sense of being a pump beam, while the second beam could be of lower intensity in the sense of being a probe beam.

The absorptivity of a two-photon process can be estimated from quantum mechanics. Without presenting the derivation, the absorptivity can be shown to be proportional to the following quantities:[3,6]

$$\delta \propto \left|\sum_k \left[\frac{< n|M|k >< k|M'|m >}{E_k - h\nu_1} + \frac{< n|M'|k >< k|M|m >}{E_k - h\nu_2}\right]\right|^2 \qquad (66)$$

Two very important features of multiphoton spectroscopy—resonance and polarization—can be understood with some appreciation of this proportionality for $\delta$. Without exploring the numerators of Equation 66 at this time, let us note what happens to the absorptivity $\delta$ as the energy of $h\nu_1$ or $h\nu_2$ approaches an intermediate stationary or virtual state $E_k$. As $h\nu_1$ or $h\nu_2$ approach $E_k$, the denominator goes to zero in one of the terms and $\delta$ tends to infinity. The rate of formation of $M^{**}$ becomes very large, the $E_m$ state becomes highly populated, and the two-photon experiment works well. Such a situation is called a *resonant absorption*. The absorptivity $\delta$ does not actually go to infinity, of course, because there is always a natural linewidth for each state that keeps the denominator finite.

A discussion of the full meaning of the numerator in Equation 66 cannot be undertaken here, but let us try to provide some connection of the equation with the idea of polarization[3]. The quantum mechanical origin for the absorption of photon energy by a molecule lies in the concept called the *transition moment*. The $<|\;|>$ terms in Equation 66 are transition moments, the quantum mechanical analogs of dipole moments. For two charges of equal magnitude and opposite sign, a classical dipole moment is defined as the product of either charge times the distance separating the charges. The dipole moment is a vector, and we should all be familiar with molecular dipole moments, either permanent or induced. It is the vector aspect of the dipole moment that ultimately leads to our ability to distinguish polarization effects in any kind of spectroscopy, but especially in multiphoton spectroscopy. If we look back at Figure 6-2, we can appreciate that the **E**-field vector will make some angle with the principal axis of the target molecule. To put this another way, the physical properties of a target molecule can be described by components of those properties relative to a convenient Cartesian axis system located somewhere in the molecule. Refer to Figure 6-2 where a Cartesian molecular axis is sketched for toluene and two incident photons. The absorption of a photon will be strongly influenced by the incident photon interacting with a permanent or induced dipole moment in the molecule that has an orientation to maximize the quantum mechanical transition moment. The point here is that the orientation, that is the polarization, of the **E**-field of the incident photon relative to the molecule has a tremendous influence on the magnitude of the transition moment and hence on the magnitude of the absorptivity $\delta$. The polarization of the incident photon can be controlled by causing the photon flux to be linearly or circularly polarized relative to directions chosen in the laboratory.

At this point we will note a significant difference in polarization effects between one- and two-photon spectroscopy. In one-photon spectroscopy, when the incident photon has some defined plane polarization, only molecules which have the correct orientation for absorbing the polarized radiation will do so. This will result in an absorption spectrum that appears identical to an unpolarized absorption spectrum. Furthermore, the same spectrum would be obtained for light of any incident plane polarization. Why is this true? Unless the molecule exists in a solid or liquid crystalline phase, there are no preferred molecular orientations. Even though the incident plane of polarization might vary, some molecules in the isotropic medium will have the correct orientation to absorb the incident light.

On the other hand, consider the two-photon experiment with each photon's polarization fixed and determined relative to the laboratory frame. Those initial polarizations can be treated as vectors, and the two vectors define a plane. Only molecules with an orientation consistent with the plane defined by the polarization vectors will absorb the two photons of light. Not all combinations of polarization vectors will match the transition moments within the molecules. Thus, no absorption of such photon pairs can occur. The two-photon experiment will yield a different absorption spectrum with each combination of polarization vectors. In fact, no absorption will be observed for many polarization vector combinations. The molecules can still rotate, but only when they achieve the correct orientation relative to the imposed "coordinate system" will absorption occur. Bottom line: Polarization effects can be observed even in molecules free to translate and rotate. Such polarization effects can be extremely important and useful in the spectroscopic study of electronic states, which we will illustrate soon with an example.

For the sake of connecting multiphoton spectroscopy with other nonlinear optical effects induced by high-intensity lasers, we will briefly discuss the other macroscopic approach based on electric susceptibility as derived from Maxwell's equations.[1,7] Recall the equations previously presented in Chapter 3 that describe the polarization induced in a sample by the presence of an electric field:

$$P = \chi^{(1)} E + \chi^{(2)} E^2 + \chi^{(3)} E^3 + \ldots \qquad (67)$$

where $P$ is the polarization, $\chi^{(i)}$ is the electric susceptibility, and $E$ is the electric field amplitude. In our earlier discussions we noted that $\chi^{(1)}$ is proportional to the sample's polarizability and measures the ease with which a dipole moment may be induced in the sample. We call $\chi^{(2)}$ the hyperpolarizability, and this measures the ease of second-harmonic generation in the medium. The ability of crystals to double a frequency depends on $\chi^{(2)}$. The third-order susceptibility $\chi^{(3)}$ is a term whose magnitude describes many effects, including Raman scattering and multiphoton spectroscopy. We remarked earlier that Raman scattering or Raman spectroscopy could be viewed as a subsection of multiphoton spectroscopy. A theoretical basis for this claim is the importance of the third-order susceptibility, or $\chi^{(3)}$ term. Physically, $\chi^{(3)}$ describes the coupling of the **E**-fields of the $n$ photons of the experiment with the dipole induced in the target molecule by the $n$ incident **E**-fields. In fact, $\chi^{(3)}$ is a tensor of rank 4, but all that means for our discussion is that it represents the magnitudes of the components of the various vectors involved along their respective laboratory (**E**-fields) and molecular axes (induced dipoles). This third-order susceptibility can also be calculated by quantum mechanics and is related to the absorptivity $\delta$. Thus the three approaches mentioned earlier to describe multiphoton effects do indeed connect with each other.

## Experimental Detection of Multiphoton Events

Suppose a sample has undergone a multiphoton absorption process. How can this be detected experimentally? A simple absorbance measurement is almost impossible. The incident photon flux is so intense and the numbers of molecules actually undergoing a multiphoton process so relatively small (recall, maybe only 1 in 100 molecules) that detecting a 0.01% to 1% change in power levels of $10^{16}$ W is virtually impossible. The two most common detection methods are based on either fluorescence or ionization.[7] Fluorescent photons are emitted from the final $E_m$ states, while ionization detection relies on yet further absorption of photons by the excited $E_m$ state to cause ionization of the molecule. Actual detection of the resulting molecular ion is accomplished by mass spectrometry.

## Preview of Case Studies

The two case studies below illustrate many of the concepts and principles that we have been discussing. The first case study features how a normal quantum mechanical selection rule can be overcome by multiphoton absorption. The second case study features how the polarization of the incident photons dramatically affects the excited states that can be reached by photon absorption.

## Case Study I: "Hidden" Electronic Transitions

### Overview of the Case Study

**Objective.** To study spectroscopically excited electronic states that cannot be achieved by normal single photon spectroscopy because of quantum mechanical selection rules.

**Laser System Employed.** A *Q*-switched high-power pulsed ruby laser driving a dye laser whose output is a tunable broad band of energies. Absorption by the target molecule is detected via fluorescence from the excited state created through absorption.

**Role of the Laser System.** To provide very high photon fluxes suitable to increase the rate of two-photon absorption.

**Useful Characteristics of the Laser Light for this Application.** High power and tunable laser light to scan the spectral range of interest.

**Principles Reviewed.** Molecular symmetries, spectroscopic selection rules, electronic states of aromatic molecules.

**Conclusions.** Two-photon spectroscopy demonstrates that certain excited electronic states do exist and provides quantitative energy data for comparison with theory.

### Electronic State Symmetries

The electronic spectra of many organic molecules have posed significant challenges to the spectroscopist interested in experimentally verifying theoretically calculated molecular energy states. Since most organic molecules contain an even number of electrons, the ground (lowest) energy state has all electrons paired and is called a singlet spin state. Moreover, the ground state is usually described as *gerade*, that is, even. A molecule will have even (*gerade*) or *ungerade* symmetry properties if a center of symmetry, also called inversion symmetry, exists in the molecular framework. Such a molecule is also called centrosymmetric. (As a quick example of a geometric figure that has a center of symmetry, consider a planar square figure.) The terms *gerade* and *ungerade* refer to the property of the wave function that describes the electronic state. If the coordinates of each electron in the molecule were inverted through the center of symmetry, which would also be the center of the coordinate system, the sign of the wave function would remain unchanged if the state were *gerade*, but would change if the state were *ungerade*. The description of a molecule as a singlet and *gerade* has great significance for spectroscopy.

Two simple rules generally help the interpretation of the electronic spectra of organic molecules. The first rule relates to changes in electronic spin. Only states of the same spin may be connected by the absorption or emission of a single photon. An all-electron-paired state, a singlet, can absorb or emit a photon to reach another singlet state. Likewise, a spin-unpaired state can reach another state whose extent of spin unpairing is the same. The second rule relates to changes in inversion symmetry. If a center of symmetry exists in a molecule, then a single photon can only be absorbed or emitted if the final electronic state has the opposite sense of *gerade* or *ungerade*. The converse of this is perhaps the easiest way to state the rule: *g* to *g* or *u* to *u* changes are not permitted. Now we can address the experimental spectroscopist's quandary. Many theoretical calculations on molecules containing $\pi$ electron systems, particularly conjugated systems, suggest the existence of singlet *gerade* excited states. Since the ground state of such molecules is also usually a singlet *gerade* state, such excited states should not be detectable by ordinary single photon spectroscopy. Thus to the single-photon spectroscopist, those excited states are "hidden." Such states are not hidden to the multiphoton spectroscopist, however, and that fact is in a large part the raison d'etre for the development of such spectroscopies.

To see that two-photon spectroscopy overcomes the selection rule prohibition against *g* to *g* transitions, consider the two-photon process as two one-photon processes. In such a case the first photon can cause a *g* to *u* change by the rule, and then the second photon can cause an allowed *u* to *g* change, with the result that all selection rules are satisfied. This approach simply requires that the virtual or stationary $E_k$ state be described by the *g* or *u* complement to the initial state.

## Two-Photon Spectra of Diphenyloctatetraene

The work of Fang et al. [8] provides one of the earliest confirmations of the ability of two-photon spectroscopy to overcome the single-photon selection rule prohibition. The centrosymmetric $\pi$ system molecule diphenyloctatetraene was studied [8] to discover the "hidden" excited states. The structure of this molecule is shown in Figure 6-3.

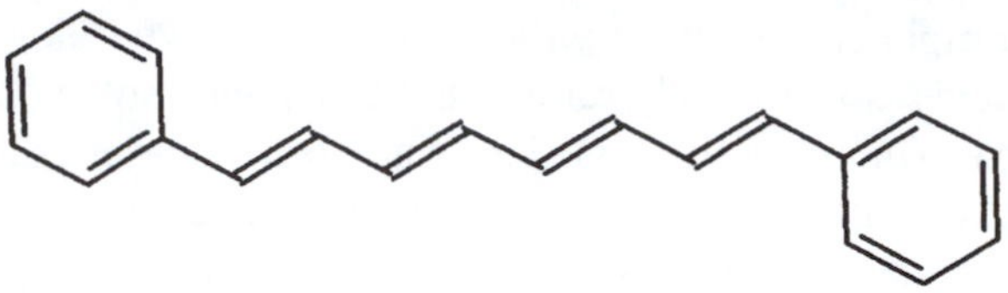

**Figure 6-3**
Structure of diphenyloctatetraene.

As might be surmised, the ground state of this molecule is a singlet and *gerade*, called $^1A_g$ in the group theoretic nomenclature of the spectroscopist, where $^1[\ ]$ denotes singlet and $[\ ]_g$ denotes *gerade*. For our purposes here, we will not discuss the meaning of the *A* or *B* labels, except to say that they denote particular electronic energy states. Theory predicts the existence of an excited electronic state with the same properties as the ground state, that is, another $^1A_g$ state. By single-photon spectroscopy, the change $^1A_g \rightarrow {}^1A_g$ is forbidden, though allowed by two-photon spectroscopy. The experimentally observed two-photon spectrum is presented in Figure 6-4.

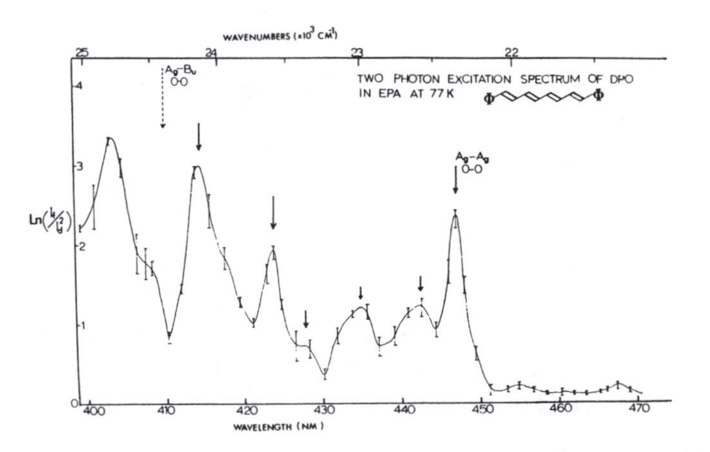


**Figure 6-4**
Two-photon excitation spectrum of diphenyloctatetraene in an EPA glass at 77 K. The arrows indicate the observed two-photon transitions from the $^1A_g$ ground state of the molecule. In particular, the $^1A_g \rightarrow {}^1A_g$ transition that is forbidden in single-photon spectroscopy is present, while the allowed single-photon transition $^1A_g \rightarrow {}^1B_u$ is absent in the two-photon spectrum. Reprinted with permission from Fang, H. L. B., Thrash, R. J., and Leroi, G. E., *J. Chem. Phys.*, 1977, 67, 3389–3391. Copyright © 1977 American Institute of Physics.

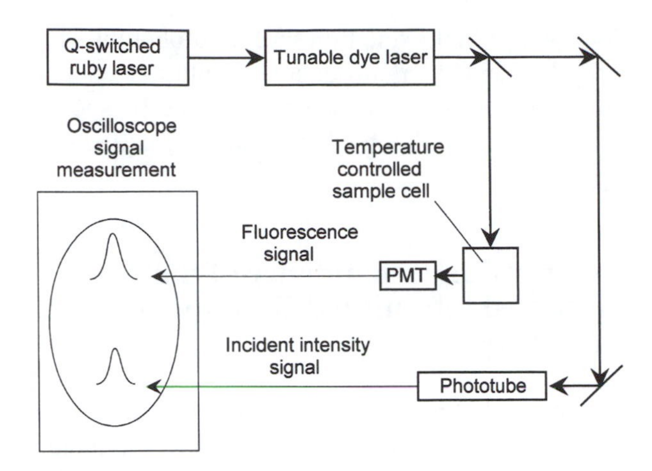


**Figure 6-5**
A two-photon excitation apparatus. A tunable dye laser is pumped by a *Q*-switched ruby laser to excite a sample with photons of the type $2\nu_1$ to induce a fluorescence signal monitored by a photomultiplier tube. As the theory for a two-photon experiment requires that the fluorescence signal be proportional to the square of the input laser intensity, the output of the dye laser is directed with a beam splitter to a phototube to monitor for this condition.

One should note, however, the absence of a peak corresponding to the single-photon allowed ${}^1A_g \rightarrow {}^1B_u$ transition. Single- and multiphoton spectroscopy together then provide a much more complete spectroscopic picture.

It might be of interest to note the experimental arrangement as shown in Figure 6-5. The dye laser used provides tuning across the wavelength range of interest, 400 - 470 nm. This experiment is a $2\nu_1$ type as opposed to a $\nu_1 + \nu_2$ type (refer to Figure 6-1 to review these notations). The beam-split dye laser is independently monitored to assure that the *intensity of the fluorescence output signal is proportional to the square of the input laser intensity*. This condition is required by the theory that describes the fluorescence signal expected from a two-photon absorption experiment.

## Case Study II: Polarization in Multiphoton Spectra

### Overview of the Case Study

**Objective.** To identify the excited electronic states of dichlorine by the study of the vibrational-rotational spectra. The excited electronic states of dichlorine are reached through multiphoton processes.

**Laser Systems Employed.** Two independently tunable dye lasers each pumped by the same high-power $N_2$ laser were used. The polarization of the incident light was controlled.

**Role of the Laser Systems.** With two tunable lasers, one laser is used to generate the excited electronic state of interest, while the second laser is used to scan the vibrational-rotational spectrum of the created excited state.

**Useful Characteristics of the Laser Light for this Application.** High power to promote two-photon absorption, controlled polarizations to provide extremely fine control over the energy states studied.

**Principles Reviewed.** Vibrational-rotational spectroscopy, general selection rules, polarized light selection rules.

**Conclusions.** Two-photon spectroscopy with polarized light provides extremely precise control over the excited states studied. The existence of ion-pair excited electronic states in dichlorine is confirmed.

### Polarization in Vibrational-Rotational Spectra

In our second example we will illustrate the power of polarization effects. In the single-photon spectroscopy of diatomic molecules, one rotational selection rule states that only those photons may be absorbed or emitted that cause a change in the energy of a rotational energy state by one or zero units of energy. With two-photon spectroscopy, the rule changes to include changes by two units as well, but eliminates changes by one unit of frequency. Thus, the complete sequence of possible rotational energy changes (and the designated branches of the observed spectrum) are 2 (*O*), 1 (*R*), 0 (*Q*), –1 (*P*), and –2 (*S*) units (the lettering scheme is a customary spectroscopic notation). In single-photon spectroscopy we observe the *P*, *R*, and *Q* branches of the vibrational-rotational spectrum. In the two-photon experiment we add the *O* (2 units) and *S* (–2 units) branches, but eliminate the *P* and *R* branches from the observed vibrational-rotational spectrum.

## Experimental Design of Spectroscopic Measurements

Ishiwata et al. [9] used multiphoton spectroscopy to identify various excited electronic states in dichlorine. Their elaborate experimental approach is shown in Figure 6-6. Their experiments required the use of three photons from two dye lasers each providing a different excitation, $\nu_1$ and $\nu_2$. The first photon $\nu_1$ was used to pump $Cl_2$ to a specific vibrational-rotational state in one of its well-known electronic states: $X^1\Sigma^+_g + h\nu_1 \rightarrow B^3\Pi_O$. The exact meaning of all the symbols is not specifically important for this discussion; however, we may view this excited state as an excited ion-pair dichlorine molecule $(Cl^{\delta+}Cl^{\delta-})^*$. The excited $Cl_2$ in the state $B^3\Pi_O$ has the possibility to be further excited by the nonresonant absorption of either: (1) two photons with energy each of $h\nu_2$ (type I spectra) or (2) two photons with energy $h\nu_1$ and energy $h\nu_2$, respectively (type II spectra). Another separate experiment employed the absorption of three photons of energy $h\nu_2$, with initial excitation with two photons of energy $h\nu_2$ and then subsequent excitation with one more photon of the same energy (type III spectra). We will focus only on what the authors called type I spectra.

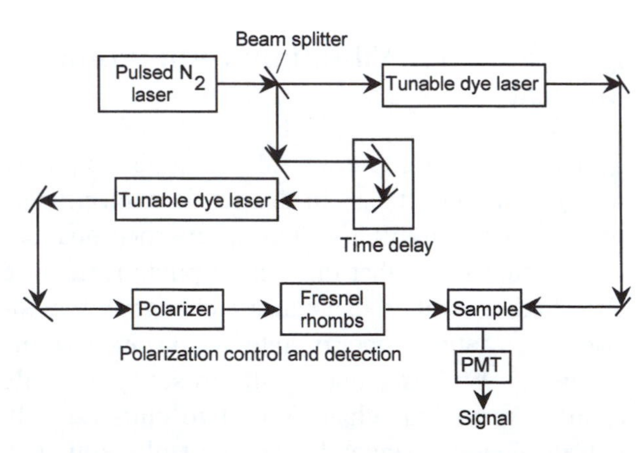


**Figure 6-6**
Schematic diagram of the apparatus for multiphoton spectroscopy of dichlorine. A pulsed nitrogen laser pumps two dye lasers, each providing a different excitation frequency, $\nu_1$ and $\nu_2$. The first photon $\nu_1$ generates the excited ion-pair dichlorine molecule $(Cl^{\delta+}Cl^{\delta-})^*$. Further multiphoton excitation of this ion-pair species can be achieved in a number of different ways, including: (1) with two photons each with energy $h\nu_2$, (2) with two photons with energy $h\nu_1$ and with energy $h\nu_2$, or (3) initially with two photons each with energy $h\nu_2$, followed by subsequent excitation with another photon of energy $h\nu_2$.

## Polarization Dependence of the Observed Vibrational-Rotational Spectrum

The multiphoton spectroscopic measurements were repeated using various polarizations of the incident photon beams. The results for circular and linear polarizations of the incident $\nu_2$ beam in the type I spectral experiment are shown in Figure 6-7. Clearly polarization causes dramatic effects. Theoretical arguments based on the calculations of the absorptiv-

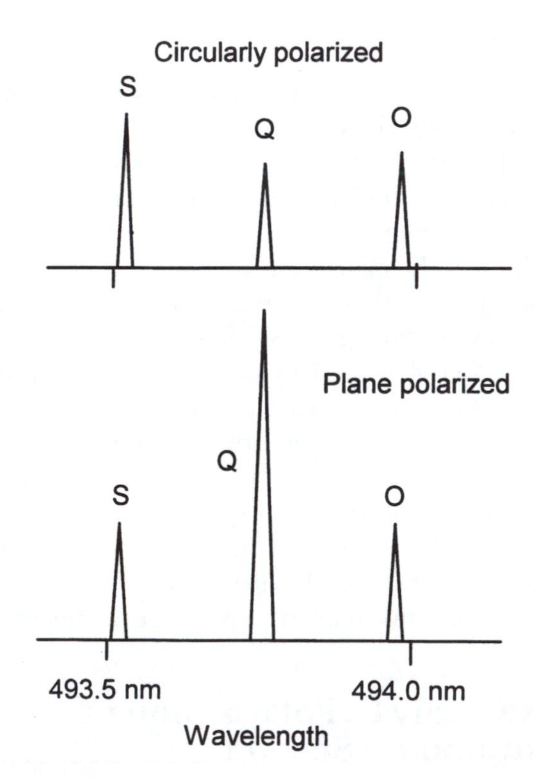


**Figure 6-7**
Polarization effects on the multiphoton spectra of dichlorine using circularly and linearly polarized light. A horizontally (linearly) polarized pump laser of energy $h\nu_1$ excites $Cl_2$ to the $v = 0$, $J = 38$ level in the $B^3\Pi_O$ state. Two-photon transitions were induced with $2h\nu_2$ photons plane polarized and circularly polarized.

ity $\delta$ predict that the ratio of the circularly (*cc*) to linearly (*ll*) polarized intensities should be:

$$\frac{I\,(cc,\,O,\text{ or }S\text{ band})}{I\,(ll,\,O,\text{ or }S\text{ band})} = \frac{3}{2} \qquad (68)$$

Indeed, the experimental results are:

| *Branch* | *I(cc)/I(ll)* |
|---|---|
| *O* | $1.35 \pm 0.15$ |
| *Q* | $0.2 \pm 0.03$ |
| *S* | $1.51 \pm 0.11$ |

Thus, to experimental precision the results agree with theory. The theory for the *Q*-branch ratio is more complicated, but the experimental value for the ratio also agrees with what would be expected theoretically, assuming the correct assignment of the spectral lines to an excited ion-pair dichlorine molecule $(Cl^{\delta+}Cl^{\delta-})^*$, the $B^3\Pi_O$ state.

## SUMMARY

Multiphoton spectroscopy is becoming more useful as lasers and associated components become more accessible and easily used. Clearly, single photon spectroscopy is not the only tool now available to spectroscopists seeking to confirm or discover new energy states in molecules.

## FOR FURTHER EXPLORATION

Have you considered the prospects for three-photon excitation? While there have been relatively few examples of simultaneous absorption of three photons ($3\,h\nu$), one exciting development is the demonstration of three-photon excitation of a tryptophan analog, N-acetyl-L-tryptophanamide (NATA).[10] Using a femtosecond mode-locked Ti:sapphire laser, researchers used the fundamental output at 840 nm to induce an emission spectrum for NATA identical to that obtained with 280 nm (one-photon) excitation. Additional experiments support the notion of a simultaneous $3\,h\nu$ absorption and not a sequential process or a generation of the third harmonic of the 840 nm light. This study demonstrates the promising possibilities for monitoring the intrinsic cellular tryptophan fluorescence in proteins without the need for UV lasers or UV optics.

## DISCUSSION QUESTIONS

**1.** Using energy diagrams discuss the principal features that distinguish single-photon spectroscopy from multiphoton spectroscopy.

**2.** Discuss whether or not CARS should be considered a multiphoton spectroscopy (see Chapter 9).

## SUGGESTED EXPERIMENTS

Suggested references to experiments that illustrate some of the principles described in the case studies:

1. Grieneisen, H. P., "The Nitrogen Laser-Pumped Dye Laser: An Ideal Light Source for College Experiments; Experiment 5—Two-Photon Sequential Excitation," Laser Science, Inc., Cambridge, MA. An experiment to demonstrate sequential two-photon excitation of iodine and analysis of the resulting fluorescence spectrum.
2. Payton, B. W., and Sieradzan, A., "Two-quantum absorption: An undergraduate's experiment in nonlinear optics," *Am. J. Phys.*, 1992, 60, 1033–1039.
3. Van Hecke, G. R., Karukstis, K. K., Underhill, J. M., "Sequential Multiphoton Excitation of Several Common Laser Dyes," *Twelve Laser Demonstrations for General Chemistry*, Harvey Mudd College Press, Claremont, Calif., 1997.

## LITERATURE CITED


1. Friedrich, D. M., "Two-photon molecular spectroscopy," *J. Chem. Educ.*, 1982, 59, 472–481.
2. Demtröder, W., "Multiphoton spectroscopy," *Laser Spectroscopy: Basic Concepts and Instrumentation*, 2nd Edition, Springer, Berlin, 1996, Chapter 7.5.
3. Lin, S. H. (Ed.), *Advances in Multiphoton Processes and Spectroscopy, Vol. 1*, World Scientific, Singapore, 1984, p. 6.
4. Andrews, D. L., "Chemical spectroscopy with lasers," *Lasers in Chemistry,* 2nd Ed., Springer-Verlag, Berlin, 1990, Chapter 4, pp. 83–137.

5. Beer's Law is $A = abC$, where $a$ is the absorptivity and is related to $\delta$, the cross section. The absorptivity $a$ of $A = abC$ is not the two-photon absorptivity $\delta$. The context should clarify which absorptivity is meant.

6. Louden, R., *The Quantum Theory of Light,"* 2nd Ed., Oxford Press, London, 1983, Chapter 9.
7. Lin, S. H., Fujimura, Y., Neusser, H. J., and Schlag, E. W. (Eds.), *Multiphoton Spectroscopy of Molecules*, Academic Press, Orlando, FL, 1984, pp. 56–67.
8. Fang, H. L. B., Thrash, R. J., and Leroi, G. E., "Observation of 'hidden' electronic states by two-photon excitation spectroscopy: Confirmation of the low energy $^1A_g$ state of diphenyloctatetraene," *J. Chem. Phys.*, 1977, 67, 3389–3391.
9. Ishiwata, T., Fujiwara, I., Shinzawa, T., and Tanaka, I., "Identification of new ion-pair states of molecular chlorine," *J. Chem. Phys.*, 1983, 79, 4779–4787.
10. Gryczynski, I., Malak, H., and Lakowicz, J. R., "Three-photon excitation of a tryptophan derivative using a fs-Ti:sapphire laser," *Biospectroscopy*, 1996, 2, 9–15.

CHAPTER 7

# Lasers as Probes: Fluorescence Spectroscopy

## Chapter Overview

With attributes such as extreme sensitivity, high selectivity, and noninvasivity, the optical technique of fluorescence spectroscopy is indeed a powerful analytical tool. Laser-induced fluorescence spectroscopy further offers the chemist a promising means of characterizing an extensive range of chemical systems and reactions. Lasers have also revolutionized a diverse range of fluorescence-based analytical methods, including the measurement of intracellular ion concentrations, the detection of trace contamination by pesticides, and the determination of DNA sequences. This chapter is devoted to examining two exciting applications of laser-induced fluorescence spectroscopy: (1) an analysis on the picosecond timescale of a sequence of ultrafast energy transfer steps in red and blue-green algae and (2) investigations involving airborne surveillance and characterization of fluorophores such as oil samples in ocean waters and terrestrial and oceanographic plants using the spectral characteristics of these substances.

## The Fate of Photogenerated Excited States

As you have explored in the previous chapter, the absorption of light of the requisite energy can lead to the generation of an electronically excited state of an atom or molecule. Chromophores with absorption spectra in the ultraviolet, visible, or near infrared range all can be activated by the output of a judiciously chosen laser. Indeed, laser-induced absorption spectroscopy is a viable technique for a vast scope of investigations. The monochromaticity and tunability of lasers enhance the resolution and sensitivity of absorption studies for such exacting applications as quantitative analyses of multicomponent samples, identification of dilute solutes or weakly absorbing species, and precise molecular structure determinations of electronically excited states.

Alternatively, the subsequent fate or outcome of the activated species may be of prime interest. In particular, information on the energetics and dynamics of photoactivated atoms and molecules can be acquired from examining the pathway of the higher-energy states. What are the temporal characteristics of the photogenerated excited state? Keep in mind that, for all but the highest intensity lasers, each molecule absorbs a single photon ($M + h\nu \rightarrow M^*$). This correspondence between the number of photons absorbed and the number of molecules undergoing activation has been previously detailed in the Law of Photochemical Equivalence. Also recall that light absorption generally promotes a molecule to a higher-order vibrational state (v′) as well as to the first excited singlet electronic state ($S_1$). The concomitant electronic and vibrational transitions arise as a consequence of the rapid timescale of absorption (on the order of $10^{-15}$ s) and the postulates of the Franck-Condon principle. The excited vibrational state is short-lived, however. A process known as *vibrational relaxation* occurs on a timescale of $10^{-12}$ s or faster to dissipate the excess vibrational energy and return the molecule to the lowest vibrational level ($v_o'$) in the first excited electronic state. It is this excited state that spectroscopists generally consider the photoactivated species, represented by $M^*$. The lifetime of this "new" and more stable excited state is also finite, on the order of $10^{-12}$ to $10^{-6}$ s for species in solution. The exact value of the lifetime depends upon the rate constants for the various reactions by which the excited electronic state is further depopulated. Let's examine the possible avenues available to the transitory excited state of a molecule, $M^*$.

### Primary Processes

The excited-state species may follow a number of competitive de-excitation pathways known as *primary processes*. These de-excitation processes are parallel reactions (i.e., occurring simultaneously) involving the excited-state species as a reactant. How do these primary processes differ? Some primary processes regenerate the ground state ($S_o$) of the excited atom or molecule. These processes may occur either in a radiative manner (e.g., via *fluorescence*) or by a nonradiative means (e.g., via *energy transfer* or *collisional quenching*). In other primary processes, the excited-state species undergoes dissociation, isomerization, or chemical reaction with other reactant molecules to either produce ground states of new molecules or create additional transient species. Table 7-1 summarizes a number of primary processes that are possible following the generation of the excited-state species $M^*$.

**Table 7-1**
Examples of Primary Processes following Photoinitiation of the Electronic Excited State of Reactant $M$, $M^*$

| | |
|---|---|
| Fluorescence | $M^* \rightarrow M + h\nu$ |
| Energy transfer | $M^* + A \rightarrow M + A^*$ |
| Collisional quenching | $M^* + Q \rightarrow M + Q$ |
| Dissociation | $M^* \rightarrow X_1 + X_2$ |
| Internal rearrangement or isomerization | $M^* \rightarrow Z$ |
| Chemical reaction with other molecules | $M^* + A \rightarrow B + C + \ldots$ |

By the attributes of a deactivation pathway, such as the rate constant of the primary process, fundamental structural and mechanistic information about either the ground state or the electronically excited species can often be revealed. The most widely characterized parameter of a primary process is its *quantum yield*, that is, the fraction of molecules that becomes deactivated by the particular pathway. Quantum yields are thus dependent upon the relative rates of all competing deactivation pathways. Fast primary processes "consume" excited-state molecules and lower the quantum yields of slower deactivation routes. Experimental parameters such as temperature, pH, choice of solvent, and ionic strength influence the kinetics of primary processes and affect their quantum yields.

### The Primary Process of Fluorescence

One particularly informative primary process is that of *fluorescence*—the emission of light by an atom or molecule in an excited state as it returns to the ground state. As a consequence of the variable rates of deactivation pathways, only about 10% of absorbing compounds are observed to fluoresce. Such chromophores may be described as *fluorophores*. What are some typical fluorophores? The amino acids tryptophan and tyrosine are common, naturally occurring fluorophores. A number of organic substances exhibit high quantum yields of fluorescence, including polynuclear aromatic hydrocarbons, such as anthracene and pyrene, and other highly conjugated ring systems such as fluorescein, rhodamine, dansyl chloride, and their derivatives. Aqueous solutions of the trivalent lanthanide ions, particularly terbium and europium, also exhibit high degrees of fluorescence.

## Advantages of Analysis Via Fluorescence Spectroscopy

The technique of fluorescence spectroscopy is clearly applicable to a more limited range of chemical systems than absorption spectroscopy, due to the limited number of fluorophores. Nevertheless, one of the major advantages of fluorescence investigations over absorption studies is a greater selectivity in the analysis of multicomponent samples. What accounts for this enhanced selectivity? The emission of each fluorophore is dictated by two characteristic wavelengths—an excitation wavelength that generates the excited state of the fluorophore and an emission wavelength at which the fluorescence signal is detected. Generally, the emission wavelength is shifted to lower energies relative to the absorption wavelength. The origin of this effect is the vibrational relaxation (i.e., the nonradiative transition to the lowest vibrational level of the excited electronic state) that occurs before fluorescence is emitted, causing the emitted photon to be "red-shifted" to longer wavelengths. In actuality, the manifold of excited vibrational states accessible upon photogeneration of the first excited electronic state broadens the range of excitation and emission wavelengths characteristic of a fluorophore. Thus, any fluorescent molecule has two characteristic spectra: an *excitation spectrum*, depicting the dependence on the wavelength of the exciting light of the fluorescence intensity at a fixed emission wavelength, and an *emission spectrum*, illustrating the dependence of the fluorescence intensity induced by a fixed excitation wavelength on the wavelength of the emitted light. Thus, two fluorescent chromophores with the same absorption spectrum can be differentiated by their emission spectra, while two fluorophores with emission at the same wavelength can be selectively studied if they have different excitation wavelengths.

The fluorescence process has additional characteristic features that enhance this primary process as a photochemical tool. One noteworthy attribute is the sensitivity of the emission properties of many fluorophores to their immediate local environment.

This spectral sensitivity is related to the differences in the electronic distribution, and therefore in the relative polarities, of the ground and excited states of the fluorescing species. In solution, a fluorophore has an optimal solvent cage configuration—a thermodynamically stable arrangement of solvated fluorophores. Upon excitation, a molecule undergoes a change in dipole moment, generally producing an excited state that is more polar than the equilibrium ground state ($\mu^* > \mu$). Such an increase in the dipole moment upon generation of the excited state leads to a reorientation of the solvent molecules surrounding the probe molecule to maintain the most stable solvent-solute arrangement.

These interactions between the solvent and fluorophore molecules affect the relative energy difference between the ground and excited states of the fluorophore and influence both emission wavelength and fluorescence quantum yield. For example, how is the energy difference between ground and excited states affected by the solvent polarity? Two pronounced trends are observed: (1) the more polar the solvent, the greater the stabilization of both the ground and excited states, and (2) the more polar the electronic state, the more stabilization observed via interactions with the solvent cage. Thus, the energy of fluorescence emission is decreased by a polar solvent (as a consequence of the greater stabilization of the excited state relative to the ground state), thereby shifting the fluorescence maximum to longer wavelengths. A variation in fluorescence quantum yield is also observed with a change in the nature of the solvation sphere of a fluorophore. Increased coupling of a solvent with the excited probe molecule often enhances the rate constants (and therefore increases the likelihood) of nonradiative transitions such as collisional quenching. Thus, increasing the polarity of the solvent typically lowers the fluorescence quantum yield.

Fluorescence lifetime measurements also enhance the use of fluorescence as a powerful diagnostic tool. By the concept of fluorescence lifetime we refer to the time duration of the vibrationally relaxed $S_1$ excited state (i.e., $v_o'$) prior to deactivation to the ground state. Numerically, the lifetime of fluorescence is defined as the reciprocal of the sum of the rate constants for all possible first-order primary processes:

$$\tau_f = \frac{1}{\Sigma k_{pp}} \quad (69)$$

In the absence of all other competitive de-excitation pathways except fluorescence:

$$\tau_f = \frac{1}{k_f} = \tau_f^{\circ} \quad (70)$$

where $\tau_f^{\circ}$ is described as the *natural fluorescence lifetime*. As additional primary processes compete with fluorescence to depopulate the excited electronic state, the fluorescence lifetime decreases. For example, with the presence of an energy acceptor $A$, the expression for the fluorescence lifetime is modified:

$$\tau_f = \frac{1}{k_f + k_t[A]} \quad (71)$$

The presence of this additional deactivation process (energy transfer) not only lowers the observed fluorescence lifetime but must also decrease the measured fluorescence intensity (by lowering the quantum yield of fluorescence). Thus, measurements of variable fluorescence lifetimes under a range of experimental conditions can often reveal the existence of alternative de-excitation mechanisms available to an excited state.

## The Role of Lasers in Fluorescence Investigations

The analytical capabilities of the fluorescence technique are further enhanced when lasers are used as excitation sources. What are some of the inherent features of lasers that have contributed to their extensive use in fluorescence applications? As one example, laser-induced fluorescence studies often rely on the wavelength tunability and narrow spectral bandwidth of dye laser systems. In particular, precise activation of fluorophores with similar or nearly coincident excitation spectra can be achieved with finely tuned laser-induced emission. For samples with either low concentrations of fluorophores, weakly absorbing chromophores, or fluorophores with low fluorescence quantum yields, the high-intensity output of certain lasers (e.g., argon ion and krypton ion lasers) is extremely beneficial to intensify the fluorescence signal to detectable levels. Pulsed laser systems also provide the capability of generating and monitoring transient excited states and the ultrafast reactions of these short-lived species. The coherence and high power of laser illumination also enable examination of the fluorescence emission of remote (i.e., distant) samples as well as fluorophores in hostile environments. Thus, lasers enable chemists to probe systems of fluorophores with an exceptional degree of spectral selectivity, sensitivity, time response, and dynamic range.

The scientific literature contains an extensive range of fluorescence investigations facilitated by the use of laser excitation sources. As a small subset of promising applications of laser-induced fluorescence in chemical and biochemical systems, the next two case studies detail inquiries aided by the extraordinary features of laser light.

## Case Study I: Energy Transfer in Photosynthetic Organisms

### Overview of the Case Study

**Objective.** To use fluorescence investigations on the picosecond timescale with laser excitation in order to delineate the pattern of ultrafast energy migration among the light-harvesting pigments of red and blue-green algae.

**Laser System Employed.** A synchronously pumped cavity-dumped dye laser and mode-locked argon ion laser to provide excitation at 540 nm and 580 nm.

**Role of the Laser System.** To provide picosecond excitation of selected pigments as well as time-correlated photon counting to allow measurement of time-resolved fluorescence spectra from intact cells with a 25 picosecond resolution.

**Useful Characteristics of the Laser Light for this Application.** Short-duration pulses on the picosecond timescale, monochromaticity, wavelength tunability.

**Principles Reviewed.** Fluorescence, energy transfer, sequential and parallel mechanisms of elementary reactions.

**Conclusions.** Aided by laser excitation on the picosecond timescale and fast detection capabilities, this study reveals predominantly sequential mechanisms of light energy transfer in red and blue-green algae and suggests some variation in the organization of the photosynthetic pigments in these organisms.

### Introduction

#### *Fluorescence Energy Transfer in Macromolecules and Membrane Systems*

Recall that we introduced energy transfer as a competitive primary process that decreases both the fluorescence quantum yield and lifetime of a fluorophore. This transfer relies on the absorption of photons by a donor molecule, the subsequent *nonradiative* transfer of the donor's excited-state energy to an acceptor fluorophore, and the ultimate fluorescence emission from the acceptor molecule. This sequence of steps can be delineated as:

$$D + h\nu \rightarrow D^*$$
$$D^* + A \rightarrow D + A^*$$
$$A^* \rightarrow A + h\nu'$$

Energy transfer measurements reveal considerable information on donor-acceptor pairs, whether positioned rigidly at a fixed separation or diffusing freely to a distance of closest approach. The rate and extent of energy transfer depend critically on both spatial and spectral considerations.

The most distinctive feature of the primary process of energy transfer is its critical dependence on the distance between donor and acceptor, varying as the inverse sixth power of the distance of separation, $R$. Generally, an interaction between donor and acceptor is possible via energy transfer with separation distances less than 50 nm. Energy transfer also requires an overlap between the emission spectrum of the energy donor molecule and the absorption spectrum of the energy acceptor molecule. Finally, a specific orientation of donor and acceptor dipole moments is required for the dipole-dipole interaction that is the mechanism for energy transfer. In particular, the rate constant for energy transfer, $k_t$, is proportional to a mathematical function relating $\theta_t$, the angle between the donor emission dipole and the acceptor absorption dipole; $\theta_d$, the angle between the donor emission dipole and the vector joining the donor and the acceptor; and $\theta_a$, the angle between the acceptor absorption dipole and the vector joining the donor and acceptor:

$$k_t \propto (\cos \theta_t - 3 \cos \theta_d \cos \theta_a)^2 \tag{72}$$

With the proper orientation of neighboring species, an electrical dipole in a donor can produce an electrical field that polarizes an acceptor molecule, creating a dipole moment by induction. It is the energy of this dipole-dipole interaction that is distance dependent and proportional to the inverse sixth power of the separation distance between donor and acceptor.

The phenomenon of energy transfer is a revealing probe of the distance between proximate chromophores on membranes or in macromolecules. Fluorescence energy transfer measurements have been used to measure distances between binding sites on proteins and between chromophores on membranes, proteins, and other rigid structures. Other exciting applications of excitation transfer have characterized the geometry of macromolecules and the lateral organization of membranes. Using measurements of time-resolved fluorescence rise and decay kinetics, processes which involve the sensitization of excitation transfer for a series of chromophores can also be monitored. The power of this technique was dramatically heightened with the availability of picosecond lasers. An excellent illustration of an investigation of fluorescence energy transfer on the picosecond timescale is the detailed characterization of the extensive sequence of fast energy transfer steps that initiate the light reactions of photosynthesis in red and blue-green algae.

Light energy transfer is central to the process of photosynthesis. As light energy transfer must be preceded by the absorption of light, we'll begin our discussion with an overview of light absorption in photosynthetic organisms.

### *Overview of Light Absorption in Photosynthetic Organelles*

Most of the pigment molecules in photosynthetic organelles serve as antennae for light gathering. Chlorophyll is the ubiquitous pigment of higher plants and algae. Carotenoids also serve as light-harvesting antennae for plants, while bacteriochlorophyll is a main constituent of green and purple bacteria. An array of additional pigments occurs in red and blue-green algae, including phycoerythrin, phycocyanin, and allophycocyanin. Why are there so many types of pigment molecules? Nature has provided a diversity of pigments to enable photosynthetic organisms to absorb a wide range of the wavelengths emitted by the sun. Large arrays of pigment complexes are also needed to ensure a sufficient production of chemical energy. Thus, the extent of absorption is directly proportional to the concentration of light-harvesting pigments. The absorption of one photon, however, generates only one excited-state pigment molecule.

### *Primary Processes of Photosynthetic Pigments*

The excited-state pigment molecule can avail itself of many pathways to release its excess energy. What primary processes are generally observed for excited-state pigment molecules in photosynthetic organisms? Two of the key processes exhibited by pigment excited states are: (1) radiative relaxation to the ground state via *fluorescence* and (2) nonradiative relaxation to the ground state via *energy transfer* to a neighboring molecule (typically another pigment molecule). Clearly, the relative quantum yield for energy transfer should far exceed that of fluorescence if an organism is to make efficient use of light energy. Both processes—fluorescence and energy transfer—occur on a slower timescale ($10^{-10}$ to $10^{-8}$ s) than light absorption.

### *Energy Transfer in Red Algae*

Thus, an essential feature of the light-harvesting or antenna pigments of photosynthetic organisms is to absorb radiation and transfer the resulting electronic excitation to a reaction center where the process of photosynthesis is initiated. The two types of reaction centers in oxygen-evolving organisms, designated photosystem I and photosystem II, are pigment-protein complexes containing series of electron donors and acceptors to conduct the critical photochemical reactions. Red and blue-green alga contain phycobilins and chlorophyll *a* as the accessory pigments serving photochemical reaction centers. The phycobiliprotein pigments, organized into supramolecular complexes called phycobilisomes—as illustrated in Figure 7-1—function by absorbing short wavelength excitation and funneling that excitation to the chlorophyll chromophores. A Forster transfer mechanism [1] via dipole-dipole interactions is proposed to accomplish energy transfer between neighboring antenna pigments.[2–4] The energy migration pattern in red algae involves three phycobiliprotein pigments: phycoerythrin (PE), phycocyanin (PC), and allophycocyanin (AP). If the energy transfer step to chlorophyll (Chl) were 100% efficient, excitation transfer would terminate with the trapping of all absorbed light by the reaction center. However, as excitation migrates within the light-harvesting pigment-protein complexes, alternative excitation decay modes, such as fluorescence, compete with reaction center trapping. The distinctive emission bands of the accessory pigments enable energy transfer to each chromophore to be resolved. Thus, organisms containing phycobilisomes provide ideal situations for monitoring excitation transfer times among an array of light-harvesting pigments. Using measurements of the kinetics of both the rise and decay of fluorescence emission at the wavelengths characteristic of each pigment, the sequence of energy transfer can be revealed.

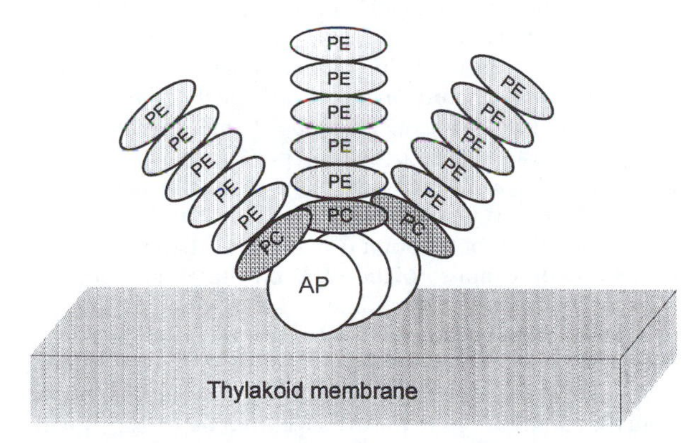


**Figure 7-1**
Schematic diagram of a cross section through a phycobilisome of a red alga. This supramolecular complex contains three phycobiliproteins or phycobilins that act as light-harvesting pigments: PE = phycoerythrin, PC = phycocyanin, and AP = allophycocyanin. The phycobiliproteins are arranged in layers and extend in a radial fashion from an allophycocyanin core to create a hemispherical structure. Light absorbed by the pigments in the phycobilisome is transferred to chlorophyll *a* in the thylakoid membrane. The phycobilisome is anchored to the thylakoid membrane by a protein shared by both structures.

What kinds of energy transfer patterns are possible in an organism with three phycobiliprotein accessory pigments and a longer wavelength chlorophyll chromophore? The two most obvious schemes are: (1) a parallel excitation transfer model and (2) a sequential excitation transfer pattern. In the parallel mechanism, each of the accessory pigments, $A$, $B$, and $C$ transfers energy directly to the chlorophyll acceptor molecule. For sequential excitation transfer, three donor-acceptor combinations are involved in succession: $A \rightarrow B$, $B \rightarrow C$, and $C \rightarrow D$.

parallel excitation transfer: $A \searrow$
$B \rightarrow D$
$C \nearrow$

sequential excitation transfer: $A \rightarrow B \rightarrow C \rightarrow D$

Variants of these two alternative schemes are also possible.

Time-resolved fluorescence measurements can be used to distinguish between these two pigment arrangements as well as other possible modes of energy transfer. How would one design a set of experiments to ascertain the energy transfer pattern? Initially, recall that each pigment is characterized by a distinct fluorescence excitation spectrum (i.e., the fluorescence intensity at a fixed wavelength as a function of the wavelength of light absorbed) and a separate fluorescence emission spectrum (i.e., the fluorescence intensity induced by excitation at a fixed wavelength as a function of the monitoring wavelength for emission). At a minimum, the excitation $\lambda_{max}$ of each phycobiliprotein $A$, $B$, and $C$ must be known as well as the emission $\lambda_{max}$ of $D$. Experiments must then focus on the time required for the fluorescence emission of pigment $D$ to reach a maximum level (i.e., the *fluorescence rise time*) following excitation of pigments $A$, $B$, and $C$ at the requisite excitation $\lambda_{max}$ values. How would the fluorescence rise time of pigment $D$ vary with light-harvesting pigment organization? For direct energy transfer from each pigment to $D$, one would anticipate similar fluorescence rise times for $D$ as light absorption by each of the pigments *A*, *B*, and *C* is induced. The similarity in fluorescence rise times would reflect comparable donor-acceptor distances. The absence of any appearance of intermediate fluorescence from the pigments not directly excited would also be consistent with a parallel excitation transfer mechanism. On the other hand, sequential energy transfer would be suggested by decreasing fluorescence rise times upon excitation of *A*, *B*, and *C*, respectively. Such a result would generally be anticipated if one observed increasing wavelengths of maximum absorbance ($\lambda_{max}$) for the pigments in the order $A < B < C$. Sequential energy transfer would also be indicated by the existence of fluorescence rise times for pigments *B* and *C* and by increasing fluorescence rise times for the pigments *B*, *C*, and *D*, respectively, upon excitation of *A*.

Early measurements of the fluorescence rise times of the antenna pigments in intact algae were limited by the time resolution of the experimental apparatus and the method of analysis of the time-resolved spectra.[2,5–12] Recent advances of picosecond excitation and time-correlated photon counting have allowed measurement of time-resolved fluorescence spectra from intact cells with a 25 picosecond resolution. Yamazaki et al.[13] proposed a sequential energy transfer pattern among the light-harvesting pigments and examined the fluorescence rise times of phycobilins and chlorophyll in the red alga *Porphyridium cruentum* and the blue-green alga *Anacystis nidulans* using a synchronously pumped, cavity-dumped dye laser. These investigations were aided by operation of the laser with very high repetition rates and low pulse energies, as well as by resolution of the fluorescence spectra into individual components by numerical solution of a linear equation system. An analysis of this time-resolved laser-initiated investigation, presented below, provides conclusive evidence of the energy transfer pattern in these complex systems.

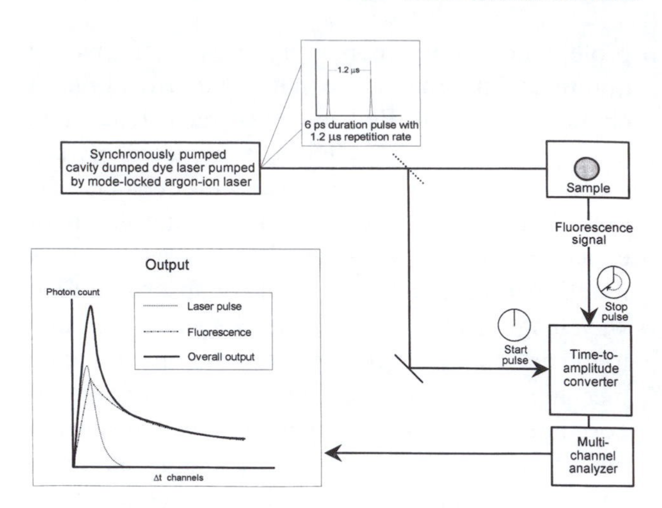


**Figure 7-2**
Schematic diagram of the picosecond laser apparatus and the time-correlated single-photon-counting system employed by Yamazaki et al. [13]. The term *single-photon-counting* underscores the sensitivity of the technique as individual fluorescence photons are detected. With a beam splitter, the output of an argon ion pumped dye laser excites the sample at a high repetition rate (800 kHz ⇒ one pulse every 1.2 $\mu$s) and triggers a reference detector (a high-speed photodiode). The reference detector starts a clock, the time-to-amplitude converter (TAC), which then charges a capacitor with a constant current. The TAC waits for a fluorescence photon to be detected by a sensitive detector, a microchannel plate photomultiplier. The detector responds to the emission with a measurable electric output pulse. The output pulse stops the clock by ceasing the charging of the capacitor. Discriminators are used to shape the electric pulses that are used for both the starting and stopping of the TAC in order to present standard electric signals to the TAC. The voltage that has accumulated in the capacitor of the TAC between the start and stop pulses is proportional to the amount of time between the excitation of the sample and the detection of a fluorescence photon. An analog-to-digital converter converts this voltage into a magnitude of time. This time is interpreted as an address or channel number in a multichannel analyzer (MCA), and the content of the channel is incremented by one. The experiment is usually repeated $10^6$ to $10^8$ times to accumulate 1000 counts or more in the peak channel. The contents of the channels of the MCA constitute a histogram of the number of counts versus time that corresponds to the probability of the occurrence of a fluorescence photon as a function of time after excitation of the sample. Deconvolution techniques are applied to separate the fluorescence decay curve from the contributions to the overall output that arise from the finite width of the excitation pulse and the response time of the detector.

## The Laser System

A schematic diagram of the picosecond laser apparatus and the time-correlated photon-counting system used by Yamazaki and co-workers appears in Figure 7-2. A synchronously pumped cavity-dumped dye laser and mode-locked argon ion laser provided excitation at 540 nm ($\lambda_{max}$ of PE) and 580 nm ($\lambda_{max}$ of PC) for *P. cruentum* and *A. nidulans*, respectively. The pulses of 6 ps duration (full-width, half-maximum) were generated at a repetition rate of 800 kHz and an intensity of $10^8$ photons/cm$^2$ at 570 nm. With a beam splitter, the dye laser pulse train functioned as both an excitation source for the sample as well as a start pulse for a time-to-amplitude converter (TAC) via a high-speed photodiode and discriminator. The fluorescence emission of the algal cells was monitored at 90° and dispersed to a microchannel plate photomultiplier using a monochromator driven in 0.625 nm steps from 600 to 750 nm. The signal was amplified by a 1 GHz preamplifier and output to a constant fraction discriminator to act as a stop pulse for the TAC. The start pulse was delayed to provide every fluorescence photon with a stop pulse. The output of the TAC was sent to a multichannel pulse height analyzer under microcomputer control. The design of the pulse height analyzer, with two microchannel plates exhibiting an extremely narrow electron transit time spread, enabled the deconvolution of fluorescence lifetimes as short as 10 ps.

## Measurement and Analysis of Time-Resolved Fluorescence Spectra

To understand the fluorescence data acquired via experiment and the subsequent spectra constructed from the primary data, it is important to distinguish among the terms *fluorescence decay curve*, *time-resolved fluorescence emission spectrum*, and *total fluorescence emission spectrum*. A *fluorescence decay curve* depicts the time dependence of the intensity of fluorescence emission induced by a fixed excitation wavelength and monitored at a specific emission wavelength. Thus, a fluorescence decay curve consists of an *xy* graph with fluorescence intensity on the y axis and time on the x axis. The decay time of fluorescence emission will vary with emission wavelength, reflecting emission from different fluorophores with distinct fluorescence spectra. A *time-resolved fluorescence emission spectrum* consists of an emission spectrum (fluorescence intensity vs. emission wavelength at a fixed excitation wavelength) acquired at a certain time interval following excitation. These time-resolved fluorescence spectra result from the contributions of all individual antenna pigments that are fluorescing at a given time. When the identity and/or number of emitting fluorophores in a sample varies over time, the shape of the fluorescence emission spectrum will also vary with time. A *total fluorescence emission spectrum* represents the sum of individual time-resolved emission spectra acquired over the entire time course of an experiment.

Now let's see how these terms apply to the data acquired and analyzed by Yamazaki and co-workers. The fluorescence decay curves of algal cells were obtained via excitation at one wavelength and subsequent monitoring of emission at wavelengths from 560 to 750 nm (*P. cruentum*) or 600 to 750 nm (*A. nidulans*) in 0.625 nm increments. Thus, fluorescence decay curves (graphs of fluorescence intensity *vs.* time) were collected for 305 emission wavelengths for *P. cruentum* and 241 emission wavelengths for *A. nidulans*. Time-resolved fluorescence emission spectra (plots of fluorescence intensity vs. emission wavelength) of *P. cruentum* and *A. nidulans* were obtained by integration of the decay curves in 12.8 ps intervals. Finally, a plot of total fluorescence intensity vs. emission wavelength for each alga was obtained by summing the integrated decay curves throughout the entire time region. Figure 7-3 summarizes these data acquisition and analysis steps.

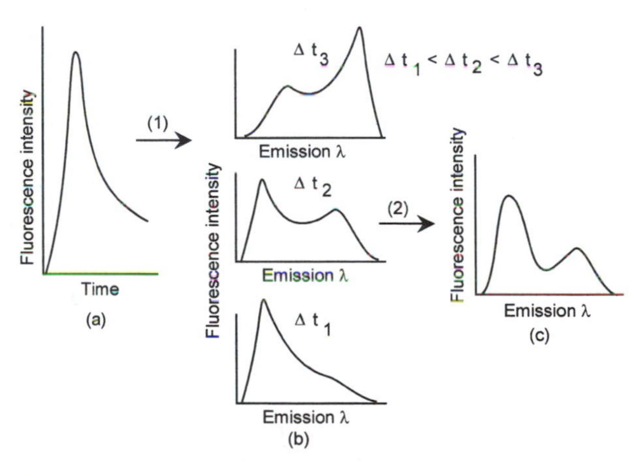


**Figure 7-3**
Summary of the data acquisition and analysis steps in the study of light-harvesting pigment energy transfer. *Fluorescence decay curves* (fluorescence intensity vs. time) as in *(a)* are acquired one at a time at a fixed excitation wavelength and a fixed emission wavelength over a range of emission wavelengths using the single-photon-counting technique. Integration of the decay curves in increments of 12.8 ps (*process 1*) yields a series of *time-resolved fluorescence emission spectra* as in *(b)*, i.e, graphs of fluorescence intensity vs. emission wavelength acquired at a fixed excitation wavelength for a fixed time interval following excitation of the sample. A plot of total fluorescence intensity as a function of emission wavelength appears in *(c)* (*total fluorescence emission spectrum*) and is obtained by summing the integrated decay curves throughout the entire time region (*process 2*).

To analyze the total fluorescence spectrum in terms of the contributions of the individual antenna pigments, the overall spectrum was assumed to consist of the fluorescence of phycoerythrin (PE) (for *P. cruentum* only), phycocyanin (PC), allophycocyanin (APC), a far-emitting allophycocyanin (fAPC), and chlorophyll *a* (Chl). Spectra of isolated PE, PC, and APC were obtained to determine the fluorescence band shape of each pigment. A red-shift of the APC fluorescence by 15 nm was assumed to obtain the band shape of fAPC fluorescence, and the total fluorescence spectrum of intact cells of *Chlorella pyrenoidosa* was taken as the band shape of Chl *a* fluorescence. The individual fluorescence bands permitted the resolution of the time-resolved fluorescence spectra into individual pigment components by a numerical solution of a system of linear equations. As the wavelength of the fluorescence maximum of individual pigments often varies with the algal species, best-fitting component spectra were permitted up to 3 nm shifts. After resolving the fluorescence spectra of the individual antenna pigments, the authors determined the time at which the fluorescence of each pigment reached its maximum value. This time is the characteristic rise time of the pigment's fluorescence.

### *Fluorescence Spectra of P. cruentum*

Figures 7-4 and 7-5 show an array of data including: (1) typical fluorescence decay curves at three monitoring wavelengths, as shown in Figure 7-4(b), (c), and (d), (2) a series of integrated time-resolved fluorescence spectra over various time domains, as shown in Figure 7-5, and (3) an overall fluorescence spectrum for *P. cruentum*, as shown in Figure 7-4(a). The decay curves exhibit a dramatic wavelength dependence, as further illustrated by the time-resolved spectra. Fluorescence from PE occurs predominantly in the 0-51 ps time region and peaks at 578 nm. The fluorescence maximum of PC at 645 nm is first resolved in the spectra obtained at 26 ps. From 51 to 128 ps the PC emission is superimposed with the APC peak at 660 nm, which dominates at later times. The Chl *a* emission is the last fluorescence spectrum to be resolved, appearing at 128 ps as a small shoulder around 683 nm. Curve fitting of the time-resolved spectra enabled the determination of the times at

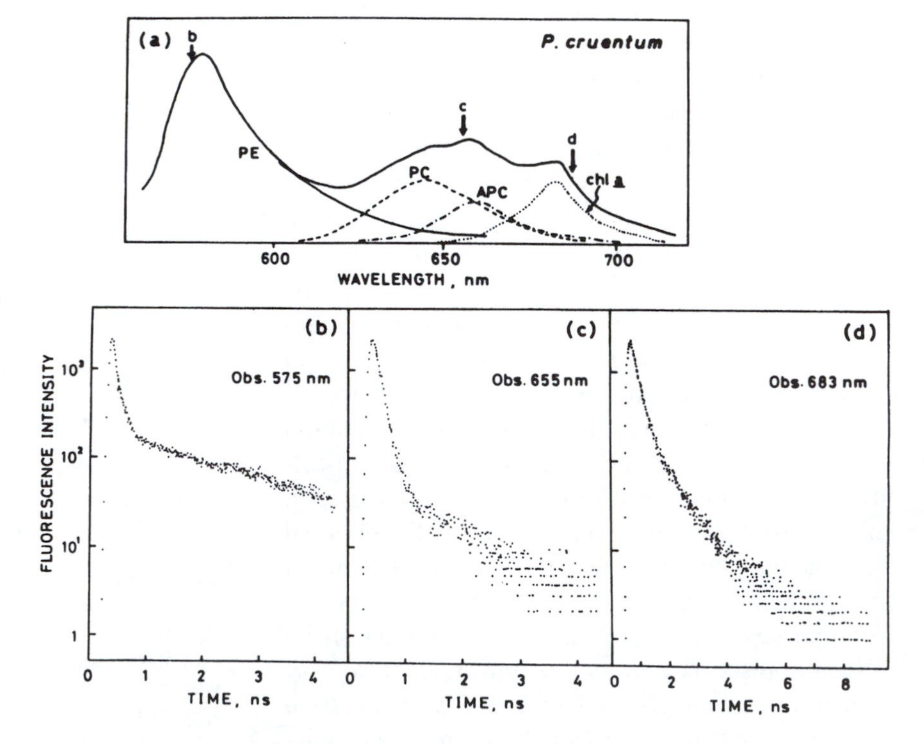


**Figure 7-4**
*(a)* Deconvolution of the overall fluorescence spectrum of *P. cruentum* into the component emission spectra of the individual light-harvesting pigments phycoerythrin (PE), phycocyanin (PC), allophycocyanin (APC), and chlorophyll *a* (Chl *a*). *(b–d)* The fluorescence decay curves of *P. cruentum* monitored at wavelengths of 575 nm (predominantly phycoerythrin emission), 655 nm (a superposition of phycocyanin and allophycocyanin emission), and 683 nm (predominantly chlorophyll *a* emission). Reprinted with permission from I. Yamazaki, et al., *Photochem. Photobiol.*, 1984, 39, 233–240. Copyright 1984 American Society for Photobiology.

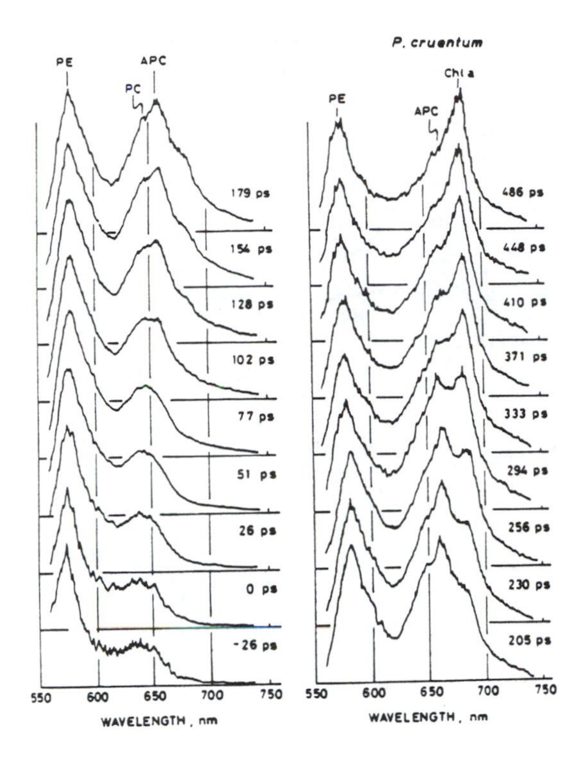


Figure 7-5
The fluorescence emission spectrum of *P. cruentum* obtained by integrating the decay curves at specific wavelengths over the desired time region. A time of 0 ps was assigned to the time at which the laser excitation pulse reached its maximum intensity. Fluorescence initially emanates from phycoerythrin (PE) at 578 nm and dominates in the data collected from 0 to 51 ps. Fluorescence emission from phycocyanin (PC) at 645 nm is apparent after 26 ps. The allophycocyanin (APC) emission at 660 nm is easily resolved after 102 ps. The small shoulder at 683 nm after 128 ps reflects chlorophyll *a* emission (Chl *a*) and is well-resolved in spectra after 256 ps. Reprinted with permission from I. Yamazaki, et al., *Photochem. Photobiol.*, 1984, 39, 233–240. Copyright 1984 American Society for Photobiology.

which the fluorescence of the individual pigments reached their maximum values. These fluorescence rise times are 0, 30, 57, and 150 ps for PE, PC, APC, and Chl *a*, respectively. Thus, sequential energy transfer from PE → PC → APC → Chl *a* is suggested by the increasing rise times. A sequential model is also consistent with the increasing $\lambda_{max}$ for the pigments PE, PC, APC, and Chl *a*, respectively.

### *Fluorescence Spectra of A. nidulans*

The fluorescence decay curves shown in Figure 7-6(b), (c), and (d); the time-resolved spectra, shown in Figure 7-7 and the overall fluorescence spectrum, shown in Figure 7-6(a) for *A. nidulans* also display a marked dependence on the emission wavelength. PC fluorescence at 645 nm appeared in the 0-51 ps spectra. The APC spectrum was red-shifted with time, appearing at about 650 nm from 51 to 300 ps and at 660 nm at later times. The broad fluorescence band of Chl *a* centered at 683 nm could be distinguished after 200 ps and increased noticeably throughout the entire time domain. Matching of theoretical and experimental fluorescence rise curves suggested the fluorescence rise times of PC, APC, and Chl *a* to be 0, 60, and 120 ps. The increasing rise times suggest a sequential pattern of energy transfer for these pigments: PC → APC → Chl *a*. The results also suggest that a special far-emitting APC pigment (fAPC) exhibits a rise time of 110 ps. If sequential energy transfer were to occur from APC → fAPC → Chl *a*, the authors reasoned that the smaller concentration of fAPC relative to APC would require a faster transfer rate from fAPC to Chl *a* than that from APC to fAPC. However, the observed comparable decay kinetics of fAPC and APC suggest that fAPC is not in the main pathway of energy transfer to Chl *a*. The pigment apparently functions as a bypass or parallel pathway of energy flow at the point of APC. Thus, a second sequential pathway for energy transfer exists in *A. nidulans*: PC → APC → fAPC.

## Summary of Results

### *P. cruentum*

Following excitation at 540 nm:

- Direct induction of PE fluorescence (no rise time).
- Increasing fluorescence rise times observed in the order PC < APC < Chl *a*.

Results consistent with a sequential model:

PE → PC → APC → Chl *a*

### *A. nidulans*

Following excitation at 580 nm:

- Direct induction of PC fluorescence (no rise time).

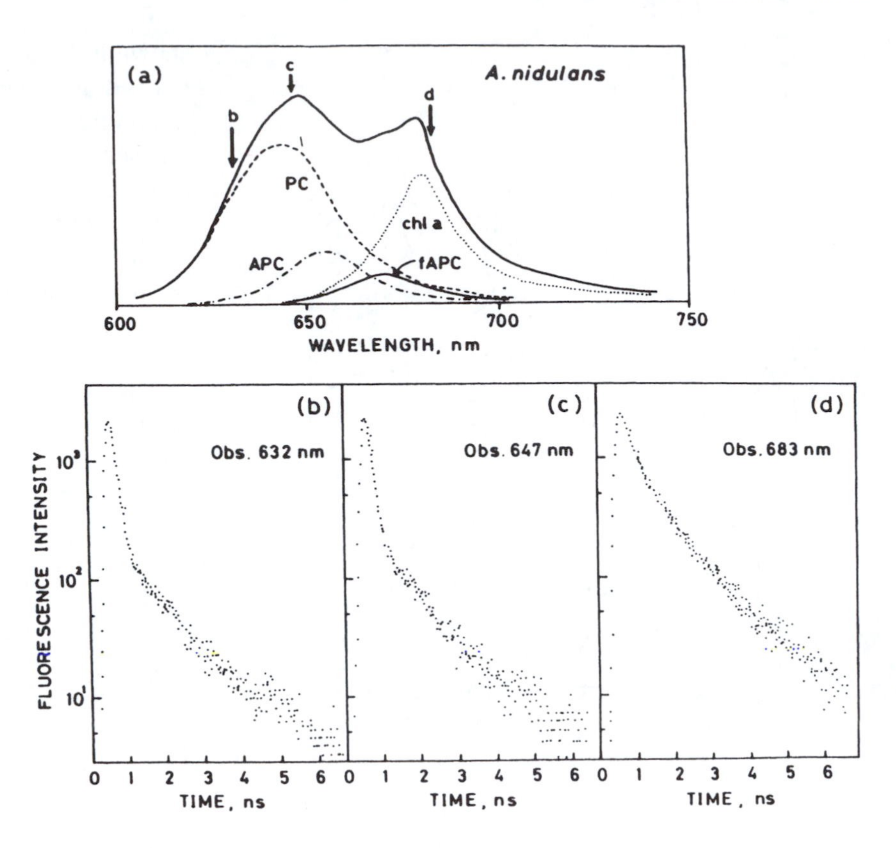


**Figure 7-6**
*(a)* Deconvolution of the overall fluorescence spectrum of *A. nidulans* into the component emission spectra of the individual light-harvesting pigments phycocyanin (PC), allophycocyanin (APC), far-emitting allophycocyanin (fAPC), and chlorophyll *a* (Chl *a*). *(b–d)* The fluorescence decay curves of *A. nidulans* monitored at wavelengths of 632 nm (predominantly phycocyanin emission), 647 nm (a superposition of phycocyanin and allophycocyanin emission), and 683 nm (predominantly chlorophyll *a* emission with a small contribution from far-emitting allophycocyanin). Reprinted with permission from I. Yamazaki, et al., *Photochem. Photobiol.*, 1984, 39, 233–240. Copyright 1984 American Society for Photobiology.

- Increasing fluorescence rise times observed in the order APC < far-APC ≈ Chl *a*.

Results consistent with a modified sequential model with a parallel pathway at APC:

$$\text{PC} \rightarrow \text{APC} \begin{array}{l} \nearrow \text{Chl } a \\ \searrow \text{far-APC} \end{array}$$

## Conclusions and Significance of the Study

This study observed the successive fluorescence emission from various pigments in the light-harvesting system of species of red and blue-green algae. In the case of the red algae *P. cruentum* the fluorescence spectra changed over the time range of 0 to 400 ps and were consistent with emission in the order phycoerythrin → phycocyanin → allophycocyanin → chlorophyll *a*, as seen in Figure 7-8. While the fluorescence spectra obtained for the blue-green alga *A. nidulans* can also be interpreted to arise from the same sequence of energy transfer upon phycocyanin excitation, the spectra changed over a significantly longer period of time from 0 to 1000 ps. These differences suggest a structural difference in the phycobilisomes of the two algae which had not previously been detected. Furthermore, the presence in *A. nidulans* of a far-emitting allophycocyanin pigment that does not direct energy flow from phycocyanin to chlorophyll *a* but instead serves as an energy sink (phycocyanin → allophycocyanin → far-emitting allophycocyanin) is a new observation of an alternative mode of energy partitioning. Thus, the ability to conduct fluorescence investigations on the picosecond timescale using laser excitation has confirmed proposed mechanisms of light energy transfer in red and blue-green algae, as well as revealed new and interesting information on the structure and organization of photosynthetic pigments in these organisms.

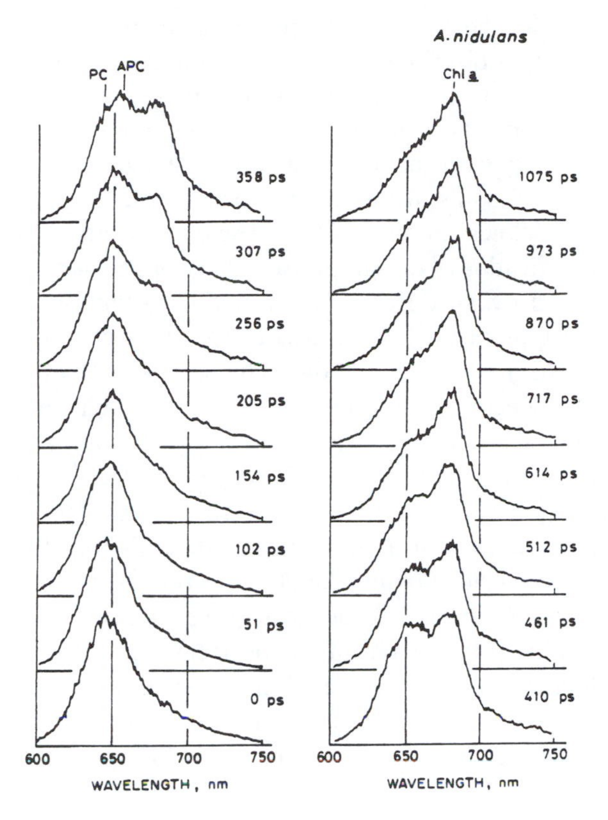


**Figure 7-7**
The fluorescence emission spectrum of *A. nidulans* obtained by integrating the decay curves at specific wavelengths over the desired time region. A time of 0 ps was assigned to the time at which the laser excitation pulse reached its maximum intensity. The initial fluorescence signal is emitted from phycocyanin (PC). The maximum intensity of the spectrum after 51 ps reflects emission from allophycocyanin (APC), with the spectrum shifting slightly to the red after 300 ps due to emission from far-emitting allophycocyanin (fAPC). The fluorescence signal from chlorophyll *a* appears as a shoulder at 683 nm after 200 ps and is increasingly resolved at later times, dominating after 400 ps. Reprinted with permission from I. Yamazaki, et al., *Photochem. Photobiol.*, 1984, 39, 233–240. Copyright 1984 American Society for Photobiology.

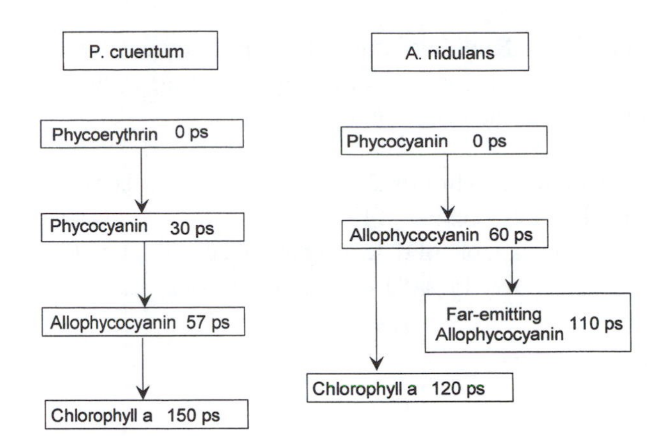


**Figure 7-8**
A schematic diagram of the sequence of energy transfer in the accessory pigments of *P. cruentum* and *A. nidulans*. The rise times of fluorescence emission from each chromophore as determined by Yamazaki et al. [13] are also presented. Following excitation at 540 nm, sequential energy transfer is observed in *P. cruentum* with direction induction of phycoerythrin (PE) and increasing fluorescence rise times for phycocyanin (PC), allophycocyanin (APC), and chlorophyll *a* (Chl *a*), respectively. Excitation of *A. nidulans* at 580 nm leads to direct induction of phycocyanin (PC) fluorescence, subsequent induction of allophycocyanin (APC) fluorescence, and excitation of either Chl *a* fluorescence (main pathway) or emission from far-emitting allophycocyanin (fAPC) (secondary pathway).

## Case Study II: Airborne Remote Sensing of Laser-Induced Fluorenscence

### Overview of the Case Study

**Objective.** To use noninvasive and remote measurements, conducted from aircraft, of laser-induced fluorescence of terrestrial and oceanographic targets. The technique of remote sensing of fluorescence signals can be utilized to detect a variety of conditions, including the presence of oil spills in marine environments and the physiological state and distribution of plant species over large areas.

**Laser Systems Employed.** Dual laser configurations consisting of either: (1) a 10 MW peak power excimer laser with a 308 nm emission wavelength and an accompanying dye laser system tuned to 450 or 533 nm or (2) a 3 MW peak power frequency doubled Nd:YAG laser tuned to 532 nm and a companion 427 nm XeCl excimer-pumped dye laser with a controlled maximum output power of 100 kW.

**Role of the Laser Systems.** To furnish high-powered and tunable excitation from high altitudes for the excitation of an array of fluorophores, including fuel oils and plant pigment molecules. Laser excitation also permits measurements to be conducted under a range of light conditions: in full sunlight, under cloud cover, or even in full darkness.

**Useful Characteristics of the Laser Light for this Application.** High power, coherence, monochromaticity, wavelength tunability, stability.

**Principles Reviewed.** Fluorescence, remote sensing, laser-induced detection and ranging systems, spectral characterization of fluorophores.

**Conclusions.** This study demonstrates that laser-induced remote sensing of fluorescence is a noninvasive technique capable of detecting and discriminating fluorophores over broad areas and in remote locales.

## Introduction

### *Remote Sensing*

An exciting new application of fluorescence as a powerful analytical tool has recently emerged. These experimental studies involve measurements, conducted from aircraft, of *laser-induced* fluorescence of a particular molecular or atomic target. This airborne *remote sensing* technique provides a sensitive and powerful means of detecting fluorophores in distant or inaccessible locations and over vast terrestrial, oceanographic, and atmospheric regions. Scientists are particularly interested in laser-based remote sensing systems, even though remote sensing of fluorescence is possible using airborne systems which utilize sunlight for excitation (i.e., *passive systems*). Systems with laser excitation sources (i.e., *active systems*) eliminate the dependence on sunlight and permit measurements to be conducted even under conditions of cloud cover or full darkness. Furthermore, the intensity and wavelength tunability of laser excitation sources enable the study of fluorescence spectral emissions from a multiplicity of fluorophores. The capabilities of remote laser-induced fluorescence investigations have been effectively exploited in two areas: the detection of maritime oil spills and discharges from ship traffic and the monitoring of the physiological state of terrestrial and oceanographic vegetation. We'll consider the nature of the information that can be acquired from remote fluorescence analysis in these two areas and then delineate the laser systems required for each application.

### *Surveillance and Monitoring of Oil in the Marine Environment*

Airborne laser-induced fluorescence measurements over ocean and coastal waters have successfully detected fluorophores with impressive sensitivity. For example, the limits of detection have stretched to substances distributed in a water column of up to 25 meters depth and with detection capabilities on the order of parts per billion.[14] In particular, remote fluorosensing is capable of the surveillance and monitoring of oil spills dispersed on a water surface with a film thickness of only 1 $\mu$m.[15] The classification and quantification of oil samples are also a practical operations. How are different classes of oils distinguished? Each type of oil (as a consequence of its chemical structure) possesses its own characteristic fluorescence spectrum—in effect, a fluorescence signature. At one extreme, light diesel fuels display high fluorescence intensities in the UV range (≈340-380 nm). In contrast, heavy crude oils exhibit markedly reduced fluorescence intensities with emission $\lambda_{max}$ values around 500 nm. Variations in fluorescence intensity and spectral shape are also used to distinguish fuel oils from less harmful substances like fish oils, which typically contain highly unsaturated fatty acids with four to six allyl groups and fluoresce at intermediate emission wavelengths (≈430 nm).

### *Noninvasive Characterizations of Photosynthetic Systems*

What kinds of information might potentially be acquired using airborne laser-induced fluorescence investigations of vegetation? Recall that the photosynthetic apparatus of plants, algae, and bacteria contain one or more organic pigments, such as chlorophyll, whose primary role is to absorb visible radiation to initiate the photochemical reactions of photosynthesis. All of these pigments are potential fluorophores. For example, isolated chlorophyll molecules emit a characteristic red fluorescence upon absorption of light in certain regions of the visible spectrum. When these chromophores are incorporated within the photosynthetic membrane, the fluorescence intensity is attenuated but detectable by sensitive instrumentation. The re-emission of absorbed light involves only a small percentage (e.g., 2–4%) of electronically excited pigment molecules, maintaining a high efficiency of photochemical conversion. Various fluorescence parameters—intensity and polarization of emission, lifetime of the excited state, quantum yield or fraction of incident light emitted as fluorescence, wavelength dependence of excitation and emission—reflect the fundamental properties of the pigment molecules and their environment. Furthermore, as the light emission from these pigment molecule fluorophores can be monitored in a nondestructive and noninvasive fashion, a direct study of a variety of physiological states is feasible. For example, laser-induced fluorescence measurements offer the potential for the detection of environmental stress in plants. Deficiencies of nutrients and water lead to decreases in photosynthetic efficiency and characteristic changes in chlorophyll fluorescence emission spectra.[16–19] Although more specific explanations of the origins of these fluorescence changes are needed, the reproducibility of these spectral changes validates the usefulness of laser-induced fluorescence measurements as a rapid, noninvasive remote technique for the detection of plant stress.

The identification of specific plants is also possible using laser-induced fluorescence investigations. Laboratory experiments using a pulsed nitrogen laser with 337 nm emission demonstrate the feasibility of the simultaneous excitation of a multiplicity of plant fluorophores.[20] These studies reveal that different plant types have characteristic fluorescence signatures, with distinct fluorescence maxima and with distinct ratios of the fluorescence intensities at the different wavelengths. For example, plants classified as monocots and dicots (i.e., having one or two seed leaves or cotyledons) possess common fluorescence maxima at 440 nm (arising from a plastoquinone electron acceptor, vitamin $K_1$), at 685 nm (arising from chlorophyll), and at 740 nm (also originating from chlorophyll). However, monocots have a significantly higher ratio of the square of the fluorescence intensity at 440 nm to the intensity at 685 nm (i.e., $I^2_{440}:I_{685}$). Conifers and hardwood trees, with common fluorescence maxima at 440 nm, at 525 nm (arising from the photoreceptor riboflavin), and at 740 nm, are distinguishable by the presence of a 685 nm fluorescence maximum in the latter. A fifth major plant type, algae, also have a characteristic fluorescence signature, with very low fluorescence at 440 nm, no emission at 525 nm, and fluorescence maxima at 685 nm and 740 nm such that the ratio of 685 nm emission to 740 nm emission is very high. Thus, the use of the laser-induced fluorescence technique for the remote identification of individual species offers considerable promise and excitement.

## The Laser Systems

The significant utility of fluorescence remote sensing for a diverse range of chemical, environmental, hydrological, and military investigations has spurred the development of what are known as *l*aser-*i*nduced *d*etection *a*nd *r*anging (*LIDAR*) systems. A basic LIDAR or (lidar) system generally consists of a high-powered pulsed laser source providing an appropriate emission wavelength for fluorescence excitation, a telescope to collect the fluorescence signal, and a spectrometer with a data acquisition system for monitoring and analyzing the data. As illustrations, schematic diagrams of the oceanographic lidar system developed by the University of Oldenburg [15] and the airborne system used by the National Aeronautics and Space Administration (NASA) for field measurements [21–23] are presented in Figures 7-9 and 7-10. These two-laser systems are some of the more advanced configurations, permitting the acquisition of fluorescence data using dual wavelength excitation. Table 7-2 compares and contrasts the main features of each system.

The design of the oceanographic lidar system was formulated to have a free field of view (by the telescope) of the water surface from an aircraft with a bottom hatch. Both the lasers and the detector system are mounted on an optical table to maintain rigid alignment. The main laser is a 10 MW peak power excimer laser with emission characterized by a wavelength of 308 nm and a pulse length of 12 ns. The excimer laser beam may be used either directly as the lidar beam (for excitation of fluorescence from oil films or plumes) or as a pump beam for a dye laser with an output wavelength of 450 nm (for excitation of chlorophyll fluorescence from phytoplankton and other naturally dissolved organic matter).

The NASA design of four major optical tiers provides flexibility for a range of applications. The excitation sources consist of a 3 MW peak power frequency-doubled Nd:YAG laser at 532 nm and a companion 427 nm XeCl excimer-pumped dye laser with a controlled maximum output power of 100 kW. These lasers furnish excitation for direct excita-

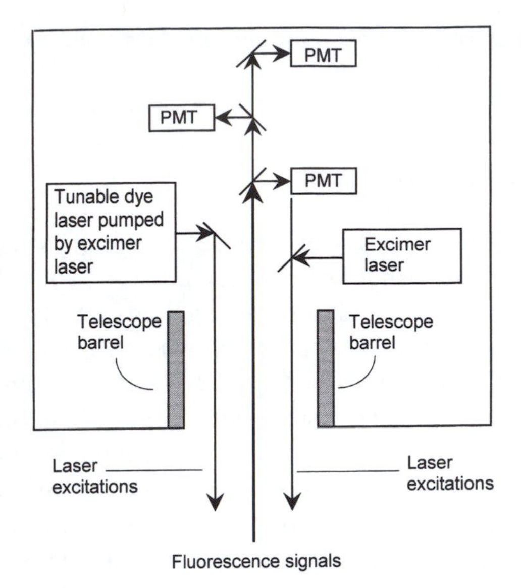


**Figure 7-9**
Schematic diagram of the oceanographic lidar system designed by the University of Oldenburg for marine surveillance [15]. The dual-laser excitation source (excimer laser + dye laser) enables fluorescence from oils and pollutants to be distinguished from the emission of naturally occurring fluorophores in seawaters. The configuration of the telescope at the bottom hatch of the aircraft provides an unobstructed field of view to the water surface. The optical axes of the telescope and the lasers are arranged at 90° to one another, with beam splitters used to direct the excitation to the ocean waters.

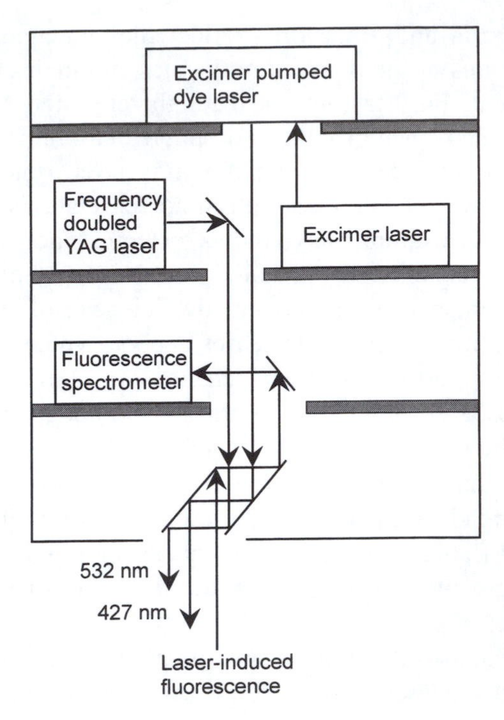


**Figure 7-10**
Schematic diagram of the airborne oceanographic lidar system used by the National Aeronautics and Space Administration (NASA) for field measurements [8–10]. The four major optical tiers consist of an excimer-pumped dye laser on the topmost tier (output at 427 nm), a Nd:YAG laser (frequency doubled for output at 532 nm) and XeCl excimer laser (308 nm output used to pump the dye laser on the top tier) on the tier below, a fluorescence spectrometer on the major tier, and various optical mirrors on the lowest level.

tion of chlorophyll molecules (427 nm) and for indirect excitation of chlorophyll via light absorption by accessory pigments (532 nm). The Nd:YAG and excimer lasers are located on the third level, while the excimer-pumped dye laser occupies the fourth or topmost tier. The major components of the second level are the telescope and fluorescence spectrometer, with the lowest level containing various optical mirrors for directing the laser beam and the returning fluorescence emission.

Let's consider the NASA system of Figure 7-10 to enumerate the key features of the lidar instrumentation. The 532 nm pulsed output of the frequency-doubled Nd:YAG laser is directed via folding mirrors to an adjustable beam divergence/collimating lens. The beam is then directed downward through folding mirrors to the angle-adjustable nutating scan mirror. The scan mirror directs the beam to the land or ocean surface. For a second excitation wavelength, the 308 nm pulsed output of the excimer laser is directed upward to the topmost tier and pulsed through a dye laser operated with stilbene-3. The 427 nm stilbene-3 dye laser output is focused through the divergence/collimating lens, folding mirrors, and scan mirror shared with the Nd:YAG laser. The lasers could be alternately pulsed, but, because of the extreme disparity in laser powers, only one excitation wavelength is generally used on each flight. To provide the start pulse timing and to monitor the output power of the lasers, a high-speed silicon photodiode views extraneously scattered radiation from the first folding mirror.

The surface fluorescence signals return through the path of scanning and folding mirrors on the low-

**Table 7-2**
Characteristics of Lidar Systems for Remote Sensing of Laser-Induced Fluorescence

| | *Oceanographic Lidar System for Monitoring Oil Spills* | |
|---|---|---|
| Lasers | Excimer | Dye |
| λ | 308 nm | 450 nm/ 533 nm |
| Peak power | 10 MW | 1 MW |
| Telescope | 40 cm receiving telescope | |
| Wavelength selection and detectors | Dichroic beam splitter assembly to deflect selected spectral ranges to interference filters and seven photomultiplier tubes. | |
| | *NASA Lidar System for Characterization of Vegetation* | |
| Lasers | Nd:YAG | XeCl excimer-pumped dye |
| λ | 532 nm | 308 nm |
| Peak power | 3 MW | 100 kW |
| Telescope | 30.5 cm receiving telescope | |
| Wavelength selection and detectors | Diffraction grating to disperse the spectral components of the return signal to 36 separate photomultiplier tubes with individual light guides. | |

est tier. The uncollimated spatial extent of the fluorescence necessitates directing the signals into a receiving telescope located on the main optical tier. The radiation is then collimated, focused, and directed via a folding mirror/beam splitter combination to the fluorosensing detector assembly. A filter in front of the fluorescence spectrometer rejects radiation below ≈ 540 nm and thus rejects the excitation wavelength component of the return signal (532 nm). A small amount of the return signal is directed through an opening in the beam splitter to a 532 nm interference filter and a focusing lens in order to generate the gate pulses for the analog-to-digital converter and to monitor altimetry information.

The fluorosensing detection assembly directs the output of the high-pass filter through a diffraction grating and a simple focusing lens to the entrance surface of 36 UVT Plexiglas light guides. These guides are optically coupled to 36 separate photomultiplier tubes (PMT). To receive spectral components over the range of 350 to 800 nm, the front faces of the light guides are physically located in the curved focal plane of the spectrometer. This design eliminates the need for physically scanning the signal as is typically the case with laboratory spectrophotometers. A high-voltage power supply allows the gain of the photomultiplier tubes to be individually adjusted to enhance detection of the fluorescence over the entire spectral range. An independent microprocessor-based data acquisition system processes and displays real-time spectral data.

## Protocol and Results of Field Experiments

To test the detection capabilities of remote sensing of oil spills, a typical airborne survey experiment would consist of multiple flights to monitor fluorescence signals at various wavelengths as a function of time and flight distance. For example, one exercise with the oceanographic lidar system monitored 60 L spills of crude oil and diesel oil of several kilometer lengths and 10-50 m widths. With laser excitation at 308 nm, fluorescence signals were measured at detection wavelengths of 366 nm and 450 nm to observe the dispersal of the diesel and crude oils, respectively. Successful detection and analysis of both types of oils could be achieved for periods within 20 minutes after spilling. Weathering (i.e., dispersal via wind conditions at the sea surface) and evaporation induce significant losses in fluorescence intensity and pose limitations in this remote sensing technique. Nevertheless, the reliability of the laser fluorosensing technique has challenged scientists to develop instruments for routine surveillance of marine and coastal waters.

To examine fluorescence data from terrestrial vegetation, the NASA lidar system uses 422 nm excitation from an excimer-pumped dye laser and detects emission at 685 nm. Recall that these wavelengths were selected on the basis of the optimal wavelengths for inducing and detecting fluorescence emission from chlorophyll pigments. A profile of the terrain elevation is also generally acquired to correct for differences in the distances of respective targets from the aircraft. This elevational profile is useful to note that considerable fluorescence is acquired from lower-height vegetation, such as grasses, as well as from the higher shrubs and trees. Abrupt breaks in the 685 nm chlorophyll signal are observed at changes in target types, such as at a boundary between a forest and a dirt road. Additional fluorescence profiles can be collected on separate flights with 532 nm excitation. Here accessory pigments absorb light and then funnel this excitation via energy transfer to neighboring chlorophyll molecules. Analyses of the results suggest that the strength of the fluorescence signals and the similarity of the response patterns over targets of the same general type provide sufficient precision to study plant pigment distributions over broad areas.

A similar correspondence of fluorescence profile to plant distribution is observed for remote sensing flights over marine waters. Profiles of the relative chlorophyll and phycoerythrin fluorescence from 532 nm excitation along a single flight line reflect the different kinds and distributions of phytoplankton in these water masses. A distinct variability in the chlorophyll and phycoerythrin fluorescence response signals indicates a distribution of these photopigments. This spatial distribution of photopigments undoubtedly reflects the variety of phytoplankton species as controlled by the physical and chemical properties of the water masses. This interpretation can be confirmed by surface investigations, demonstrating the feasibility of airborne mapping of oceanographic plant distributions using laser-induced fluorescence.

As we have seen, various aspects of the physiological state of vegetation can be determined from the level of chlorophyll fluorescence induced by airborne lidar systems. The chlorophyll fluorescence spectrum, that is, the dependence of fluorescence intensity on emission wavelength, is also sensitive to a number of characteristics of the vegetation, including the plant species, the chlorophyll content, the age and seasonal period of the plant, and the structure and location of the leaf. Depending on the excitation wavelength employed, other fluorophores may contribute to the observed emission spectrum, including flavin adenine dinucleotide, nicotinamide adenine dinucleotide phosphate, coumarin, and light-harvesting pigments such as carotenes. As a consequence, the ability to resolve emission at discrete wavelengths from a lidar signal would vastly enhance the analysis of the acquired spectral information. A system to perform multicolor imaging of vegetation fluorescence has been designed [24] using frequency-doubled or -tripled emission from a Nd:YAG laser and both a lidar telescope and a split-mirror Cassegrainian telescope.

## Further Contributions of Lasers to Remote Sensing

From an environmental vantage point, the considerable significance of these applications is sufficient to ensure the continued development of remote sensing techniques. But the astounding diverse range of terrestrial, marine, and atmospheric investigations involving remote sensing is a clear indication of the technological power of this technique. For example, while we've mentioned the airborne detection of laser-induced fluorescence from ocean waters as a means of studying oil spills and vegetation, the method is also useful for detecting groundwater contaminants.[14,25] In particular, small aromatic compounds which absorb far-UV radiation and possess measurable fluorescence quantum yields are readily monitored. These pollutants include phenol, toluene, o-cresol, xylenes, and various chloro- and nitro-substituted phenols. Another exciting oceanographic application involves the remote detection of submerged submarines from another search submarine through the monitoring of subsurface water temperatures.[26] By directing a pulsed laser beam into the water, the subsurface ocean waves created by the turbulence of a submarine's wake are detectable. The subsurface temperature-depth profile of the water is determined from an analysis of the resulting Brillouin and Rayleigh backscattering components.

Atmospheric processes over large areas are conveniently monitored by remote sensing techniques. A recent study reports that airborne and satellite infrared instrumentation have been successfully used to characterize the type and intensity of large scale forest and brush fires and to identify the products of such fires.[27] These techniques are valuable from an environmental viewpoint to study atmospheric pollution and from a forestry management viewpoint to detect wildfires and to evaluate fire dynamics. Real-time data acquisition with transmission to a ground station recently allowed for the planning of both fire-fighting strategies and resource allocations during a fire in the California Ojai Forest.[28] Furthermore, such remote sensing has important meteorological and military applications, for recent studies have recognized the possibility of added possible short-term climatic effects arising from fires ignited during a nuclear exchange. Ground-based infrared absorption measurements using a tunable diode laser have also been conducted to measure methane concentration in the stratosphere.[29] Similar infrared remote sensing techniques will even be used to study the Martian atmosphere aboard the Mars Observer spacecraft.[30]

The development of more complex lidar systems has enabled the investigation of atmospheric conditions over large areas and a range of altitudes, despite the presence of interferences from pollutants varying in concentration and identity. In particular, the vital role of ozone in the gas-phase chemistry of the troposphere demands accurate estimates of both its temporal and spatial distributions. Ground-based and airborne lidar systems have been developed to demonstrate that such accurate measurements of ozone distributions can be achieved using a differential absorption lidar (DIAL) technique. This method requires two fixed-wavelength laser lines to be emitted simultaneously from a single laser source, with one line strongly absorbed by ozone and the other line serving as an atmospheric reference as it is not absorbed by ozone. Solid-state Nd:YAG lasers with doubling crystals and KrF excimer lasers provide suitable lines in the UV region for both day- and night-time operation. For infrared operation DIAL systems have been designed using high-energy multiwavelength $CO_2$ lasers. In addition to atmospheric ozone depletion studies, DIAL systems have been used to detect halogenated hydrocarbon emissions during mobile ground operation, trace gases such as $N_2O$ and CO in the upper atmosphere, natural and anthropogenic aerosols in the troposphere, and chemical and biological warfare agents on the battlefield.[31–33]

Satellite-based remote sensing systems have improved the ability to acquire information on cloud properties for climatological and meteorological purposes.[34] The absorption, scattering, and reflection properties of clouds have been used to quantify a range of cloud parameters including cloud height, thickness, phase, droplet size, and particle distribution. Attenuation of the laser beam by the dense lower atmosphere is eliminated by conducting laser remote sensing from an upper atmosphere satellite, improving the accuracy of measurements. However, satellite-borne lasers must satisfy additional requirements imposed by the rigors of launching and the adversity of the space environment. Diode-pumped solid-state lasers (particularly Nd:YAG and Nd:YLF lasers) offer the reliability and durability needed for satellite operation for extended periods of time.

This brief compendium serves as an illustration of the exciting directions in which scientists are applying methods of remote detection. Whether the application monitors a sample's absorption, emission, or scattering properties from land, sea, air, or space, the laser excitation is instrumental in the production of signals for remote detection.

## FOR FURTHER EXPLORATION

The technique of laser-induced fluorescence spectroscopy is currently playing a key role in deciphering the dynamics of the protein folding process. A protein is able to perform its biological functions only when the string of amino acids of which it's composed is folded into a compact three-dimensional structure. For many years scientists have sought to understand the process by which protein folding occurs. In particular, one major goal of studying the reversible folding and unfolding of proteins is to determine whether the mechanism of protein folding involves intermediates. Currently lasers are facilitating the characterization of the kinetics and mechanism of protein folding. In one application, scientists at the National Institutes of Health are using a Nd:YAG laser to examine the refolding of the protein cytochrome c.[35] While the 1064 nm fundamental output of the Nd:YAG laser is capable of inducing denaturation (i.e., unfolding) of the protein by heating the surrounding solvent (an increase on the order of 50 K), unfolding is accomplished chemically using a denaturant such as urea. Refolding is induced through a rapid mixing of the denatured cytochrome c with a pH 7 phosphate buffer. The fourth harmonic of the Nd:YAG output (266 nm) serves as an excitation wavelength for an intrinsic tryptophan fluorophore, Trp 59, the sole tryptophan residue in the protein. Refolding of the protein to its native structure is monitored by the subsequent variation in the wavelength and intensity of Trp emission. The emission wavelength maximum typically shifts during the folding process (from about 345-350 nm to 325-330 nm) to reflect the removal of the tryptophan from an environment where it is solvated with a polar solvent to a nonpolar region buried within the interior of the folded protein. Also as the protein folds, the proximity of Trp 59 to the covalently-bound heme within the cytochrome c protein increases, permitting nonradiative energy transfer from tryptophan to the heme. The efficiency of energy transfer is revealed by the extent of quenching of the tryptophan fluorescence. Processes occurring within 50-100 $\mu$s can be detected, well within the time needed to observe local structure formation ("secondary structure") such as $\alpha$-helices and $\beta$-sheets.

## DISCUSSION QUESTIONS

**1.** What fluorescence experiments should be conducted to either verify or disprove a mixed energy transfer pattern characterized by: two accessory pigments ($A$ and $B$) that serve a long wavelength chromophore $D$ in a sequential excitation transfer and a third accessory pigment ($C$) that is involved in a parallel excitation transfer to $D$? The energy transfer scheme is represented by:

$$\begin{array}{ccc} A \rightarrow B \rightarrow D \\ \nearrow \\ C \end{array}$$

**2.** Rather than focusing on the decay of the excited state to monitor energy transfer, what other technique could be used to follow the transfer of energy among pigment molecules? Would the same laser apparatus be suitable for this alternative technique? If not, suggest an appropriate laser system for these new investigations.

**3.** Photosynthetic organisms often respond to different growth conditions by synthesizing different relative amounts of light-harvesting pigments. How would such changes alter the fluorescence results obtained?

**4.** Case Study II describes one advantage of remote sensing of emission—the ability to probe a sample that is spread over a large area. What other situations are ideal settings for remote sensing of laser-induced fluorescence?

**5.** In addition to identifying vegetation, atmospheric constituents, and ocean contents, we've noted that remote sensing is also useful in obtaining data to evaluate long-range changes in environmental conditions. What kinds of temporal studies might be of interest?

## LITERATURE CITED


1. Forster, T., "Experimental and theoretical investigation of intermolecular energy transfer of electron activation energy," *Z. Naturforsch.*, 1949, 4a, 321–327.
2. Tomita, G., and Rabinowitch, E., "Excitation energy transfer between pigments in photosynthetic cells," *Biophys. J.*, 1962, 2, 483–499.
3. Grabowski, J., and Gantt, E., "Photophysical properties of phycobiliproteins from phycobilisomes: Fluorescence lifetimes, quantum yields, and polarization spectra," *Photochem. Photobiol.*, 1978, 28, 39–45.
4. Grabowski, J., and Gantt, E., "Excitation energy migration in phycobilisomes: Comparison of experimental results and theoretical predictions," *Photochem. Photobiol.*, 1978, 28, 47–54.
5. Brody, S. S., and Rabinowitch, E., "Excitation lifetime of photosynthetic pigments in vitro and in vivo," *Science*, 1957, 125, 555.
6. Brody, S. S., Porter, G., Tredwell, C. J., and Barber, J., "Picosecond energy transfer in *Anacystis nidulans*," *Photobiochem. Photobiophys.*, 1981, 2, 11–14.
7. Brody, S. S., Tredwell, C. J., and Barber, J., "Picosecond energy transfer in *Porphyridium cruentum* and *Anacystis nidulans*," *Biophys. J.*, 1981, 34, 439–449.
8. Mar, T., Govindjee, Shinghal, G. S., and Merkelo, H., "Lifetime of the excited state in vivo. 1. Chlorophyll *a* in algae, at room temperature and at liquid nitrogen temperatures; rate constants of radiationless deactivation and trapping," *Biophys. J.*, 1972, 12, 797–808.
9. Nicholson, W. J., and Fortoul, J. I., "Measurement of the fluorescent lifetime of chlorella and porphyridium in weak light," *Biochem. Biophys. Acta*, 1967, 143, 577–582.
10. Singhal, G. S., and Rabinowitch, E., "Measurement of the fluorescence lifetime of chlorophyll a in vivo," *Biophys. J.*, 1969, 9, 586–591.

11. Porter, G., Tredwell, C. J., Searle, G. F. W., and Barber, J., "Picosecond time-resolved energy transfer in *Porphyridium cruentum*," *Biochem. Biophys. Acta*, 1978, 501, 232–245.

12. Hader, D. P., and Tevini, M., "Photosynthesis," *General Photobiology*, Pergamon Press, Oxford, 1987, Chapter 9, pp. 112–165.

13. Yamazaki, I., Mimuro, M., Murao, T., Yamazaki, T., Yoshihara, K., and Fujita, Y., "Excitation energy transfer in the light-harvesting antenna system of the red alga *Porphyridium cruentum* and the blue-green alga *Anacystis nidulans*: Analysis of time-resolved fluorescence spectra," *Photochem. Photobiol.*, 1984, 39, 233–240.

14. Chudyk, W. A., Carrabba, M. M., and Kenny, J. E., "Remote detection of groundwater contaminants using far-ultraviolet laser-induced fluorescence," *Anal. Chem.*, 1985, 57, 1237.

15. Hengestermann, T., and Reuter, R., "Lidar fluorosensing of mineral oil spills on the sea surface," *Applied Optics*, 1990, 29, 3218–3227.

16. Chappelle, E. W., Wood, Jr., F. M., McMurtrey III, J. E., and Newcomb, W. W., "Laser-induced fluorescence of green plants. 1. A technique for the remote detection of plant stress and species differentiation," *Applied Optics*, 1984, 23, 134–138.

17. Chappelle, E. W., McMurtrey III, J. E., Wood, Jr., F. M., and Newcomb, W. W., "Laser-induced fluorescence of green plants. 2. LIF caused by nutrient deficiencies in corn," *Applied Optics*, 1984, 23, 139–142.

18. Lichtenthaler, H. K., "Remote sensing of chlorophyll fluorescence in oceanography and in terrestrial vegetation: an introduction," *Applications of Chlorophyll Fluorescence* (H.K. Lichtenthaler, Ed.), Kluwer Academic, Dordrecht, 1988, 287–297.

19. Karukstis, K. K., "Chlorophyll fluorescence as a physiological probe of the photosynthetic apparatus," *Chlorophylls* (H. Scheer, Ed.), CRC Press, Boca Raton, 1991, 769–795.

20. Chappelle, E. W., Wood, Jr., F. M., Newcomb, W. W., and McMurtrey III, J. E., "Laser-induced fluorescence of green plants. 3. LIF spectral signatures of five major plant types," *Applied Optics*, 1985, 24, 74–80.

21. Hoge, F. E., and Swift, R. N., "Airborne simultaneous spectroscopic detection of laser-induced water Raman backscatter and fluorescence from chlorophyll *a* and other naturally occurring pigments," *Applied Optics*, 1981, 20, 3197–3205.

22. Hoge, F. E., and Swift, R. N., "Airborne dual laser excitation and mapping of phytoplankton pigments in a Gulf Stream warm core ring," *Applied Optics*, 1983, 22, 2272–2281.

23. Hoge, F. E., Swift, R. N., and Yungel, J. K., "Feasibility of airborne detection of laser-induced fluorescence emissions from green terrestrial plants," *Applied Optics*, 1983, 22, 2991–3000.

24. Edner, H., Johansson, J., Svanberg, S., and Wallinder, E., "Fluorescence lidar multicolor imaging of vegetation," *Applied Optics*, 1994, 33, 2471–2479.

25. Kenny, J. E., Jarvis, G. B., Chudyk, W. A., and Pohlig, K. O., "Groundwater monitoring using remote laser-induced fluorescence," *Luminescence Applications in Biological, Chemical, Environmental, and Hydrological Sciences*, ACS Symposium Series 383 (M. C. Goldberg, Ed.), American Chemical Society, Washington, D. C., 1989, Chapter 14.

26. Leonard, D. A., and Sweeney, H. E., "Method of remotely detecting submarines using a laser," *Applied Optics*, 1990, 29, 2656.

27. Stearns, J. R., Zahniser, M. S., Kolb, C. E., and Sanford, B. P., "Airborne infrared observations and analyses of a large forest fire," *Applied Optics*, 1986, 25, 2554–2562.

28. Ory, T. R., "Photonics in earth sciences: Protecting the environment," *Photonics Spectra*, 1990, 24(9), 115.

29. Koide, M., Taguchi, M., and Fukunishi, H., "Ground-based remote sensing of methane height profiles with a tunable diode laser heterodyne spectrometer," *Geophys. Res. Lett.*, 1995, 22, 401–404.

30. McCleese, D. J., "Infrared remote sensing of the Martian atmosphere," *Applied Optics*, 1989, 28, 1909.

31. Sachse, G. W., and Browell, E. V., "Airborne lasers accurately measure greenhouse gases," *Laser Focus World*, 1992, 28(4), 73.

32. Barnes, N. P., "Lidar systems shed light on environmental studies," *Laser Focus World*, 1995, 31(4), 87.

33. Leonelli, J., "Lidar technology: Measuring the atmosphere," *Photonics Spectra*, 1995, 29(6), 97–106.

34. Lindner, B. L., and Isaacs, R. G., "Remote sensing of clouds by multispectral sensors," *Applied Optics*, 1993, 32, 2744.

35. Anonymous, "Laser probes mystery of protein folding," *Biophotonics International*, 1995, 2(6), 21–22.

CHAPTER 8

# Separation and Analysis of Mixtures Using Capillary Electrophoresis and Laser-Induced Fluorescence Detection

## Chapter Overview

Electrophoresis is a transport phenomenon describing the movement of particles through a solvent under the influence of an electrical field. In this chapter we will examine a specialized form of electrophoresis—*capillary electrophoresis*—which has developed into a powerful technique for the separation and analysis of both charged and neutral substances. When capillary electrophoresis is coupled with laser-induced fluorescence as a detection method, the technique is an indispensable qualitative and quantitative tool for the separation and analysis of multicomponent systems.

## Physical Separation Techniques for Multicomponent Systems

To analyze a complex heterogeneous system effectively, initial separation of the constituent components is often required. Techniques that separate assemblies of particles on the basis of distinct physical properties include such diverse methods as sedimentation (ultracentrifugation), chromatography, and electrophoresis. *Sedimentation* describes the motion of a macromolecule in a centrifugal field. Relationships between the sedimentation coefficient of a molecule and its molecular weight, density, and shape have been established. The various forms of *chromatography* accomplish separation on the basis of different affinities of the analytes for mobile and stationary phases in a tube, capillary, or column apparatus. Chromatographic behavior can be sensitive to such parameters as molecular weight, molecular charge or charge density, molecular size, and chemical structure. *Electrophoresis*, the migration of charged particles in solutions under the influence of an applied electric field, characterizes macromolecules by their rate of movement in that electric field. This property not only provides information to determine molecular weight, but also to quantify a molecule's net charge, to distinguish a molecule's shape or conformation, and to assay a sample's purity. When electrophoretic separation is conducted in a liquid medium, even neutral molecules can be separated as a consequence of *electroosmosis*, a phenomenon leading to the bulk flow of liquid solvent due to the effect of the electric field on counterions adjacent to the wall of the sample chamber.

## Principles of Electrophoretic Separation

The migration in solution of electrically charged compounds under the influence of an electric field constitutes the phenomenon of electrophoresis. Complete separation of solutes is possible for small sample sizes. What factors influence the movement of a chemical species through a solvent in response to an electric field? In general, the components of a mixture migrate with velocities reflecting the charge density of the particles. Let's explore this idea for the most common contemporary mode of electrophoresis, *zone electrophoresis*.

At the core of a typical zone electrophoresis system is a reservoir containing an electrolyte which conducts electric current and provides a buffering capacity. Into this continuous buffer system a sample containing a mixture of cations, anions, and neutral molecules is introduced as a sharp initial band of analyte or a "zone." Under the influence of an electric field, the ionic species contained within both the sample and the carrier electrolyte migrate to the corresponding electrode. Cations move toward the cathode, and anions migrate toward the anode. Distinct ionic solutes differentially migrate in a homogeneous buffer to provide discrete, moving zones, as illustrated in Figure 8-1. The characteristic velocity v of a charged particle (charge = $q$) in an insulating medium or an electrolyte is determined by the balance between the electrical force ($E \cdot q$) exerted by the applied electrical field $E$ and the viscous drag $f \cdot \mathrm{v}$ of the particle:

$$E \cdot q = f \cdot \mathrm{v}$$

The translational frictional coefficient $f$ is a function of the physical properties of the particle. Generally, the movement of a macromolecule in a charged field is expressed in terms of the particle's *mobility*, $\mu$, defined as the velocity per unit field:

$$\mu = \mathrm{v}/E = q/f$$

As $\mu$ is a function of $q$ and $f$, the mobility should give information about the charge, size, shape, molecular

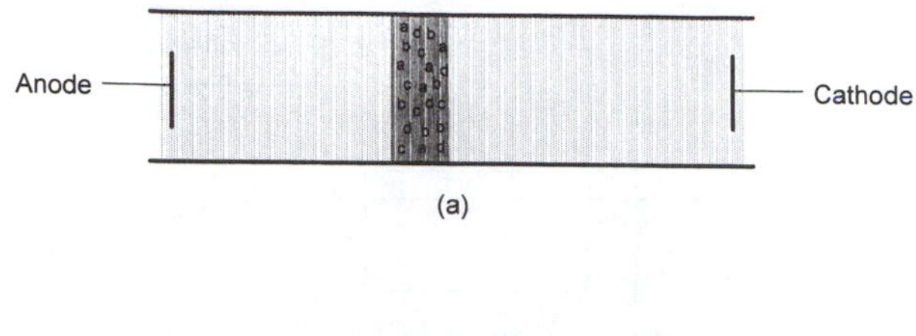


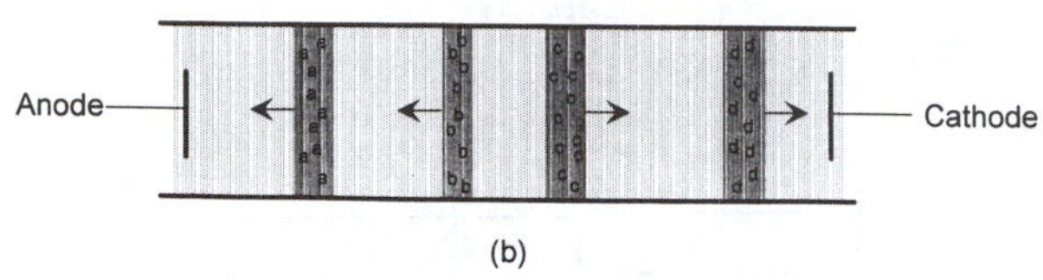


**Figure 8-1**
Principles of zone electrophoresis: *(a)* sharp initial band of analyte, *(b)* differential migration of the components into distinctly resolved and moving zones with migration times for the components increasing in the order $d < c < b < a$.

weight, and solvation of the molecule. With a suitable detection device, the passage of each solute zone may be recorded, yielding a two-dimensional *electropherogram*—a graph of detector response as a function of time. Migration time and peak area are two quantitative parameters extracted from an electropherogram.

Zone electrophoresis can be conducted in free solution or with the use of a supporting medium, such as paper or a gel network in contact with the electrolyte solution. The solid (e.g., paper) or semisolid (e.g., gel network) support medium can enhance (as well as reduce) resolution by acting as a molecular sieve to partition components as a consequence of differences in diffusion through or adsorption to the supporting medium. Thus, separation of components is governed by both electrophoretic migration and an adsorption or molecular sieving effect. Alternatively, open-tube fused silica capillaries can serve as the separation chamber in an electrophoretic separation. This mode of electrophoresis, known as *capillary electrophoresis*, offers numerous advantages despite the absence of a supporting substance. Let's evaluate the advantages offered by capillary electrophoresis that have contributed to the rapid development of the field.

## Capillary Electrophoresis—Merits and Drawbacks

The advantages of *capillary electrophoresis (CE)* arise intrinsically from the use of capillary tubing with small inside diameters (20 - 200 $\mu$m). What specific benefits are provided by these separation chambers? Clearly, the large surface area provided by the inner walls of the capillary furnishes an effective means of dissipating the heat generated by a current flow ("joule heating"). Thus, the position and width of the detection zone is not altered by temperature effects ("zone dispersion" or "band broadening"). As a further consequence, high electric fields can be applied along small diameter capillaries without a significant increase in temperature. Since the electrophoretic velocity of the charged species is proportional to the applied field, CE with high-strength electric fields can achieve rapid, high-resolution separation. The large surface area of the capillary wall can limit the extent of recovery of a sample, however, and thus surface adsorption must be minimized. Some approaches to eliminate or reduce adsorption of the analyte on the inner walls of the capillary include: (1) adsorption of polymers on the surface walls prior to the introduction of the analyte into the separation chamber and (2) for fused silica capillaries, addition of a chemical derivatizing agent to covalently block the silanol groups on the capillary surface.

As a third advantage, CE offers the ability to manipulate conveniently small sample volumes, typically on the order of nanoliters. Two possible drawbacks of small sample sizes must be recognized, however. CE is generally practical only as an analytical technique, not as a preparative technique to produce large quantities of an isolated component (except in the case of fine chemicals such as pharmaceuticals). To enhance the quantity of material analyzed, systems have been designed to include arrays of parallel capillaries rather than longer capillaries which lengthen migration time. A second limitation of small sample volumes is the need for extremely sensitive detection systems. One of the most sensitive detection modes that has emerged for CE is on-column fluorescence detection, particularly involving laser-induced fluorescence. We will explore this application further, later in this chapter. Table 8-1 summarizes the advantages and limitations of capillary electrophoresis.

Thus, a generic capillary electrophoresis CE system could be divided into five main units: (1) an injection system to introduce the analyte mixture into the separation chamber; (2) a separation system consisting of a temperature-regulated compartment,

**Table 8-1**
Advantages and Limitations of Capillary Electrophoresis

| | |
|---|---|
| Advantages of CE | 1. Heat dissipation by surface area of capillary walls limits zone dispersion. |
| | 2. Heat dissipation by surface area of capillary walls permits use of high electric fields to decrease migration time. |
| | 3. Only small sample volumes are required. |
| Disadvantages of CE | 1. Low sample recovery if surface adsorption occurs. |
| | 2. Small sample volumes limit quantity that can be analyzed practically. |
| | 3. Small sample volumes necessitate sensitive detection systems. |

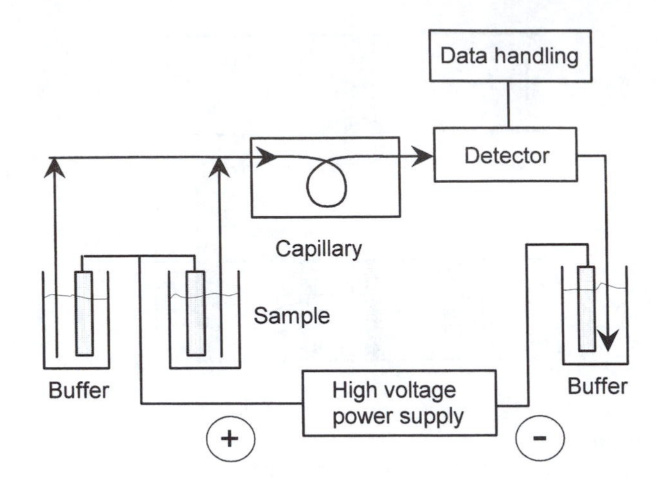


**Figure 8-2**
A generic *capillary electrophoresis (CE)* system consisting of: (1) an injection system to introduce the analyte mixture into the separation chamber; (2) a separation system consisting of a temperature-regulated compartment, capillary tubing (constructed of fused silica, borosilicate glass, or Teflon), and buffer reservoirs; (3) a detector; (4) a high-voltage power supply and controller to apply the electric field; and (5) a data processing system.

capillary tubing (constructed of fused silica, borosilicate glass, or Teflon), and buffer reservoirs; (3) a detector; (4) a high-voltage power supply and controller to apply the electric field; and (5) a data processing system. The detection system used in the majority of commercial capillary electrophoresis instruments consists of a UV-visible absorbance or fluorescence-based detector that has been adapted optically for the very short light path lengths across the capillary. Figure 8-2 presents a schematic diagram of a typical capillary electrophoresis system.

## The Phenomenon of Electroosmosis

Before measuring the electrophoretic mobility of a particle, one must first take into account the influence of electroosmosis on the measurements. Electroosmosis (EO) is a basic phenomenon in all electrophoretic separation processes. Electroosmotic flow is the bulk flow of liquid due to the effect of the electric field on counterions adjacent to a negatively charged capillary wall. The magnitude of the flow and its direction depend on the composition of the capillary and the components within the buffer reservoir. What effect introduces the negative charge on capillary walls? Let's examine the case of a fused silica capillary and an aqueous-based buffer system to answer this question.

Silica in contact with an aqueous solution hydrolyzes to form silanol surface groups, SiOH. The pH of the surrounding medium dictates the charge on these groups. At pH values generally below 2.5, the surface silanols behave as weak bases to form positively charged groups, $SiOH_2^+$. The glass surface is essentially neutral (i.e., SiOH) at pH 2.5, and, as pH increases, the surface silanols behave as weak acids and dissociate to form $SiO^-$ ions. Complete titration is accomplished by pH 7–8. To balance the negative surface charge present in most pH-neutral systems, counterions in the buffer medium adsorb onto the silica wall via electrostatic attractions. When an electric field is applied to induce the electrophoretic separation of the analyte, the layer of adsorbed charge is drawn towards the corresponding electrode. As the charges are solvated by water molecules, a bulk flow of solvent (electroosmotic flow) is induced as the charges migrate. Because of its extremely high dielectric constant, water is usually positively polarized in comparison to a fused silica surface. Hence, when an electric field is applied across the fused capillary tube, the mobile ions of the solution migrate with their hydrated water molecules towards the cathode. The flow profile of electroosmotic flow is uniform across the capillary tube, not parabolic as in laminar flow. This characteristic is optimal, as the flat flow profile of electroosmosis will add the same velocity component to all solutes, regardless of their radial position in the capillary. (In other words, a particle near the capillary wall will experience the same electroosmotic velocity as a particle positioned at the center of the tube.) Thus, electroosmosis does not introduce any significant dispersion of the zone.

In summary, under the influence of an electric field, two electrokinetic actions occur in a silica capillary: (1) electrophoresis of the ions and (2) electroosmosis of the solution. The apparent mobility, $\mu_{app}$, contains both an electrophoretic component, $\mu$, and an electroosmotic component, $\mu_{os}$:

$$\mu_{app} = \mu \pm \mu_{os}$$

If the electrophoresis and the electroosmosis both act in the same direction (as is the case for cations in a neutral aqueous medium), $\mu_{app}$ is given by the sum of $\mu$ and $\mu_{os}$. When the two electrokinetic actions work in opposite directions (e.g., for anions in a neutral aqueous solution), $\mu_{app}$ is determined by the difference in these two terms. Hence, particles that are positively or negatively charged or electrically neutral can be separated in aqueous buffer systems as long as the electroosmotic flow velocity is higher than the velocities of all anionic components.

The schematic electropherogram in Figure 8-3 illustrates the relative migration velocities of charged and neutral components in a system. The net migration velocity of component $i$ is the sum of the electrophoretic velocity (denoted $v^+$ for cations, and $v^-$ for anions, and equal to 0 for neutral species) and the electroosmotic flow velocity, $v_{eo}$. Cations will elute first as the electrophoretic velocity ($v^+$) and $v_{eo}$ add. Neutral species have no electrophoretic component to the migration velocity and thus move with $v_{eo}$ only. Anions, with electrophoretic velocities with negative signs due to the movement in the direction toward the anode (i.e., $v^- < 0$), are observed to migrate with the slowest velocities equal to $v^- + v_{eo}$. Adjustments of parameters in the buffer reservoir (e.g., pH, ionic strength, composition) can control the relative magnitudes of $|v_{eo}|$ and $|v^-|$ to ensure anion elution.

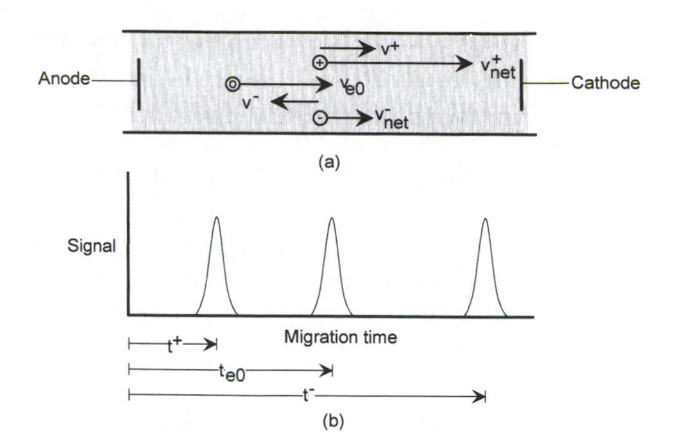


**Figure 8-3**
*(a)* A schematic electropherogram to illustrate the relative migration velocities of positively charged, neutral, and negatively charged components. The observed net migration velocity of each component is the sum of its electrophoretic velocity and the electroosmotic flow velocity of the system, $v_{eo}$. The electrophoretic velocities for cations and anions are denoted $v^+$ and $v^-$, respectively; for neutral species, the electrophoretic velocity is equal to zero. *(b)* The net migration times for cationic, neutral, and anionic species are denoted $t^+$, $t_{eo}$, and $t^-$, respectively. Cations elute first as both components of the net migration velocity add to reinforce one another, $v^+ + v_{eo}$. Neutral components appear next, moving with $v_{eo}$ only. Anions, with a negative electrophoretic velocity (i.e., moving toward the anode), migrate with the slowest net velocity, equal to the vector addition of $v^- + v_{eo}$.

## Capillary Electrophoresis Using Laser-Induced Fluorescence

### Advantages of Detection via Laser-Induced Fluorescence

We have previously enumerated the contributions of lasers as excitation sources in the study of fluorescent systems (Chapter 7). These aforementioned capabilities of lasers include enhancements to the spectral selectivity, sensitivity, time response, and dynamic range attainable in investigations of fluorophores. In addition, lasers offer further advantages as excitation sources for fluorescence detection of the components separated by capillary electrophoresis. The high spatial coherence of a laser beam allows for highly efficient excitation of the analyte within small capillaries. Most of the emission from arc lamps, in contrast, is wasted even with attempts to focus the beam onto capillaries. When the fluorescence signal is weak, particularly in the case of many intrinsic fluorophores, excitation efficiency is critical. Furthermore, high laser power facilitates the detection of small concentrations (typically, femtomole or attomole levels) of fluorophores. The higher the laser power, the greater the fluorescence signal, in general. However, care to avoid optical saturation or photodegradation is required when working with laser excitation sources as a consequence of too high a number of photons absorbed per second. Parameters such as laser irradiance (W per $cm^2$), power (W), and illumination time must be controlled to minimize these effects.

Helium-cadmium and argon lasers are the most common excitation sources, with output at 325, 354, and 442 nm for the He/Cd laser and at 514, 488, 476, or 350–360 nm for the argon ion laser. Laser-induced fluorescence with diode laser wavelengths in the red (>630 nm) and near-infrared regions has also proved to be quite successful.[1] Visible diode laser-induced fluorescence is especially promising due to the stability, small size, and inexpensive cost of such lasers. Furthermore, interferences from Raman scattering are reduced significantly (i.e., 50–fold) with excitation in the red or near-IR compared to the UV.[2]

### Fluorescent Labeling via Derivatization

The excitation wavelength of the intrinsic fluorescence of a fluorophore clearly dictates the choice of laser source. While lasers are tunable and highly monochromatic, the use of lasers cannot guarantee the observation of intrinsic fluorescence of separated components in a chemical sample. In fact, very few molecules exhibit significant levels of intrinsic fluorescence. Electron delocalization via extended conjugation or aromaticity is often present in fluorophores, but the presence of these structural features does not insure a high quantum yield of fluorescence. Furthermore, these characteristics often lower the stability of the compound to light. As an additional deterrent to high fluorescence intensities, the electrolyte concentrations, pH, and solvent properties within the separation chamber of a capillary electrophoretic system can influence fluorescence attributes dramatically.

Do these obstacles to the observation of intrinsic fluorescence preclude the analytical chemist's use of laser-induced fluorescence detection in capillary electrophoresis? Not necessarily. The need for a correspondence between the laser emission wavelength and an absorbance band in the analyte is often circumvented by the design of fluorophore derivatives to match available laser lines. Table 8-2 illustrates some common derivatizing agents and the typical classes of reagents that they modify. Pre-column or post-column derivatization may be performed, i.e., derivatization prior to or after electrophoretic separation, respectively. Does pre-column derivatization offer any advantages over post-column modification? Labelling the sample with fluorescent tags before undergoing electrophoretic separation eliminates the need to modify the electrophoresis instrumentation to perform chemical derivatization on the separated components prior to detection. Nevertheless, pre-column derivatization procedures may lead to changes in sample stability, mobility, and purity that could affect the resolution of the separation. Thus, the choice to use native fluorescence or tagging procedures and the selection of pre-column or post-column derivatization methods must be made for each individual system.

The following case studies illustrate the extensive capabilities of laser-induced fluorescence capillary electrophoresis. The first case study examines the successful and rapid resolution of racemic mixtures of amino acids[3]. The second study demonstrates the extreme sensitivity of the technique to low concentrations of analytes through an innovative enzymatic assay procedure[4].

## Case Study I: Resolution of a Racemic Mixture of Amino Acids

### Overview of the Case Study

**Objective.** To resolve racemic mixtures of derivatized amino acids into their optically active (D and L) component enantiomers.

**Laser System Employed.** Helium-cadmium laser with 325 nm output.

**Role of the Laser System.** To induce on-column fluorescence of the dansyl fluorophore in both derivatized amino acids and in the diastereomeric complexes formed between the amino acids and the copper(II) complex of L-histidine.

**Table 8-2**
Derivatizing Agents and their Substrates for LIF-CE

| *Substrate* | *Derivatizing Agents* | *Excitation and Emission Wavelengths* | *Laser Excitation Source* |
|---|---|---|---|
| Primary amines $RNH_2$ | Dansyl chloride | 325 nm; 600 nm | He/Cd 325 nm |
| | 3-(4-Carboxybenzoyl)-2-quinoline carboxaldehyde | 442 nm; 550 nm | He/Cd 442 nm |
| | fluorescein isothiocyanate | 494 nm; 525 nm | Ar 488 nm |
| | 6-iodoacetamidofluorescein | 493 nm; 518 nm | Ar 488 nm |
| | CY5.11a-OSuc (dicarbocyanine, succinimidyl ester) | 667 nm; 689 nm | Near-IR diode 670 nm |
| Thiols RSH | pyrene maleimide | 339 nm; 376 and 396 nm | He/Cd 325 nm or Ar 351 nm |
| Aldehydes RCHO and | fluorescein thiosemicarbazide | 492 nm; 516 nm | Ar 488 nm |
| ketones RCOR′ | dansyl hydrazide | 336 nm; 531 nm | He/Cd 325 nm or Ar 351 nm |
| Carboxylic acids | N-cyclohexyl-N′-(4-dimethyl-aminonaphthyl)carbodiimide | 329 nm; 415 nm | He/Cd 325 nm |

**Useful Characteristics of the Laser Light for this Application.** Wavelength tunability for direct excitation of the dansyl fluorophore.

**Principles Reviewed.** Chiral compounds, enantiomers, racemic mixtures, diastereomers.

**Conclusions.** Femtomole amounts of racemic mixtures of derivatized amino acids can be resolved and analyzed rapidly by means of a high-voltage capillary zone electrophoresis system using capillary columns containing a chiral support electrolyte and employing laser-induced fluorescence detection.

## Chirality and Enantiomers

A molecule is said to be *chiral* if it exhibits the property of handedness; in other words, a chiral molecule cannot be superimposed on its mirror image. A chiral molecule and its mirror image are called *enantiomers*, configurational isomers that differ in the arrangement of atoms or groups of atoms bonded to a chiral center. (Recall that configurational isomerism is one type of stereoisomerism whereby atom connectivities are identical, but different arrangements of atoms in space are present. Conformational isomerism—involving rotation about a single bond—is another type of stereoisomerism.) Enantiomers have identical achiral properties, such as melting point, boiling point, density, solubility in a given solvent, etc., yet exhibit distinct chiral properties, such as the direction of rotation of plane polarized light. Chiral molecules are said to be *optically active* substances.

A 50:50 mixture of enantiomers constitutes a *racemic mixture,* and the separation of a racemic mixture into its constituent enantiomers is termed *resolution.* By what means can enantiomers be separated? Enantiomers cannot be separated on the basis of achiral properties, or by methods that rely on achiral properties, such as recrystallization or distillation. An alternate approach is necessary. In particular, one can form *diastereomers* (stereoisomers that are not mirror images of one another) by reacting the racemic mixture with a chiral reagent. Diastereomers may behave as distinct substances and differ in all properties. As a consequence, ordinary physical methods used to separate achiral compounds will enable the separation of diastereomers. After separation, subsequent reactions can be conducted to regenerate the original (separated) enantiomers.

## Electrophoresis of 1:1 Mixtures of D- and L-Amino Acids

In this study, the principle of generating diastereomers from a racemic mixture of enantiomers is coupled with capillary electrophoresis as a separation means. The enantiomers of interest in this investigation are the $\alpha$-amino acids, R–CH($NH_2$)(COOH). With the exception of glycine (where R = H), amino acids are optically active molecules as a consequence of the four different substituent groups at the $\alpha$-carbon atom. The two mirror-image enantiomers of an amino acid are designated the D- and L-isomers. The D- (or dextrorotatory) isomer rotates the plane of polarization of plane polarized light to the right, while the L- (or levorotatory) isomer rotates the polarization plane to the left. Although naturally occurring amino acids have the L-configuration, it is possible to synthesize both the D- and L-forms.

The investigation focuses on the resolution of a 1:1 mixture of D- and L-amino acids that have been labelled with the fluorescent dansyl (5-dimethylaminonaphthalene-1-sulfonyl) group. Reaction of nonfluorescent dansyl chloride with amines produces fluorescent sulfonamides ($\lambda_{em} \approx$ 515-600 nm). The basis for the resolution is the differential interaction of the D- and L-forms with a chiral support electrolyte [copper(II) complex of L-histidine] present in the capillary column. Thus, a reaction of a chiral reagent with each of the enantiomers occurs within the capillary column. A mixed chelate complexation occurs to form two diasteromeric ternary [Cu(II) + L-histidine + amino acid] complexes. The amino acids to be separated include several with neutral side chains—tyrosine, methionine, phenylalanine, serine, valine, and cysteine—and two with negatively charged side chains—aspartate and glutamate. Amino acids bound to Cu(II) L-histidine acquire a positive charge from the histidine at the pH conditions in the electrolyte medium (pH 8.0). Thus, amino acids bound to Cu(II) L-histidine (cations) migrate faster than free amino acids (which have either no net charge or are negatively charged). The more strongly bound an enantiomer, the lower the fluorescence signal exhibited because of fluorescence quenching by the copper ion. Resolution of the D- and L-optical isomers at femtomole levels is achieved for all cases except DL-serine. The migration order of the amino acids varies, however. For example, the D-forms of tyrosine, phenylalanine, and aspartate bind more strongly to the chiral support electrolyte and migrate first, while the D-form of glutamate forms a weaker association with Cu(II) L-histidine and exhibits a slower migration velocity. Replacement of L-histidine with D-histidine in the electrolyte reverses the migration order of the DL-amino acids, while use of a 1:1 mixture of D- and L-histidine eliminates resolution of the enantiomeric amino acids. Figure 8-4 illustrates electropherograms

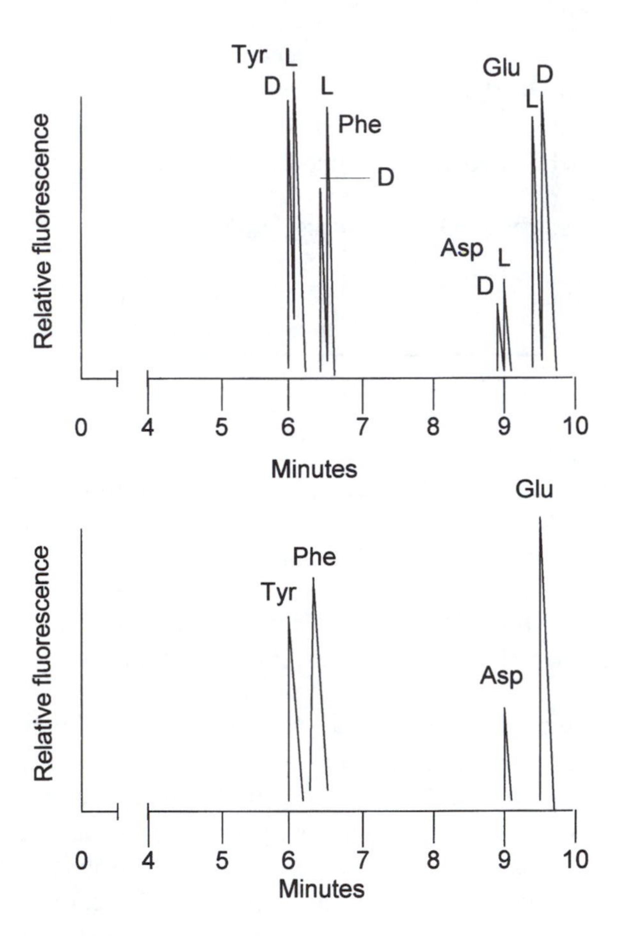


**Figure 8-4**
Schematic of electropherograms of racemic mixtures of D- and L-amino acids obtained with: (1) Cu(II) L-histidine support electrolyte and (2) a 1:1 mixture of Cu(II) D- and L-histidine. With the chiral support electrolyte in (1), resolution of the D- and L-forms of the amino acids tyrosine (Tyr), phenylalanine (Phe), aspartate (Asp), and glutamate (Glu) is obtained. For Tyr, Phe, and Asp, the D-form migrates faster as a consequence of a tighter association with the support electrolyte. With a 1:1 racemic mixture of the D- and L-enantiomers of Cu(II) histidine in the capillary column, resolution of the D- and L-forms of the four amino acids is lost. In each case the amino acids with neutral side chains, Tyr and Phe, migrate faster than the amino acids with negatively charged side chains, Asp and Glu.

obtained with: (1) Cu(II) L-histidine support electrolyte and (2) a 1:1 mixture of Cu(II) D- and L-histidine. Such rapid and effective resolution of a racemic mixture is an analytical achievement facilitated by both the separation technique of capillary electrophoresis and the laser-induced detection method.

## Case Study II: Enzymatic Assays

### Overview of the Case Study

**Objective.** To quantify the amounts of enzymes in picoliter volumes of cell extracts using an enzymatic assay that generates multiple fluorescent products.

**Laser System Employed.** An argon ion laser tuned to 488 nm (10 mW).

**Role of the Laser System.** To enable ultrasensitive off-column laser-induced fluorescence detection of species eluting from a fused silica capillary column.

**Useful Characteristics of the Laser Light for this Application.** Wavelength tunability and high power for direct excitation of a fluorophore present at low concentration.

**Principles Reviewed.** Enzyme reactions, enzyme activity.

**Conclusions.** Lower limits of detection than are attainable by conventional enzymatic assays are achieved through the coupling of the separation capabilities of capillary electrophoresis and the sensitivity of detection provided by laser-induced fluorescence.

### Enzymatic Assays

An enzymatic assay is a quantitative measurement of enzyme concentration and activity. Due to the small amounts of enzymes in cell extracts or preparations, sensitive enzymatic assays with exceptional limits of detection are required. Conventional enzymatic assays generally require long incubation times in order to generate detectable amounts of enzymatic products. A sensitive detection technique, such as laser-induced fluorescence, can dramatically improve the accuracy and efficiency of enzyme quantification. The combination of efficient electrophoretic separation of substrates and products with highly sensitive detection of a fluorophore (usually a fluorescent product) is a powerful approach to the measurement of enzyme activity.

### Analysis of β-Galactosidase

A fluorometric assay to measure enzyme concentration often involves the first-order reaction of the enzyme with a nonfluorescent species to produce a fluorophore. As a process that exhibits first-order kinetics with respect to the enzyme, the rate of the overall reaction is proportional to the enzyme concentration. Thus, the rate of change of the concentration of the fluorescent product with time, and therefore the rate of change of fluorescence with time, is also proportional to the enzyme concentration. This technique has been used[5,6] to determine the concentration of the enzyme β-galactosidase, a catalyst for the hydrolysis of lactose in *E. coli* to galactose and glucose. The overall reaction for the enzymatic assay combines β-galactosidase (β-gal) with the nonfluorescent substrate fluorescein-diβ-D-galactopyranoside (FDG) to produce the fluorescent product fluorescein. The reaction mechanism describes a two-step hydrolysis. FDG first combines with β-gal to produce fluorescein-mono-β-D-galactopyranoside (FMG) and galactose. After sufficient accumulation of FMG to compete with FDG for β-gal, fluorescein is generated:

$$\text{FDG} \xrightarrow{\beta\text{-gal}} \text{FMG} + \text{galactose}$$

$$\text{FMG} \xrightarrow{\beta\text{-gal}} \text{fluorescein} + \text{galactose}$$

In this investigation, a 60-minute incubation of β-gal ($3.1 \times 10^{-12}$ M) and FDG (200 μM) was followed by analysis of 40 picoliters of the enzymatic mixture via capillary electrophoresis. Why was a powerful separation technique like CE employed prior to laser-induced fluorescence measurements? Depending on the kinetics of the reaction pathway, the simultaneous presence of two fluorescent materials (FMG and fluorescein) is possible. Furthermore, these species contain the same chromophore and display quite similar fluorescence characteristics; FMG has an excitation $\lambda_{max}$ = 452 nm and an emission $\lambda_{max}$ = 518 nm, while fluorescein has excitation and emission $\lambda_{max}$ values of 490 and 514 nm, respectively [7]. When small concentrations of $\beta$-gal are employed ($< 3.1 \times 10^{-11}$ M), the concentration of FMG does not build up significantly during a one-hour incubation and, as a consequence, no fluorescein is detected. The first-order initial step of the mechanism is thus monitored. With a 1 $\mu$L reaction volume a limit of detection of $6.5 \times 10^{-14}$ M (35 pg/mL) was reported for FMG, an exceptionally low level. Numerous potential interferences were controlled or tested for, including accounting for the presence of FMG as a contaminant in commercially obtained FDG, quantifying the extent of nonspecific hydrolysis of FDG without the use of $\beta$-gal catalyst, investigating the impact of dilution on $\beta$-gal activity, and determining the effect of FDG concentration on $\beta$-gal activity. While the assay requires careful analysis to accurately quantify enzyme concentrations, the exceptional limits of detection that can be achieved warrant the meticulous evaluation necessary.

## Future Directions

Applications involving capillary electrophoresis and laser-induced fluorescence detection are quite promising and an active area of research. Detection limits in the zeytomole ($10^{-21}$) range have been achieved[8] for derivatized peptides, enzymes, and carbohydrates using laser-induced fluorescence. The small sample requirements and extraordinary resolution afforded by capillary electrophoresis have led to peptide mapping schemes for amino acid residues with native fluorescence (tryptophan, tyrosine, and phenylalanine) and for those residues requiring derivatization (e.g., arginine). These studies have been aided by the increasing availability of lasers with deep UV emission, permitting the direct excitation of intrinsic fluorescence of molecules containing tryptophan, tyrosine, and phenylalanine. Extensions to the analyses of cellular contents at the single-cell level are the ultimate microanalyses attainable by capillary electrophoresis and laser-induced fluorescence. Such quantitative, multicomponent analyses on a cell-by-cell basis will have enormous impact in the biological and medical sciences. Numerous advances are possible from progress in single-cell analytical technologies, including the development of diagnostic and therapeutic applications vital to the study of diseases, as well as the elucidation of the chemical bases of such phenomena as cellular differentiation and cell-cell communication. With advances in lasers, electrophoretic systems, and fluorescence detection systems, the full importance of capillary electrophoresis as a sensitive multicomponent analysis tool has yet to be realized.

### FOR FURTHER EXPLORATION

Laser-induced fluorescence detection coupled with capillary electrophoresis is playing an important role in the Human Genome Project[9]. With a goal of identifying all three billion nucleotide base pairs of the human genome (i.e., the totality of the DNA sequences in humans), international teams of researchers are collaborating to acquire and analyze the sequence data. To accelerate the sequencing methods, numerous new approaches and technologies have been devised. One such method recently developed by scientists is *capillary array electrophoresis*, whereby numerous separations occur simultaneously using a bundle of parallel silica capillaries with fluorescence detection of labelled oligonucleotides using the 488 nm line of an argon laser[10]. Further miniaturization of the electrophoretic process has been accomplished by using microfabricated capillaries on glass chips[11]. Other DNA sequencing developments include using 488 nm excitation to induce near-infrared fluorescence of bases separated in multiple capillary columns filled with linear polyacrylamide matrices[12]. A commercial device employs an argon ion laser with a sequencer containing 96 capillaries filled with a fluid-gel matrix of polyethylene oxide to increase fragment resolution and speed[13]. Solid-state diode lasers (e.g., the 795 nm emission of an argon ion-pumped Ti:sapphire laser) are also being used for near-IR fluorescence detection of labelled oligonucleotides with single-lane capillaries[14]. With many near-IR dyes having absorption maxima matching the fundamental lasing lines of inexpensive semiconductor diode lasers, the prospects for reducing the costs of the Human

Genome Project are quite favorable. These studies represent exciting examples of the use of laser technology to increase the efficiency and effectiveness of traditional DNA sequencing methods.

## DISCUSSION QUESTIONS

1. Suppose a mixture of three fluorescein-labelled amino acids was analyzed using capillary electrophoresis and laser-induced fluorescence detection. Using a 10 mM phosphate buffer at pH 6.8, in what order would the following fluorescein-tagged amino acids be detected: arginine, glutamic acid, and glycine? Given the amino acid parent structure R–CH–$COO^-$–$NH_3^+$, at pH 6.8 the R groups are $CH_2CH_2CH_2NHC(=NH_2^+)NH_2$ for arginine, $CH_2CH_2COO^-$ for glutamic acid, and H for glycine.
2. Mixtures of structurally similar proteins are often separated and identified using capillary electrophoresis. Suppose the wild-type (i.e., the native form) enzyme ribonuclease T1 and three of its mutants are present in aqueous solution. The mutations involve variations in the primary sequence at either Glutamine–25 (Gln–25) and/or Glutamate-58 (Glu–58). The substitutions are: Mutant I, Gln–25 → Lysine; Mutant II, Glu–58 → Alanine; and Mutant III, Gln–25 → Lysine *and* Glu-58 → Alanine. At pH levels above 3.0, the wild-type enzyme exhibits an electrophoretic mobility away from the detector (i.e., the enzyme moves toward the anode) due to its net negative charge. Using pH levels above 7.0, the enzyme moves toward the detector as a result of the countercurrent electroosmotic flow.
   a. In what order would you expect the enzymes to reach the detector?
   b. If all enzymes are resolved in their native conformations, what information is revealed about the extent of solvent accessibility of the enzymes?
   c. If migration times suggest that Mutant I has a more positive net charge than Mutant II, which amino acid—Gln-25 or Glu-58—exhibits the greater solvent accessibility?

   (Note: The side-chain structures of the amino acids above pH 7.0 are: glutamine: $-CH_2CH_2CONH_2$; glutamate: $-CH_2CH_2COO^-$; lysine: $-CH_2CH_2CH_2CH_2NH_3^+$; and alanine: $-CH_3$.)
3. If the components of a mixture are intrinsically fluorescent, what are the advantages and drawbacks of derivatizing the samples post-column for laser-induced fluorescent detection using the parameters of the fluorescent tag?

## LITERATURE CITED


1. Mank, A. J. G., and Yeung, E. S., "Diode laser-induced fluorescence detection in capillary electrophoresis after pre-column derivatization of amino acids and small peptides," *J. Chromatogr. A*, 1995, 708, 309–321.
2. Mank, A. J. G., van der Laan, H. T. C., Lingeman, H., Gooijer, C., Brinkman, U. A. Th., and Velthorst, N. H., "Diode laser-induced fluorescence detection in capillary electrophoresis after pre-column derivatization of amino acids and small peptides," *Anal. Chem.*, 1995, 67, 1742–1748.
3. Gassmann, E., Kuo, J. E., and Zare, R. N., "Electrokinetic separation of chiral compounds," *Science*, 1985, 230, 813–814.
4. Craig, D., Arriaga, E. A., Banks, P., Zhang, Y., Renborg, A., Palcic, M. M., and Dovichi, N. J., "Fluorescence-based enzymatic assay by capillary electrophoresis laser-induced fluorescence detection for the determination of a few $\beta$-galactosidase molecules," *Anal. Biochem.*, 1995, 226, 147–153.
5. Rotman, B., "Measurement of activity of single molecules of $\beta$-D-galactosidase," *Proc. Natl. Acad. Sci. USA*, 1961, 47, 1–6.
6. Huang, Z., "Kinetic assay of fluorescein mono-$\beta$-D-galactoside hydrolysis by $\beta$-galactosidase: A front-face measurement for strongly absorbing fluorogenic substrates," *Biochemistry*, 1991, 30, 8530–8534.
7. Haugland, R. P., "Phycobiliproteins and their conjugates," in *Molecular Probes: Handbook of Fluorescent Probes and Research Chemicals*, Molecular Probes, Inc., Eugene, OR, 1992, 77–98.
8. Monnig, C. A., and Kennedy, R. T., "Capillary Electrophoresis," *Anal. Chem.*, 1994, 66, 280R–314R.
9. Fitzgerald, M. E., "Photonics speeds up DNA research," *Photonics Spectra*, 1996, 30(4), 70–77.

10. Mathies, R. A., and Huang, X. C., "Capillary array electrophoresis: an approach to high-speed, high-throughput DNA sequencing," *Nature*, 1992, 359, 167–169.

11. Woolley, A. T., and Mathies, R. A., "Ultra -high-speed DNA sequencing using capillary electrophoresis chips," *Analytical Chemistry*, 1995, 67, 3676–3680.

12. Ruiz-Martinez, M. C., Berka, J., Belenkii, A., Foret, F., Miller, A. W., and Karger, B. L., "DNA sequencing by capillary electrophoresis with replaceable linear polyacrylamide and laser-induced fluorescence detection," *Analytical Chemistry*, 1993, 65, 2851–2858.

13. Anonymous, "Speeding up DNA sequencing," *Biophotonics International*, 1996, 2(6), 20.

14. Williams, D. C., and Soper, S. A., "Ultrasensitive near-IR fluorescence detection for capillary gel electrophoresis and DNA sequencing applications," *Analytical Chemistry,* 1995, 67, 3427–3432.

## GENERAL REFERENCES

### Capillary Electrophoresis

Kuhn, R., and Hoffstetter-Kuhn, S., *Capillary Electrophoresis: Principles and Practice*, Springer-Verlag, Berlin, 1993.

### Fluorescent Labelling via Derivatization

Haugland, R. P., *Molecular Probes: Handbook of Fluorescent Probes and Research Chemicals*, Molecular Probes, Inc., Eugene, OR, 1992.

CHAPTER 9

# Laser Light Scattering

## Chapter Overview

We have noted many times in our discussions that light can be viewed as an oscillating electric field. When the frequency of oscillation of a light wave matches a frequency associated with an energy change in a molecule, we realize that the light can be absorbed by the molecule. What happens when the frequency of a light wave does not match an energy change in a molecule? Instead of being absorbed, the light wave is *scattered*. This chapter discusses two uses of scattered light. One use involves elastically scattered light to probe for the existence of intermolecular forces between two components in a liquid mixture. The second application involves the use of inelastically scattered light to detect the presence of specific molecules in a hostile environment such as a fire. Since scattering is only a weak effect whose magnitude depends on the intensity of the incident light, it should be no surprise that laser light finds significant uses in scattering experiments.

## Light Scattering

### Elastic and Inelastic Scattering of Light

Light interacting with matter is either absorbed or scattered. We might classify the subsequent possibilities for the absorbed or scattered light with the presentation in Table 9-1. With respect to scattering, we need to appreciate the difference between elastic and inelastic interactions. As might be surmised, *elastic* refers to an interaction of the incident light with the substance of interest which leads to no change in the frequency (energy) of the light wave. *Inelastic* must then refer to the situation where the scattered light wave has a frequency different than the incident light wave. The frequency of the inelastically scattered light wave might be higher or lower, depending on whether the substance loses or gains some energy, respectively. We can study elastic or inelastic scattering either as a static or dynamic process. A static scattering experiment does not follow the evolution of the scattered light wave as a function of time. A dynamic experiment looks to appreciate how the intensity of the scattered light develops in time after exposure of the sample to the incident light wave.

**Table 9-1**
Absorbed Light, Scattered Light

| *Initial Process* | *Fate of Incident Light* | | |
|---|---|---|---|
| Spontaneous absorption | Radiative | | Photon Emission |
| | Nonradiative | | Energy transfer to lattice, possible photon emission |
| Scattering | Elastic—static or dynamic | Rayleigh | No photon energy loss |
| | | Brillouin | No photon energy loss |
| | Inelastic—static or dynamic | Raman | Partial loss or gain of energy |
| | | Nonlinear | Partial loss or gain of energy |

Static elastic light scattering has been used for a long time to measure particle and molecular sizes. Dynamic elastic light scattering is still a developing field, largely made possible by fast-pulsed lasers to initiate scattering and fast-response detectors to follow the intensity of the resultant scattered light as a function of time. An important physical property measurable by dynamic light scattering is the speed of molecular diffusion. Dynamic inelastic processes are studied to ascertain how fast energy can be put into or removed from molecules. However, most of the new spectroscopies based on inelastic scattering are static processes such as the Raman effect and the CARS example that we will discuss here.

### The Mathematics of Induced Dipoles

Light scattering is highly directional. If it were not, we would not see red sunsets or blue skies! Directionality arises from the relative orientation of the electric field of the incident photon with the target molecule. The basic scattering picture is illustrated in Figure 9-1. The interaction of light with matter is described to a good approximation as the interaction of the oscillating electric field of the light beam with the charge cloud of a molecular target, the substance of interest. In effect, the oscillating

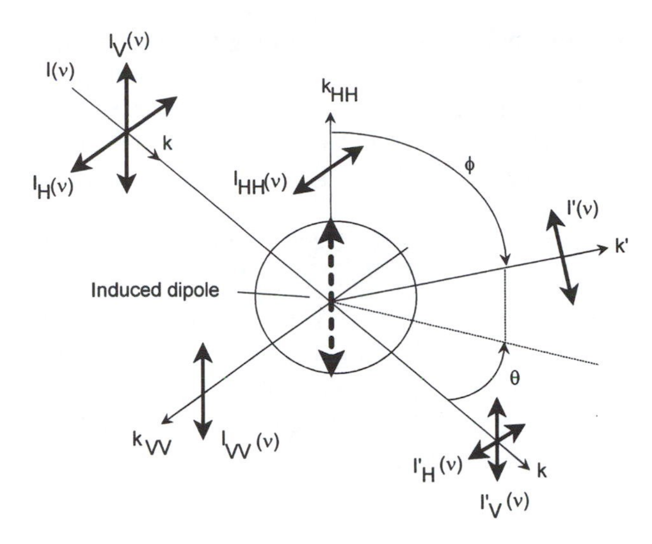


**Figure 9-1**
Incident light intensity $I_{incid}$ induces a dipole in a target substance (molecule) which in turn generates scattered light. The frequency of the incident light does not match any energy transition frequency in the substance. When the frequency of the scattered light is the same as that of the incident light, $\nu = \nu_o$, the scattering is elastic; inelastic otherwise. Scattering occurs in all directions, but not with equal intensity. The angles θ,ϕ define the direction of the scattered light wave, that is, the direction of the photon's wave vector *k*. If vertically polarized light $I_v$ is incident on the sample, the resultant scattered light viewed at 90° from incident is $I_{vv}$. Incident light horizontally polarized and viewed at 90° through a vertical polarizer is $I_{vh}$ ( = $I_{hv}$ if θ = 90°).

electric field of the photon induces a momentary electric dipole moment in the charge cloud of the target molecule. If the frequency of the electric field of the light beam matches one of the natural frequencies of the molecular target, then the photon may be absorbed, that is, captured by the target, causing the target to make a transition from one energy state to another. However, if the frequency of the electric field differs from the natural frequencies of the molecular target, then the photons are scattered, often so that they emerge from the target at the same frequency and possibly in a different direction.

Key to elastic or inelastic scattering is the induction of an oscillating dipole in the target molecule by the oscillating electric field of the photon. Quantitatively, the induced dipole μ can be related to the perturbing electric field strength **E** by:

$$\mu = \alpha E + \beta E^2 + \chi E^3 + \ldots \qquad (73)$$

where $\alpha$ is the *normal polarizability* of the molecule, $\beta$ is called the *hyperpolarizability*, and $\chi$ is sometimes referred to as the *second hyperpolarizability* or the *third-order susceptibility*. In our discussion of nonlinear optical behavior in Chapter 3, we saw an alternative version of Equation 73 written in terms of polarization. Whether one speaks of an induced dipole, as we are here, or an induced polarization, as we did in Chapter 3, is a matter of convenience, for the concepts are essentially the same. What is of more critical importance is to note that the magnitude of the induced effect, dipole or polarization, depends on powers of the incident light wave's electric field. The normal polarizability $\alpha$ describes ordinary elastic and inelastic scattering that arises in response to a linear electric field strength **E**. The higher-order polarizabilities $\beta$ and $\chi$ are important for the so-called nonlinear Raman spectroscopies which are possible as the magnitude of **E** increases. To give some sense of the magnitude of these effects, $\alpha$ is typically $10^{-40}$ $Cm^2/V$ (coulomb meter²/volt) while $\beta$ is often about $10^{-50}$ $Cm^3/V^2$. A typical gas laser outputs photons whose **E** is about $10^4$ to $10^5$ V/m. When the electric field strength **E** becomes $10^{10}$ to $10^{12}$ V/m, as is possible with very high energy pulsed lasers, the magnitudes of the polarizability terms contributing to the induced dipole $\mu$ become comparable. Thus it is possible to study nonlinear spectroscopies with high-power lasers that provide very high electric field strengths.

For the present, let us focus on the magnitude of the induced dipole that results from the linear polarizability $\mu = \alpha \mathbf{E}$. Imagine that the target molecule is fixed in space so that all of its atoms have known spatial coordinates relative to some established coordinate system. A photon incident to the target molecule will also have some specific orientation of its **E**-field with respect to this established coordinate system. This implies that the electric field of the photon will have in general some *x*, *y*, and *z* components relative to the coordinate system, since it is unlikely that the photon's electric field will be exactly coincident with any of the *x*, *y*, or *z* axes. Moreover, the dipole induced in the target molecule will not typically lie exactly along the *x*, *y*, or *z* axes of the coordinate system; the dipole will therefore have *x*, *y*, and *z* components. The mathematical relationship between the photon's electric field *x*, *y*, and *z* components and the components of the induced dipole can be given by Equation 73 expressed in matrix form as:

$$\begin{bmatrix} \mu_x \\ \mu_y \\ \mu_z \end{bmatrix} = \begin{bmatrix} \alpha_{xx} & \alpha_{xy} & \alpha_{xz} \\ \alpha_{yx} & \alpha_{yy} & \alpha_{yz} \\ \alpha_{zx} & \alpha_{zy} & \alpha_{zz} \end{bmatrix} \begin{bmatrix} E_x \\ E_y \\ E_z \end{bmatrix} \quad (74)$$

In this expression, the polarizability $\alpha$ is a matrix whose elements describe how the photon induces each component of the dipole. For example, $\alpha_{xy}$ describes how much $E_y$ of the incident photon's electric field contributes to the formation of $\mu_x$. One important property of $\alpha$ is its symmetric character. Mathematically this symmetry is interpreted to mean $\alpha_{xy} = \alpha_{yx}$, and physically the symmetry implies that we cannot tell whether the dipole is up or down but only the direction along which it must lie. A second characteristic of $\alpha$ is that the symmetry of the target molecule might dictate that some of the elements of the matrix be rigorously zero. For example, if the target molecule is "spherical," as is the case for methane $CH_4$, all of the diagonal elements of the matrix are equal and nonzero (that is, $\alpha_{xx} = \alpha_{yy} = \alpha_{zz} \neq 0$), and all of the off-diagonal elements are equal to zero (e.g., $\alpha_{xy} = \alpha_{yz} = 0$, etc.). The symmetry requirement that some polarizability elements be zero has important, far-reaching consequences for investigations known as depolarization studies.

Thus far, the target molecule has been viewed as an electron cloud distortable by the perturbing photon's electric field. We know, however, that the molecule actually consists of atoms joined together by bonds, each of which could be viewed as a localized electron cloud with a specific orientation with respect to the molecule, and therefore, with respect to the coordinate system. It should not be difficult to imagine that as the molecule vibrates (via bond stretching, bond bending, etc.), the polarizability of the electron cloud can change. If the change in a molecular motion can be considered harmonic (regularly periodic in time), then the amount that the harmonic molecular motion can affect a general element of the polarizability matrix $\alpha$ is given by:

$$\alpha_{ij} = (\alpha_{ij})_{\text{eq}} + \left(\frac{\partial \alpha_{ij}}{\partial Q_k}\right)_{\text{eq}} Q_k \quad (75)$$

The origin of Equation 75 assumes that the molecular motion can be described by the coordinate $Q_k$ and that for small displacements of the $Q_k$ from its equilibrium value $Q_{k,\text{eq}}$, the change in polarizability element $\alpha_{ij}$ depends on the rate of change of $\alpha_{ij}$ with the coordinate $Q_k$. To a good approximation, the change in $Q_k$ is harmonic. We will write:

$$Q_k = Q_{k,\text{eq}} \cos(2\pi\nu_k t) \quad (76)$$

which assumes a simple cosine time dependence for the harmonic change in $Q_k$ whose frequency of harmonic motion is $\nu_k$. Substituting Equation 76 into 75 introduces a time dependence to the general polarizability element given by:

$$\alpha_{ij} = (\alpha_{ij})_{\text{eq}} + \left(\frac{\partial \alpha_{ij}}{\partial Q_k}\right)[Q_{k,\text{eq}} \cos(2\pi\nu_k t)] \quad (77)$$

What is the consequence of this harmonic molecular motion on the induced dipole, especially if we now take into account the fact that the electric field of the photon is periodic with a frequency $\nu_o$? Let us write for the photon's electric field: $E = E_o \cos(2\pi\nu_o t)$. Now if this expression for $E$ and Equation 77 are substituted into Equation 74, Equation 78 results:

$$\begin{bmatrix} \mu_x \\ \mu_y \\ \mu_z \end{bmatrix} = \begin{bmatrix} (\alpha_{xx})_{\text{eq}} + \left(\frac{\partial \alpha_{xx}}{\partial Q_k}\right)_{\text{eq}} [Q_{k,\text{eq}} \cos 2\pi\nu_k t] & \alpha_{xy} & \alpha_{xz} \\ \alpha_{yx} & \alpha_{yy} & \alpha_{yz} \\ \alpha_{zx} & \alpha_{zy} & \alpha_{zz} \end{bmatrix} \begin{bmatrix} E_{ox} \cos 2\pi\nu_o t \\ E_{oy} \cos 2\pi\nu_o t \\ E_{oz} \cos 2\pi\nu_o t \end{bmatrix} \quad (78)$$

Here only one polarizability element $\alpha_{xx}$ is written out to save space and sanity. Each $\alpha_{ij}$ could be written in an analogous manner. When the matrix multiplication implied in Equation 78 is actually carried out, terms involving $Q_{k,\text{eq}} \cos(2\pi\nu_k t) \cdot E_{ok} \cos(2\pi\nu_o t)$ will appear. A trigonometric multiple angle identity yields the result:

$$2\cos(2\pi\nu_k t)\cos(2\pi\nu_o t) = \cos 2\pi(\nu_o - \nu_k)t + \cos 2\pi(\nu_o + \nu_k)t \quad (79)$$

What does this mean with respect to the induced dipole? Let us look at the term that results for the $(\,)_{xx}$ term contributing to $\mu_x$, for the result is crucial to our understanding of Rayleigh scattering and the spontaneous Raman experiment. The complete $(\,)_{xx}$ term is:

$$()_{xx} = (\alpha_{xx})_{eq} E_{ox}\cos(2\pi\nu_o t)$$
$$+\left(\frac{\partial\alpha_{xx}}{\partial Q_k}\right)_{eq} Q_{k,eq}E_{ox}[\frac{1}{2}\cos 2\pi(\nu_o - \nu_k)t]$$
$$+\left(\frac{\partial\alpha_{xx}}{\partial Q_k}\right)_{eq} Q_{k,eq}E_{ox}[\frac{1}{2}\cos 2\pi(\nu_o + \nu_k)t] \quad (80)$$

Now in fact each $(\,)_{ij}$ term would look the same with only the specific $ij$ component(s) changing. The upshot of this result is that there will be a dipole induced in the molecule whose frequency is just that of the incident photon's frequency $\nu_o$, as well as two new dipoles induced whose frequencies are $(\nu_o + \nu_k)$ and $(\nu_o - \nu_k)$. The term whose frequency is just $\nu_o$ is the result of a purely elastic collision and is the origin of the so-called *Rayleigh elastic scattering*. The new frequencies quantify the inelastic scattering resulting from the collision between the molecule and incident photon and describe the *Raman effect*. The $(\nu_o - \nu_k)$ term leads to the *Stokes photon* and the *Stokes Raman spectral line*. The $(\nu_o + \nu_k)$ term leads to the *anti-Stokes photon* and the *anti-Stokes Raman spectral line*. These two new, additional spectral lines are the basis of normal—or what is also being called *spontaneous—Raman spectroscopy*. These ideas are summarized in Table 9-2. Just as Rayleigh scattering occurs all of the time, so does spontaneous Raman scattering. However, it can be estimated that only 1 in $10^{15}$ photons undergoes a Stokes inelastic collision, and even fewer photons undergo an anti-Stokes collision. The Raman effect is very weak, and thus only the advent of the laser with its inherent high intensity has sparked the Raman renaissance.

## Rayleigh Scattering

### Scattering Intensities and Rayleigh Ratios

Let's first focus on the Rayleigh term and return to the Raman effect later. What is the intensity of the light wave generated by the induced dipole whose frequency is $\nu_o$? The intensity we seek is just that of the elastically scattered light wave or photon. The magnitude of the electric field of the scattered light wave can be written as:

$$E_{scat} = \frac{1}{c^2} f(\theta, \phi)\left(\frac{d^2(\alpha E_o\cos(2\pi\nu_o t))}{dt^2}\right) \quad (81)$$

Here, the $f(\theta,\phi)$ term relates to the geometry of scattering, that is, the angles that the propagation wave vector makes with the induced dipole (see Figure 9-1). The time derivative describes how the induced dipole depends on time and how its time dependence creates the electric field of the scattered light wave. Look now at the ratio of the intensity of the scattered light wave to the intensity of the incident light wave, recalling that $I \propto E^2$:

$$\frac{I(\nu_o)_{scat}}{I(\nu_o)_{incid}} =$$
$$\frac{\left(\frac{2\pi\nu_o}{c}\right)^4 \alpha^2 E_o^2\cos^2(2\pi\nu_o t)(f(\theta,\phi))^2}{E_o^2\cos^2(2\pi\nu_o t)} \quad (82)$$

**Table 9-2**
Scattering Consequences of an Incident Photon of Frequency $\nu_o$

| *Frequency of Dipole Induced in Molecular Target* | *Nature of Collision* | *Type of Scattering* | *Scattering Probability upon Interaction with Target* |
|---|---|---|---|
| $\nu_o$ | Elastic | Rayleigh | Nearly 100% |
| $\nu_o - \nu_k$ | Inelastic | Stokes Raman | 1 in $10^{15}$ |
| $\nu_o + \nu_k$ | Inelastic | Anti-Stokes Raman | $<<$ 1 in $10^{15}$ |

Since the frequency of the scattered light is the same as that of the incident light, Equation 82 describes elastically scattered light and can be simplified to:

$$\frac{I(\nu_o)_{scat}}{I(\nu_o)_{incid}} = \frac{16\pi^4}{\lambda_o^4}\alpha^2(f(\theta,\phi))^2 \qquad (83)$$

Equation 83 describes Rayleigh scattered light as dependent on $\lambda^{-4}$ with at least one simple consequence: Blue light (short $\lambda$) scatters more than red light (long $\lambda$); skies are blue and sunsets are red.

Rayleigh recognized that the measurement of absolute light intensities was experimentally very difficult and proposed a rearrangement of Equation 83 to:

$$\frac{I(\nu_o)_{scat}}{I(\nu_o)_{incid}}(f(\theta,\phi))^{-2} \equiv R(\theta,\phi) = \frac{16\pi^4}{\lambda_o^4}\alpha^2 \qquad (84)$$

where $R(\theta,\phi)$ is called the Rayleigh ratio. The Rayleigh ratio is experimentally realized by measuring, at a fixed distance and angle from the sample, signals that are proportional to each intensity and then calculating their ratio.

## The Correspondence of Rayleigh Ratios to Molecular Structure

What does $R(\theta,\phi)$ have to do with the substance causing the scattering? The amount of scattering depends on the substance's polarizability $\alpha$—the ability of a molecule to form a temporary dipole moment under the action of an oscillating electric field. Thus, $R(\theta,\phi)$ is quite substance dependent, as the light-scattering strength of a molecule varies with its shape as well as its constituent atoms.

The above comments are perfectly applicable to isotropic substances, of which spherical molecules represent one class. For nonspherical, anisotropic molecules the total scattered light intensity in any given direction arises from the inherent anisotropies of the molecules. The total scattered intensity at a viewing angle $\theta$ from incident, but in the plane formed by the source, substance, and detector, is:

$$I(\theta) = I_{is}(\theta) + I_{an}(\theta) \qquad (85)$$

where the subscripts *is* and *an* refer to isotropic and anisotropic, respectively. [When viewed in the plane defined above, $R(\theta,\phi)$ becomes a function of $\theta$ only, $R(\theta)$.] To place scattering measurements on a relative but easily compared basis, we use Rayleigh ratios. Equation 86 defines three pertinent Rayleigh ratios:

$$R_{tot}(\theta) = (f(\theta))^2\frac{I(\theta)}{I_{incid}}$$

$$= \frac{(f(\theta))^2}{I_{incid}}(I_{is}(\theta) + I_{an}(\theta)) \qquad (86)$$

or

$$R_{tot}(\theta) = R_{is}(\theta) + R_{an}(\theta)$$

where $R_{tot}(\theta)$ is the total ratio which is actually measured, $R_{is}(\theta)$ is the isotropic ratio, and $R_{an}(\theta)$ is the anisotropic ratio. The latter two arise from the isotropic and anisotropic scattering mechanisms. We have made the point that scattering depends on the polarizability of the target molecule, which in turn depends on molecular shape. Without details, it should seem reasonable that we can separate the polarizability effect into: (1) a term which assumes $\alpha$ is the same in all directions leading to $R_{is}(\theta)$, and then, correcting for that assumption, (2) a term $R_{an}(\theta)$ specifically to account for shape. Equation 86 applies when the viewer of the scattered light is a fixed distance from the sample and in the horizontal plane defined above. Moreover, the incident light noted in Equation 86 is nonpolarized. (We will discuss the effects of incident polarized light shortly.) While scattered light is a function of $\theta$, using an angle of 90° provides many conveniences and all further discussion will be based on 90° observations. For ease of notation, $R_k(\theta = 90°)$ will be denoted by just $R_k$ where $k = is$ or $an$.

To separate $R_{is}$ and $R_{an}$ from the measured $R_{tot}$, Cabannes suggested taking into account the polarizations of the scattered light:[1]

$$R_{is} = R_{tot}\left[\frac{3I_{VV} - 4I_{VH}}{3I_{VV} + 9I_{VH}}\right] \qquad (87)$$

Equation 87 shows how $R_{is}$ is obtained from $R_{tot}$ by measuring the scattered light intensities defined in Figure 9-1.[2,3] $I_{VV}$ refers to light intensity measured at the detector when the incident light is vertically polarized and only vertically polarized scattered light is allowed to reach the detector. When the incident light is vertically polarized and only horizontally polarized light is allowed to reach the detector, the intensity will be called $I_{VH}$. If no polarization effects occur, that is, if a photon is scattered from the target molecule with its plane of polarization the same as that of the initial orientation, $I_{VH}$ would be zero (there would be no horizontal component to the scattered light), and $R_{is}$ would be identical to $R_{tot}$. If the molecule is anisotropic, the photon will have its plane of polarization rotated from the initial orientation, and $I_{VH}$ will not be zero. The Cabannes factor (the square bracketed term in Equation 87) then estimates the extent of that rotation.

How is $R_{tot}$ actually obtained? In Equation 87, $R_{tot}$ refers to the total Rayleigh ratio measured using unpolarized light. Use of polarized laser light necessitates some extra experimental steps relative to more conventional procedures with unpolarized light sources. Defining total Rayleigh ratios for vertically and horizontally polarized incident light, $R_V$ and $R_H$, respectively, the following is true:[4]

$$R_{tot} = \frac{1}{2}(R_V + 3R_H) \tag{88}$$

While Rayleigh ratios depend on the specific geometry of the measuring apparatus, in practice determination of those factors is avoided by calibration with substances whose absolute Rayleigh ratios are known. The idea is:

$$R_{sample} = R_{standard}\left(\frac{I_{sample}}{I_{standard}}\right) \tag{89}$$

Using incident vertically polarized laser light, the necessary total Rayleigh ratio is given by Equation 90:

$$R_{tot} = \frac{1}{2}\left[R_{V,std}\left(\frac{I_{VV}}{I_{VV,std}}\right) + 3R_{H,std}\left(\frac{I_{VH}}{I_{VH,std}}\right)\right] \tag{90}$$

which is obtained via Equations 88 and 89 in terms of the standard vertical and horizontal Rayleigh ratios. Strictly, the horizontal term should be $I_{HH}/I_{HH,std}$ but for 90° scattering $I_{HH} = I_{VH}$. Thus, the necessary experimental measurements are just $I_{VV}$ and $I_{VH}$ for the samples and standard(s).

## The Correspondence of Rayleigh Ratios to Intermolecular Forces

How can microscopic photons be used to probe the properties of matter? In any phase, deviations from the macroscopic average value of any physical property in that phase are always occurring. These microscopic deviations are called *fluctuations*. Density and concentration fluctuations play a major role in scattering phenomena because they affect the bulk polarizability of the collection of target molecules. When a large number of molecules, such as a mole ($6 \times 10^{23}$), are viewed, a substance's density and concentration (if a solution) appear quite uniform. However, if any 100 molecules are observed, the density and concentration of these 100 would fluctuate about the macroscopic average values. For a pure substance there are no concentration fluctuations, only density fluctuations, which means that the actual number of molecules in a specific volume will vary slightly over time, but the types of molecules will remain the same. In a small volume of a solution, the number of molecules fluctuates with time, as does the number of each type of particle. In a collection of molecules of two kinds, at any given instant an incident photon would interact in some particular way based on the polarizability of the collection. Now, if in another instant, the density or concentration of those molecules differed, an incident photon would react in a different way. These slight differences are influenced by the nature of the intermolecular forces occurring in the mixture. Thus, studying fluctuations by light scattering provides a direct means of studying intermolecular forces.

Although we have discussed how $R_{tot}$ and $R_{is}$ are obtained, $R_{is}$ has not been connected to solution properties. The connection is made in the following way. Recall that the Rayleigh ratio is really a "normalized" intensity of scattered light. The scattered light originates from the oscillating dipole created in the electron cloud of the molecule by the incident electric field of the laser photon. Whatever affects the induced oscillating dipole also affects the production of the scattered light. As implied earlier, the polarizability of the collection of molecules measures the collection's response to the photon's oscillating electric field. The practical way to estimate polarizability, however, is the dielectric constant of the collection. A given molecule exists in a dielectric medium formed by its neighboring molecules. Fluctuations of density and, if appropriate, concentration, affect the dielectric constant, and in turn, the scattered intensity. Thus:

$$R \propto < (\delta\varepsilon)^2 > \tag{91}$$

sets the Rayleigh ratio proportional to the average of the square of the fluctuations of the dielectric constant of the sample (solution).

Specific uses of light scattering measurements will depend on estimating fluctuations in the dielectric constant of the substance or medium under study. Table 9-3 presents a brief compilation of how Rayleigh scattering has been used to probe microscopic properties of matter. The case study presented later in this chapter (Case Study I) deals with how we can elucidate the nature of the intermolecular forces existing between molecules in simple binary liquid mixtures. Before turning to that case study, however, let's explore ramifications of the $(\nu_o + \nu_k)$ and $(\nu_o - \nu_k)$ terms in Equation 80.

## Raman Effect

As mentioned earlier in the Raman experiment, photons which are inelastically scattered with less energy than incident are observed as spectral signals called the *Stokes lines*. Photons that are inelastically scattered with more energy than incident give rise to the *anti-Stokes spectral lines* in the Raman experiment.

At first thought it would seem that each vibrational motion of the molecule with its associated coordinate $Q_k$ and frequency $\nu_k$, when exposed to incident photons of frequency $\nu_o$, should yield Raman spectral lines due to inelastic scattering. Not all molecular vibrations will scatter light inelastically, however, and thus we have to consider in a little more detail the origin of Raman lines. The intensity of a spontaneous Raman spectral line is given by:

$$I = K(\nu_o \pm \nu_k)^4 \left(\frac{\partial \alpha_{ij}}{\partial Q_k}\right) E_o^2 \tag{92}$$

where $K$ is a collection of physical constants and apparatus parameters. Three important conclusions can be drawn from Equation 92:

1. If there is no change in polarizability as a consequence of a particular molecular motion, that is, if $(\partial \alpha_{ij}/\partial Q_k) = 0$, the Raman effect due to that molecular motion is rigorously zero and is not observed.
2. The magnitude of the Raman effect depends on the frequency sum or difference to the fourth power, a very marked frequency dependence. In fact, since $h\nu_o$ is usually UV or visible energy while $h\nu_k$ is IR, we find $\nu_k << \nu_o$ and thus $\nu_o \pm \nu_k \approx \nu_o$.
3. The intensity depends on the square of the incident photon's electric field. Thus the laser with its high intensity has made the spontaneous Raman technique viable.

A schematic Raman spectrum showing Rayleigh and Raman lines is shown in Figure 9-2. The spectral lines occur at $\nu_o$ and $(\nu_o \pm \nu_k)$ with an important intensity difference between the Stokes $(\nu_o - \nu_k)$ and anti-Stokes $(\nu_o + \nu_k)$ lines. The Stokes lines are often 100 to 1000 times more intense than the anti-Stokes lines, but to understand why we need to appreciate some aspects of molecular energy levels and the Boltzmann distribution.

Figure 9-3(a) shows the Raman effect qualitatively in terms of molecular energy levels. The process of inducing the dipole in the target molecule is akin to the molecule first absorbing the incident photon and being raised to a "virtual" excited electronic state. Then in a very short time, on the order of femtoseconds, a photon is reemitted whose energy is either proportional to $\nu_o$, $(\nu_o - \nu_k)$, or $(\nu_o + \nu_k)$, which are the Rayleigh, Stokes Raman, and anti-Stokes Raman lines, respectively. The Stokes Raman line

**Table 9-3**
Uses of Static Rayleigh Light Scattering

| *Physical Property* | *System Studied* |
|---|---|
| Sizes | Particles, molecular aggregates<br>Molecules |
| Shapes | Particles, molecular aggregates<br>Molecules |
| Intermolecular forces | Binary liquid mixtures,<br>Ionic solutions |

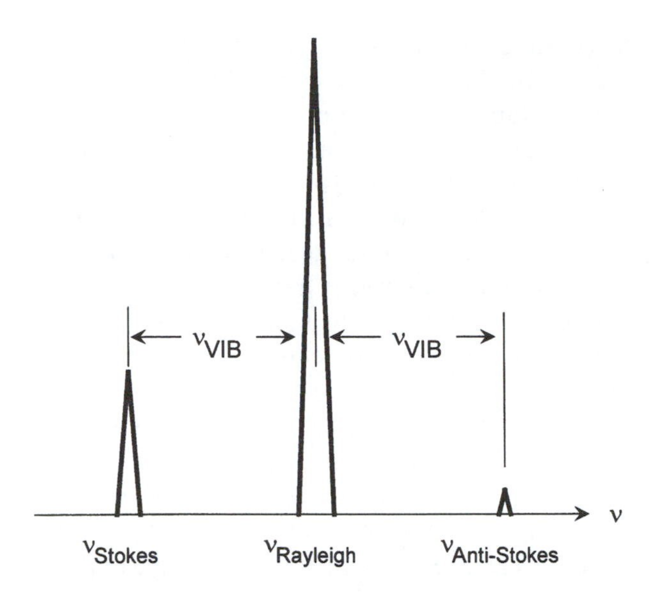


Figure 9-2
A schematic spontaneous Raman spectrum. Photons which are elastically scattered with a frequency equal to the incident photon's frequency $\nu_o$ give rise to the Rayleigh spectral lines. Inelastically scattered photons give rise to the Stokes and anti-Stokes spectral lines, depending on whether the scattered photons have less or more energy than the incident photon, respectively. The Stokes line ($\nu_o - \nu_{vib}$) is often 100–1000 times more intense than the anti-Stokes line ($\nu_o + \nu_{vib}$).

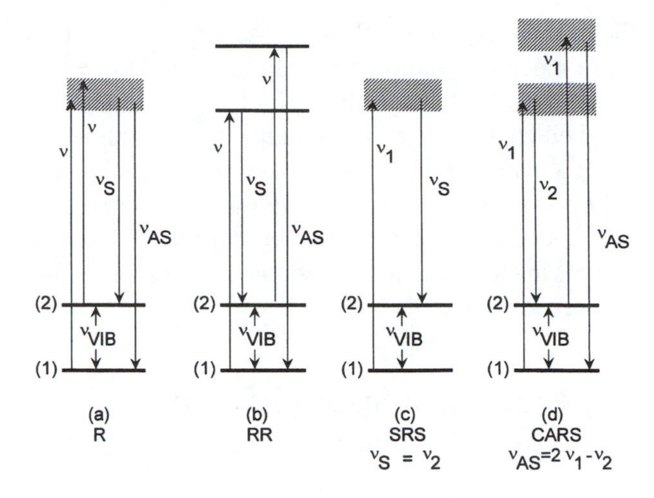


Figure 9-3
A qualitative energy level diagram illustrating the energy changes involved in (*a*) spontaneous Raman spectroscopy, *R*, (*b*) resonance Raman spectroscopy, *RR*, (*c*) stimulated Raman spectroscopy, *SRS*, and (*d*) coherent anti-Stokes Raman spectroscopy, *CARS*.

starts when a molecule in a low energy state absorbs the energy $h\nu_o$, raising its total energy by $h\nu_o$, as shown in Figure 9-3(a). The anti-Stokes Raman line starts when a molecule in a higher vibrational energy state absorbs the energy $h\nu_o$, also raising it to a virtual state. The intensity of any spectral line depends on the number of molecules with the correct initial energy. The number of molecules with a given initial energy depends on the magnitude of the given energy and the temperature as described by the Boltzmann distribution:

$$N_v = N_o e^{-\varepsilon_v / kT} \tag{93}$$

For a population of $N_o$ molecules at a constant temperature, there are fewer molecules with higher energies. Since the anti-Stokes Raman lines start with molecules of "higher" energy, there are fewer of them initially and hence the intensity of the anti-Stokes line should be less than the Stokes line. Note this is a valid observation when the same vibrational motion is involved, that is, when the energy transitions in the molecule are from vibrational energy levels marked (1) and (2) in Figure 9-3(a). On this point you should keep in mind that the Stokes line describes an increase in molecular vibrational energy while the anti-Stokes line describes a decrease in molecular vibrational energy. The ratio of intensities of the anti-Stokes and Stokes lines for the molecular vibration whose frequency is $\nu_k$ is:

$$\frac{I(\text{anti - Stokes})}{I(\text{Stokes})} = \left(\frac{\nu_o - \nu_k}{\nu_o + \nu_k}\right)^4 e^{h\nu_o / kT} \tag{94}$$

and the magnitude of this ratio is commonly 0.01 to 0.001. As alluded to earlier, the spontaneous Raman Stokes effect is small at best, but even less probable for anti-Stokes lines.

## Resonance Raman

The spontaneous Raman effect can be initiated by a photon with sufficient energy to raise a molecule to a virtual state which exists long enough to emit the Stokes or anti-Stokes photon in an inelastic manner. What happens if the incident light photon's energy matches the energy necessary to reach an excited but stable electronic state of the molecule? For such a condition, the Raman effect is enhanced $10^2$-fold to $10^4$-fold, and the process is called *resonant Raman* (RR). The basis for RR lies in the fact that the probability for spontaneous absorption becomes large when the photon's energy is at or close to the energy level difference between the ground and excited electronic state. The condition of the photon's energy matching the energy level difference is often called *resonance* and provides the name "resonant Raman." The RR effect has been used to observe signals in cases where the spontaneous Raman effect is just too weak.

## Experimental Spontaneous Raman Spectroscopy

How are Raman spectra obtained? Figure 9-4 shows a schematic for a spontaneous Raman experiment. As may be surmised at this point, the experiment starts with a laser light source which needs to be capable of providing output spectral lines sufficient to excite molecular electronic transitions. Argon ion and helium-neon lasers have been popular sources. Near-IR Raman is a growing field of interest in which higher-power diode lasers will find ready application. The output of the laser is collimated and passed through the sample, which might be a solid, liquid, or gas. In fact, one of the advantages of spontaneous Raman is the wide range of types of samples that can be studied with comparative ease.

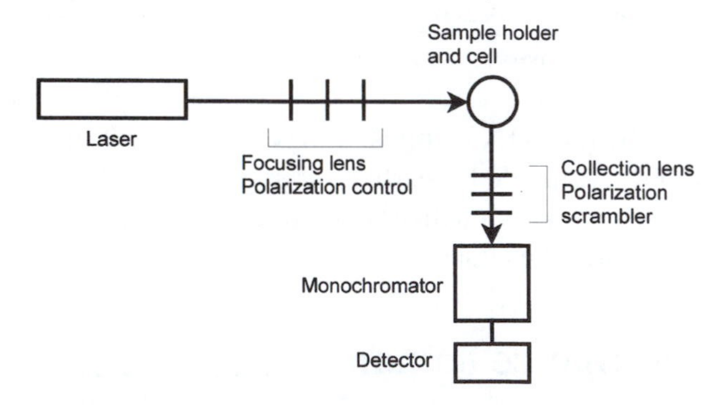


**Figure 9-4**
A schematic of the instrumental setup to perform a spontaneous Raman spectroscopy experiment. Collimated emission from the laser source, often an argon ion laser or a helium-neon laser, is used to induce molecular electronic transitions in the sample. Scattering is detected at 90° from the incident light. As scattered light is often depolarized, the extent of the depolarization is measured by directing the scattering through a polarization analyzer. A polarization scrambler is often placed just before the monochromator to obtain an accurate measure of the scattered light at the detector, as the response of the monochromator is often sensitive to polarization.

As Raman is a scattering spectroscopic technique, sample viewing is achieved off-axis from the incident photon beam, typically at 90° for spontaneous Raman. The light is collected by focusing optics, possibly passed through a polarization analyzer, and then directed through a polarization scrambler before passing through the monochromator where it is ultimately detected on exit by a photomultiplier tube (PMT).

Some comments are pertinent to the function of the polarization analyzer and the polarization scrambler. Many lasers output inherently polarized light. Scattering and absorption of a photon by a target molecule can depend markedly on the angle of the incident plane of polarization—the plane of the photon's electric field—as the photon strikes the target molecule. For example, it should not be hard to imagine that no difference in scattering or absorption would be observed if the target molecule were spherical, such as methane $CH_4$, regardless of the angle made by the photon's electric field. In contrast, it should be easy to imagine significant differences for scattering or absorption if the plane of the photon's electric field approached in the plane of the benzene ring or perpendicular to it. Moreover, when plane polarized light strikes a target molecule, the shape of the molecule can cause the scattered light to be depolarized. In other words, the plane of the scattered Stokes or anti-Stokes photon's electric field can be rotated away from its initial angle of incidence. This rotation is called *depolarization* and is quantified by measuring the vertical and horizontal components of the scattered Raman photons using the polarization analyzer. For vertically polarized light incident on a molecule, if no depolarization occurs, then all scattered photons are vertically polarized and the measured intensity when the analyzer is horizontal would be zero. Each Raman line, Stokes and anti-Stokes, has some degree of depolarization, and the depolarization can be used as an additional characteristic of the molecular motion that gives rise to the Raman line.

The function of the polarization scrambler is in one sense curious, but important. The optical surfaces in the monochromator might respond differently to light of different polarizations. This could mean, for example, that if vertically polarized light entered the monochromator, the PMT signal would be different than if horizontally polarized light entered. The scrambler destroys the polarization effects and allows the PMT to sense the true intensity of the light and not an intensity modified by passage through the optical surfaces of the monochromator.

## Nonlinear Oscillators

No discussion of laser-induced spontaneous Raman spectroscopy could be complete without at least some mention of the nonlinear Raman techniques made possible with the advent of the laser. The names of these techniques virtually describe an alphabet soup, which is deciphered in Table 9-4. While the acronyms CARS, RIKES, HORSES, etc., stand for fascinating, even exotic types of spectroscopy, it is well beyond our scope here to discuss all of these spectroscopies. These spectroscopies fall under the rubric of Raman because they involve studying the photons whose emission depends on nonlinear, third-order susceptibilities. Of the entries in Table 9-4 other than Raman, CARS is probably the most often used. Coherent anti-Stokes Raman spectroscopy depends on temporal and spatial coherence of two or more laser beams incident on a sample. The coherence requirement is specified with the adjective added to the name. These techniques are also referred to as *two-*, *three-*, or *four-color mixing* or *two-*, *three-*, or *four-wave mixing experiments*. Stimulated Raman spectroscopy (SRS) is a two-color mixing experiment, while CARS is a four-wave (or four-color) mixing experiment.

**Table 9-4**
Raman Spectroscopies Revealed

R = spontaneous Raman = normal Raman
RR = Resonant Raman
PARS = Photoacoustic Raman Spectroscopy
SRS = Stimulated Raman Spectroscopy
  SRGS = Stimulated Raman Gain Spectroscopy
  IRS = Inverse Raman Spectroscopy = TIRE
  TIRE = The Inverse Raman Effect
RIKES = Raman-Induced Kerr Effect Spectroscopy
CARS = Coherent Anti-Stokes Raman Spectroscopy
CSRS = Coherent Stokes Raman Spectroscopy
TRIKE = The Raman-Induced Kerr Effect
HORSES = Higher-Order Stokes Effect Scattering
HORAS = Higher-Order Anti-Stokes Scattering

### Stimulated Raman Spectroscopy

Spontaneous Raman allows incident light of frequency $\nu_1$ to create inelastically scattered light of frequency $\nu_S$ or $\nu_{AS}$. Consider a sample simultaneously irradiated by $\nu_1$ and $\nu_2$ light, such that $\nu_1 - \nu_2 = \nu_{VIB}$ as shown in Figure 9-3(c). In essence, $\nu_1$ creates a population inversion which $\nu_2$ stimulates to emit additional $\nu_2$ photons. Signal detection occurs by measuring the intensity gain for the $\nu_2$ light or by measuring the intensity loss of $\nu_1$ light. If for fixed pump $\nu_1$ light, when $I(\nu_1)$ decreases or $I(\nu_2)$ increases while changing $\nu_2$, then for $\nu_2$ such that $\nu_1 - \nu_2 = \nu_{VIB}$, a vibrational spectral line is detected. Measuring the $I(\nu_2)$ gain is often called *stimulated Raman gain spectroscopy* (SRGS), while measuring $I(\nu_1)$ loss is often called *inverse Raman spectroscopy* (IRS). Both SRGS and IRS are varieties of stimulated Raman spectroscopy (SRS). The possibility of increased signal strength for ease of detection and the lack of a required monochromator are two important driving forces for stimulated Raman spectroscopic experiments. Both of these motivations are also important in the application of the CARS technique, our next topic.

### Coherent Anti-Stokes Raman Spectroscopy

In contrast to spontaneous Raman anti-Stokes spectroscopy where the target molecule changes its vibrational energy level, coherent anti-Stokes Raman spectroscopy, CARS, causes no net effect in the target molecule. The emitted signal, however, is an anti-Stokes line. Figure 9-3(d) illustrates the energetic bases for CARS. By exposing the target molecule to two oscillating fields of amplitude $E_1$ and $E_2$ with frequencies $\nu_1$ and $\nu_2$, the target molecule tries to respond to these two frequencies. When the difference in the incident frequencies equals the frequency of a molecular motion, that is, when $\nu_k = (\nu_1 - \nu_2)$, a forced oscillation with a frequency of $(\nu_1 + \nu_k)$ is created that results in the emission of an anti-Stokes photon. The net result is $(2\nu_2 - \nu_1) = \nu_{AS}$, where $\nu_{AS}$ is the frequency of the scattered anti-Stokes photon that is detected.

Alternatively, one may view the process in the following manner. By exposing the sample to $h\nu_1$, the sample can be raised to some virtual intermediate level. By exposing the sample to $h\nu_2$, the excited intermediate level can be stimulated to return to an excited level. By exposing the sample in the excited level again to $h\nu_1$, a new virtual level can be reached that emits a Raman photon of energy $h\nu_{AS}$. The energy $h\nu_k$ of the excited level can be determined in the following way. Referring again to Figure 9-3(d):

$$\begin{aligned} \nu_{VIB} &= \nu_1 - \nu_2 & \text{(a)} \\ \nu_{VIB} &= \nu_{AS} - \nu_1 & \text{(b)} \\ \nu_{AS} &= 2\nu_1 - \nu_2 & \text{(c)} \end{aligned} \quad (95)$$

Experimentally, $\nu_1$ and $\nu_2$ are known, $\nu_{AS}$ is measured, and $\nu_k$ is determined from Equation 95(c). The experiment depends on being able to tune $\nu_1$ or $\nu_2$ (usually $\nu_2$) through the energy necessary to satisfy Equation 95(a). The CARS experiment is a four-color experiment because three input frequencies are used ($\nu_1$ twice and $\nu_2$) and one output frequency is obtained ($\nu_{AS}$).

The intensity of the CARS signal is given by:

$$I_{CARS} = K\chi_{CARS} I^2(\nu_1) I(\nu_2) L^2 \times \left[ \frac{\sin\left(\frac{L}{2}(2k_1 - k_2 - k_{AS})\right)}{\left(\frac{L}{2}(2k_1 - k_2 - k_{AS})\right)} \right] \quad (96)$$

The factor $L$ is the length over which the wave mixing can occur in the sample. Typically, the mixing region is very small. Since the intensity depends on the magnitude of the third-order susceptibility $\chi_{CARS}$, CARS is a nonlinear process. The [ ] term has a maximum equal to 1 when the wave vectors describing the spatial phase of each laser beam satisfy the condition $2k_1 - k_2 - k_{AS} = 0$. Recall that the direction of a light wave is described by its wave vector, which also relates the phase of one light wave to another. Geometrically, two ways of adding the wave vectors are illustrated in Figure 9-5 where the vectors are either collinear or crossed. The vector addition, which is an equivalent way to describe the required phase matching condition, can be written mathematically as $2k_1 = k_2 + k_{AS}$.

The anti-Stokes photons emitted via the CARS experiment are orders of magnitude more intense than those that result from the spontaneous Raman process. There are two main reasons for the greatly increased intensity. One reason has its origin in the spatial coherence of the Raman photons. For spontaneous Raman, photons are scattered randomly in any direction of the $4\pi$ solid angle of a sphere. For CARS photons, the scattered angle is restricted to either collinear or to the cone defined in Figure 9-5(b). Assuming that roughly the same number of photons emitted in the same time, and assuming that the experimental solid angle viewed is close to the cone established by $2k_1 = k_2 + k_{AS}$, the signal enhancement would depend on the ratio ($4\pi$/solid angle of the cone). A second reason for the significantly increased intensities of the anti-Stokes photons can be seen from Equation 97. Spontaneous Raman depends linearly on incident power, say $I(\nu_1)$. Thus, the ratio of $I_{CARS}$ to $I_{RAM}$ (the spontaneous Raman intensity) for a gaseous sample of pressure $p$(torr) is given by:

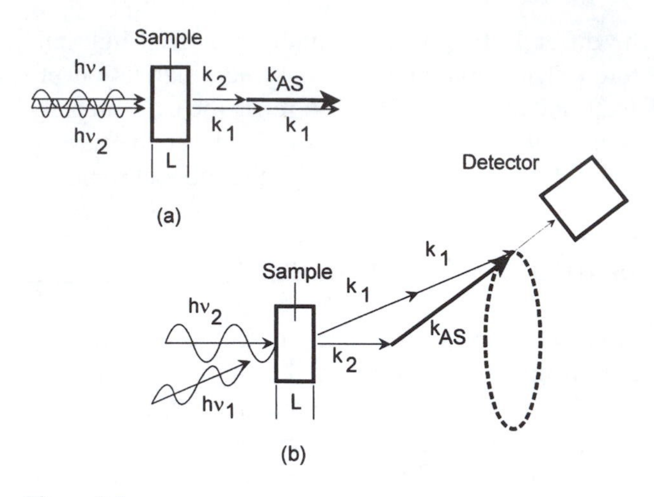


**Figure 9-5**
Schematic of spatial phase matching; wave vector additions for the coherent anti-Stokes Raman experiment (CARS). Not only is temporal coherence necessary for frequency matching, but spatial coherence determines the direction of the scattered output Raman photon. (*a*) is collinear wave vector addition. (*b*) is crossed planar wave vector addition.

$$\frac{I_{CARS}}{I_{RAM}} \approx 10^{-5} I(\nu_1) I(\nu_2) p(\text{torr}) \quad (97)$$

Since it is easy to achieve high pump powers of $I(\nu_1)I(\nu_2) \approx 10^{10}$, the CARS signal can be $10^5$ or greater in intensity than the spontaneous Raman signal. In fact, the CARS signal can achieve a significant fraction of incident pump power.

Because of the wave vector addition of the input laser beams, the CARS output signal has a specific direction which allows the signal to be readily separated from interfering signals (for example, other scattered or fluoresced radiation) without the need for a monochromator. The spatial coherence wave vector matching requirement also places a restriction on the volume of sample that is "probed," that is $L^3$, the volume from which the mixing process and scattered light originate. This scattering volume can be extremely small, 0.1 $nm^3$ for example. If broadband probing lasers are used and the output signal is studied with a monochromator, a spectrum of the sample may be obtained. Of course, spectra can be used as fingerprints of molecules in the sample.

The family of Raman spectroscopies has quickly developed into a very useful and powerful resource for many types of chemical problems. Case Study II in this chapter is an example of such an application.

# Case Study I: Thermodynamic Properties Measured by Light Scattering

## Overview of the Case Study

**Objective.** To use light scattering measurements of nonideal methanol-water solutions to calculate the composition dependence of the excess Gibbs potential of the binary mixture and relate the extent of nonideality to intermolecular forces.

**Laser System Employed.** An argon ion laser tuned to 488 nm.

**Role of the Laser System.** As an incident light source for scattering measurements.

**Useful Characteristics of the Laser Light for this Application.** High-intensity light enables detection of the small fluctuations in solution density and concentration that arise from the intermolecular forces in binary mixtures.

**Principles Reviewed.** Light scattering, excess functions, anisotropy, nonideal solution behavior.

**Conclusions.** Light scattering measurements offer an alternative means of measuring thermodynamic properties of binary solutions. In particular, light scattering measurements successfully yield derivations of excess functions that characterize the extent of nonideality and reveal the nature of the intermolecular forces present in such solutions.

## Introduction—Thermodynamics of Mixtures

A reasonable view of the world would assume that the whole of an object is the sum of its parts. This is, however, a naive perception of the chemical world. If one mixes 50 mL of water and 50 mL of methanol, the total volume is not 100 mL but something less. When the whole is not the sum of its parts, we ascribe the difference between what was observed and what was expected to the nonideality of the system. From a microscopic viewpoint, nonideality for water-methanol mixtures can be understood in terms of the differences in the intermolecular forces between water molecules, between methanol molecules, and between water and methanol molecules. These forces are reflections of the hydrogen bonding in water-methanol mixtures. Nonideal effects are very common and are manifest in a number of experimental phenomena. For solutions, two common physical properties readily reveal nonideality: (1) the total volume of a mixture and (2) the pressure of a vapor phase in equilibrium with a liquid mixture. When measuring the latter property, composition dependent vapor pressure behavior in accord with Raoult's Law defines an ideal liquid-vapor system. The appearance of an azeotrope, however, signifies a nonideal system.

To quantify the nonideality occurring in a mixture, we introduce new thermodynamic variables called *excess functions*. Thus, the total observed volume will equal the expected ideal volume plus the excess volume (a negative, positive, or zero quantity) and can be expressed by:

$$V_{\text{tot}} = V_{\text{ideal}} + V^E \tag{98}$$

Because it has proven almost impossible to predict theoretically the excess volume $V^E$ for any given system, excess volumes are determined experimentally by the difference $V_{\text{tot}} - V_{\text{ideal}}$, where $V_{\text{ideal}}$ is calculated assuming simple additivity of the volume of each component.

We will look at the excess Gibbs potential $G^E$ of the mixture as defined by:

$$G_{\text{tot}} = G_{\text{ideal}} + G^E \tag{99}$$

The Gibbs potential is of interest because, under isobaric (constant pressure) conditions, the sign and magnitude of Gibbs potential differences determine phase (and chemical) equilibrium. The object of this case study is to quantify the excess Gibbs potential and, from that quantification, draw conclusions about the nature of the intermolecular forces between the molecules in the solutions under study. The fascinating aspect of the quantification is that it can be achieved by studying scattered laser light.

## Introduction—Light Scattering of Mixtures

For liquid mixtures the term $R_{\text{is}}$ (introduced earlier) is further divided into contributions from density fluctuations, $R_d$ and concentration fluctuations, $R_c$:

$$R_{\text{is}} = R_d + R_c \tag{100}$$

Noting how fluctuations in density affect the fluctuations of the dielectric constant yields Equation 101 below:[4,5]

$$R_d = \left(\frac{\pi^2}{2\lambda^4}\right) kT\kappa_T \left[ f\left(\frac{(n^2-1)(n^2+2)}{3}\right)\right]^2 \tag{101}$$

Here $\lambda$ is the wavelength of incident light, $\kappa_T$ is the solution isothermal compressibility, $n$ is the refractive index of the solution, and $kT$ is the Boltzmann constant times the temperature. The factor $f$ requires some comment. The model used to derive Equation 101 is applicable to gases where significant intermolecular effects are absent or small. When dealing with condensed phases, such as liquids, intermolecular effects are not negligible. The factor $f$ corrects the gas-phase refractive index expression for use in condensed phases. The correction is small and generally around $1.0 \pm 0.1$.[6] Note that since $\kappa_T$ and $n$ are functions of composition measurable for each solution studied, $R_d$ is readily obtained for each value of $T$ and $\lambda$ used in the experiment.

The concentration Rayleigh ratio $R_c$ is really the term of interest and is obtained experimentally by subtracting $R_d$ from $R_{\text{is}}$. The theoretical expression for $R_c$ is given by:

$$R_c = \left(\frac{2\pi^2}{\lambda^4}\right) kTV(1-x_2) \left[\frac{n^2\left(\frac{\partial n}{\partial x_2}\right)^2}{\left(\frac{\partial \mu_2}{\partial x_2}\right)}\right] \tag{102}$$

where the new terms are $V$, the molar volume of the mixture, and $x_2$, the mole fraction of component 2, usually called the solute. Equation 102 shows that $R_c$ depends on the derivative of the chemical potential $\mu_2$ with composition, $\partial\mu_2/\partial x_2$. It is this derivative that we need to calculate in order to extract important thermodynamic information.

For an ideal solution, $\partial\mu_2/\partial x_2 = N_A kT/x_2$ which allows us to reduce $R_c$ to what is called $R_{\text{id}}$, the ideal Rayleigh ratio, given by:

$$R_{\text{id}} = \left(\frac{2\pi^2}{\lambda^4}\right)\left(\frac{V(1-x_2)x_2}{N_A}\right)\left[n^2\left(\frac{\partial n}{\partial x_2}\right)^2\right] \tag{103}$$

where $N_A$ is Avogadro's number. For nonideal solutions we can solve for $\partial\mu_2/\partial x_2$ (many steps omitted) to obtain:

$$\frac{\partial \mu_2}{\partial x_2} = \frac{N_A kT}{x_2} - (1-x_2)\left(\frac{\partial^2 G^E}{\partial x_2^2}\right) \tag{104}$$

Substituting Equations 103 and 104 into Equation 102, we obtain after some rearrangement:

$$\left[\frac{R_{id} - R_c}{R_c}\right] = \frac{x_2(1-x_2)}{N_A kT}\left(\frac{\partial^2 G^E}{\partial x_2^2}\right) \tag{105}$$

This equation connects thermodynamics, here the excess Gibbs potential, with laser-generated photon probes of the fluctuations. Since $R_{\text{id}}$ and $R_c$ are obtainable independently, Equation 105 relates the isothermal composition dependence of the scattering (left-hand side of the equation) to a composition dependence of the excess Gibbs potential. Since $G^E$ is zero for an ideal mixture, clearly any deviation from ideality appears immediately as a nonzero numerator of the left-hand side of Equation 105, and this nonideality reflects the magnitude of the excess Gibbs potential.

To proceed further, some functional form is required for $G^E$. The Redlich-Kister polynomial is often used and is given by:[7]

$$G^E = x_2(1 - x_2)\sum_{j=1}^{n} A_j(1 - 2x_2)^{(j-1)} \qquad (106)$$

Here the $A_j$ are parameters independent of composition to be determined by data fitting. The calculation at this point proceeds by choosing a polynomial, taking its second derivative with respect to composition, substituting this into Equation 105, and then solving for the unknown coefficients $A_j$ by some least squares technique.

Generally, the composition dependence of $[(R_{id} - R_c)/R_c]$ is complex. In the methanol-water system discussed below, a sixth-order polynomial ($n = 4$) was used to fit the data. In this case, plots of $[(R_{id} - R_c)/R_c]$ versus composition were fit to Equation 106 written for $G^E$ as a sixth-order polynomial in $x_2$.

## Experimental

In this case study, the results of the above procedure are applied to solutions of methanol-water, a system well known for its nonideal behavior because it exhibits a minimum boiling point vapor-liquid azeotrope and negative excess molar volumes of mixing. The block diagram of a typical apparatus illustrated in Figure 9-6 shows that the experimental apparatus is basically quite simple. A laser light source, such as a 488 nm argon ion laser line (which is excellent for this purpose), is directed toward the thermostatted sample, and the scattered light intensity is detected with a photomultiplier tube. The scattered intensity is directly proportional to the photomultiplier's output current, which becomes then the primary data signal. The scattering measurements are usually made at a 90° angle from incident.

## Extent of Intermolecular Forces

Typical excess Gibbs potential data as a function of composition for room temperature are shown in Figure 9-7. Figure 9-7 is the payoff picture presenting the total excess Gibbs potential versus composition and temperature.[8] A major result of the experiment is that the excess Gibbs potential reaches a maximum at concentrations of methanol slightly in excess of water. This behavior parallels the excess volume data where an asymmetry in the minimum in $V^E$ exists for water-rich solutions. One reasonable interpretation of the maximum in $G^E$ on the water-rich side is that greater nonideal intermolecular forces exist in the $x_{MeOH} < 0.5$ region. What would be the origin of these intermolecular interactions? If only simple binary molecular interactions occur in a mixture, then the excess functions might be expected to be a simple parabola. Because there is asymmetry in each of the

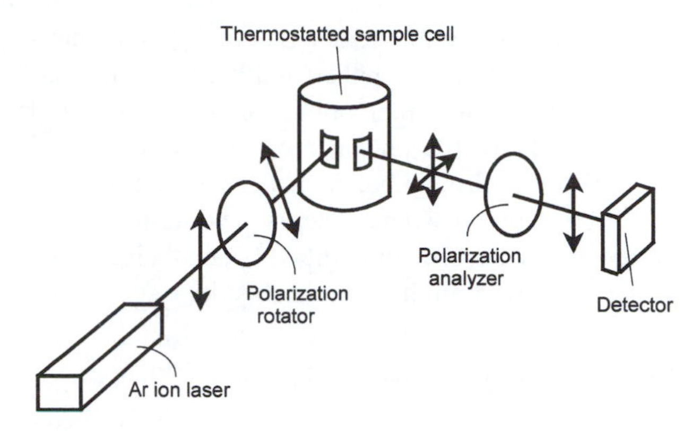


**Figure 9-6**
Typical light scattering apparatus defining polarizations of incident and scattered light. An argon ion laser is directed toward a thermostatted sample. The intensity of the scattered light is measured at 90° to the incident beam.

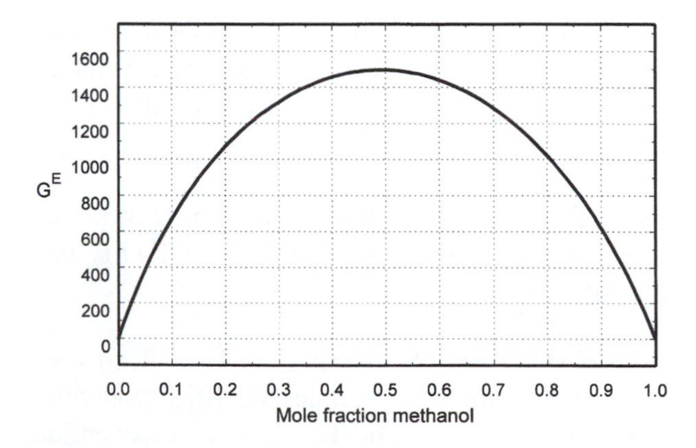


**Figure 9-7**
Excess Gibbs potentials for methanol-water mixtures as a function of composition at 10°C. The asymmetric shape of the data (skewed to the water-rich side) is indicative of greater nonideal intermolecular forces in the water-rich region.

excess functions, we conclude that many-body interactions, three-dimensional networks, exist in addition to the binary interactions in water-methanol solutions. Thus, the presence of three-dimensional networks of hydrogen bonds that are more extensive in binary solutions where water is the dominant component is in accord with the $G^E$ data being more strongly skewed on the water-rich side.

### Significance of the Study

A major objective for this investigation is to better understand intermolecular forces, especially as a function of composition in solution. The connection between the microscopic fluctuations and the macroscopic Gibbs potential is crucial to the success of the experiment, and that connection is possible by laser-generated scattered photons. Traditional thermodynamic measurements do not directly measure Gibbs potentials, but must derive them from other measurements. However, since measuring the scattered light due to fluctuations is easy, a new, more direct method of measuring Gibbs potentials and hence the effects of intermolecular forces is in hand.

## Case Study II: Coherent Anti-Stokes Raman Spectroscopy: A New Way to Measure Temperature

### Overview of the Case Study

**Objective.** To use coherent anti-Stokes Raman spectroscopy to better understand the process of combustion, a reaction often too dangerous and difficult to access by normal means.

**Laser Systems Employed.** A high-energy pulsed laser is used as a pump beam, and a dye laser is used as a probe beam.

**Role of the Laser Systems.** The laser beams are mixed in the sample to generate a CARS signal which reflects the presence and temperature of chemical constituents of the flame.

**Useful Characteristics of the Laser Light for this Application.** The high-intensity laser pump beam in conjunction with the tunable dye laser probe beam initiates the nonlinear Raman effect known as CARS. The energy and intensity of the resulting anti-Stokes photons carry the desired chemical analysis information.

**Principles Reviewed.** Vibrational-rotational spectroscopy, anti-Stokes Raman lines, four-color mixing, temperature dependent spectral line intensities.

**Conclusions.** Use of CARS allows the temperature of flames to be measured and profiled. These flame temperature profiles help in understanding the chemical reactions that occur in combustion.

### Introduction

As described earlier, CARS depends on the mixing of at least two laser beams in a sample and the subsequent analysis of the resultant anti-Stokes beam. Two very important practical aspects of CARS, however, are that: (1) the required laser beams can be brought together some distance from the lasers and their associated optics and (2) the anti-Stokes line signal(s) can also be detected some distance away from the mixing region. As a consequence, the domain of CARS applications far exceeds the typical sphere of conventional spectroscopic experiments. The ability to spectroscopically sample at a safe distance has made CARS a very important tool for studying flames and other hostile environments such as nucleating explosions. One important objective of CARS investigations is to measure flame temperature.

### Vibrational-Rotational Basis for CARS

The CARS technique normally studies the vibrational (or under enough spectral resolution, vibrational-rotational) energy levels of molecules in their ground electronic state. Most molecules exist in their lowest vibrational state at room temperature and need to become quite hot before a significant fraction of molecules is vibrationally excited. Even at room temperature, however, many molecules might exist in excited rotational states. The fraction of molecules which exists in vibrationally (or rotationally) excited states can be estimated by the Boltzmann distribution (Equation 93) with $\varepsilon_v = (v + 1/2)h\nu_{VIB}$, where $\nu_{VIB}$ is the fundamental vibrational frequency of the molecule of interest. For any flame produced by burning a fuel in air, dinitrogen gas $N_2$ will be present. The intensities of the vibrational-rotational spectrum of $N_2$ gas will depend on the temperature of the system as a consequence of the temperature dependence of the vibrational level populations. Thus, from an analysis of the intensities of the vibrational-rotational spectrum, the flame temperature at a particular location can be determined.

The fundamental vibrational frequency for $N_2$ is about 2345 $cm^{-1}$. Table 9-5 gives the fractions of molecules in the first and second excited vibrational state relative to the number in the ground vibrational state at several temperatures. Note that the intensity of the vibrational anti-Stokes lines increases with temperature, as we expect when we recall that the intensity of the anti-Stokes lines depends on the population of excited states and that increasing temperature enhances the excited state populations by Equation 93.

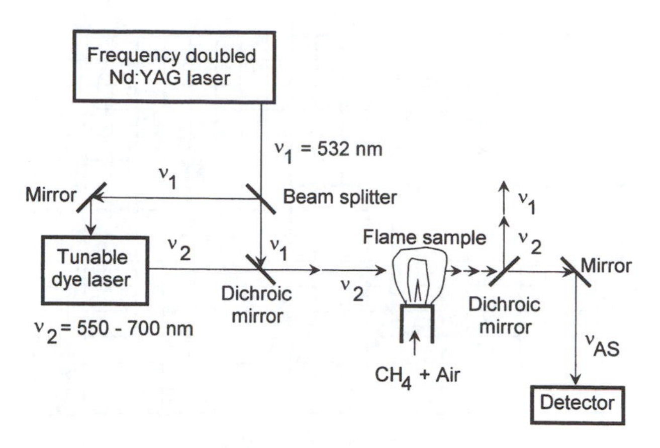


**Figure 9-8**
A schematic illustrating a coherent Raman spectroscopic system used to determine the temperature of a flame. A frequency-doubled YAG pulsed laser serves as both the primary pump beam $\nu_1$ as well as the optical pump for a tunable dye laser. The broad output of the dye laser $\nu_2$, when combined with the primary pump beam, facilitates generation of anti-Stokes lines over a range of wavelengths. The dichroic mirror before the sample reflects $\nu_1$ and transmits $\nu_2$, while the dichroic mirror positioned after the flame enables the output at $\nu_{AS}$ to be separated from $\nu_1$ and $\nu_2$.

## Experimental

How is the CARS experiment performed? Verdieck et al.[9] report measuring the temperature of a methane-air flame using CARS. A schematic of their apparatus is shown in Figure 9-8. A frequency-doubled YAG pulsed laser whose output is 532 nm is the primary pump beam $\nu_1$. Part of the YAG laser's output is used (via the beam splitter) to drive a dye laser whose output is tunable over a narrow or broad wavelength range, depending on the dye used. The YAG pulse width is on the order of 10 ns with repetition rates of 10 Hz. The peak power is about $10^6$ W with the average power about 2 W. The wave mixing arrangement is collinear with the separation of the output beams accomplished by clever optical mirrors known as *dichroic mirrors*. Such mirrors can be made to reflect certain frequencies while transmitting others. The dye laser is used with a dye that provides a broadband laser light output to generate anti-Stokes lines over a wide energy range. In effect, each laser pulse generates a spectrum of anti-Stokes lines.

**Table 9-5**
Fractions of Dinitrogen Molecules in Excited Vibrational States for Several Temperatures

| *Vibrational State Quantum Number (v)* | *Temperature (K)* | | |
|---|---|---|---|
| | 300 | 1000 | 2100 |
| 1 | $\approx 10^{-5}$ | 0.03 | 0.2 |
| 2 | $\approx 10^{-10}$ | $\approx 10^{-3}$ | 0.04 |

## Results

The broadband probe laser output may be viewed as centered about the pump laser line $\nu_1$ at 532 nm or 18,800 $cm^{-1}$. Thus, the main $\nu_{AS}$ anti-Stokes line, which corresponds to the $v = 1$ to $v = 0$ transition in the ground electronic state of $N_2$, should occur around 18,800 + 2300 = 21,100 $cm^{-1}$. Figure 9-9 shows the experimentally obtained flame spectrum for methane-air with the strongest signal around 21,100 $cm^{-1}$ and with vibrational-rotational lines occurring at lower $\nu_{AS}$ values.[10] This spectrum was theoretically fit to a temperature of 2104 K with excellent agreement of the calculated and experimental spectra.

The details of the spectrum arise from the $\Delta J = 0$ rotational lines associated with the $Q$ branch of the v = 1 to v = 0 branch of the vibrational transition. In addition, a sizable amount of the so-called "hot band" (i.e., a transition between two excited states) v = 2 to v = 1 transition with its own vibrational-rotational structure is observed. The flame spectrum shows the

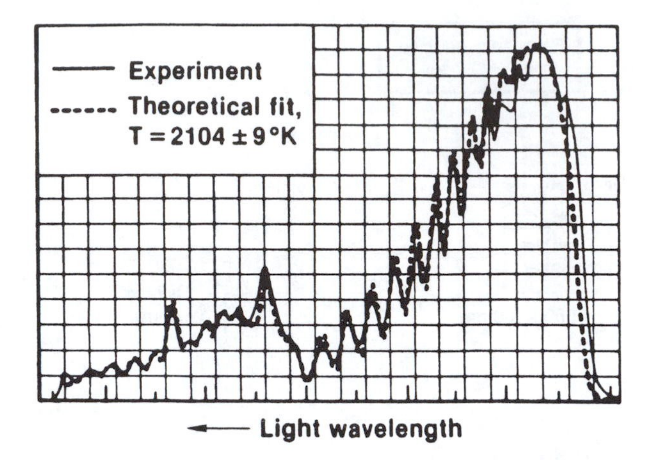


**Figure 9-9**
Experimental and theoretical CARS spectrum of $N_2$ found in methane-air flame at 2104 K. The strongest portion of the spectrum reflects the anti-Stokes transition at around 21,000 $cm^{-1}$. Reprinted with permission from Verdieck, J. F., Hall, R. F., Shirley, J. A., and Eckbreth, A. C., *J. Chem. Educ.*, 1982, 59, 495–503. Copyright 1982 Division of Chemical Education, American Chemical Society.

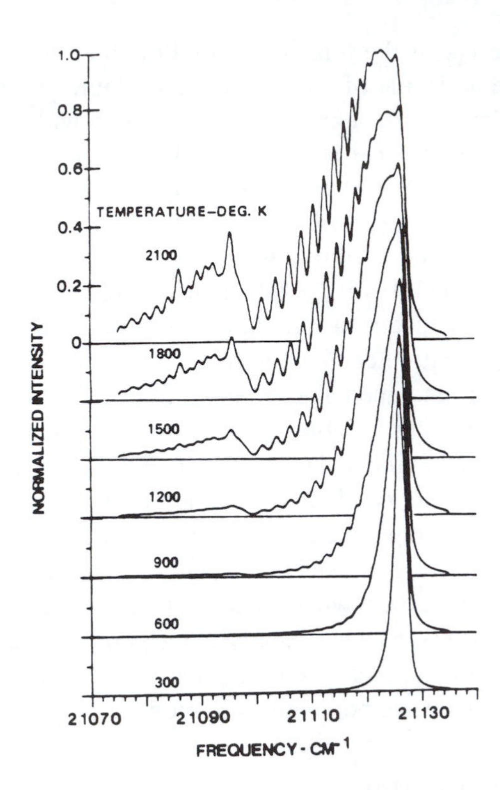


**Figure 9-10**
Theoretical vibrational-rotational anti-Stokes lines for the $\Delta J = 0$ $Q$ branch of $N_2$ for several temperatures. Reprinted with permission from Verdieck, J. F., Hall, R. F., Shirley, J. A., and Eckbreth, A. C., *J. Chem. Educ.*, 1982, 59, 495–503. Copyright 1982 Division of Chemical Education, American Chemical Society.

hot band intensity to be about 40%[9] of the fundamental v = 1 to v = 0 band, which is in very rough agreement with the calculation in Table 9-5 showing the ratio $N_2/N_1 \approx 0.04/0.2 \approx 0.2 \Rightarrow 20\%$. Again without the details of the calculation, Figure 9-10 shows a progression of theoretical spectra calculated at various temperatures.[11] These spectra reflect transitions between the rotational states of the first excited vibrational state (v = 1) to rotational levels in the ground vibrational state (v = 0). At higher temperatures, similar anti-Stokes transitions occur from the rotational states of the second vibrational state (v = 2) to various rotational states of the first excited vibrational state (v = 1). The agreement between experiment and theory in the calculation of the flame's temperature is outstanding. It is not clear that any other approach to measuring the flame's temperature could be as accurate.

## Why CARS?

Why use CARS? The first reason might be the magnitude of the $\nu_{AS}$ signal. Given the powers possible from the laser system, the $\nu_{AS}$ signal is easily detected. Background noise in this case is negligible and the exciting line $\nu_1$, which in spontaneous Raman would give rise to an immensely intense Rayleigh line that generally makes the recording of an anti-Stokes difficult, is easily separated by means of the dichroic mirror.

A very fundamental reason for the use of CARS to analyze $N_2$ centers on what is called the *exclusion rule*. Since $N_2$ is a symmetric diatomic molecule, its single stretching vibration does not change or induce a molecular electronic dipole (the basis for normal infrared spectroscopy). The stretching does change the molecular polarizability, however. What this means is that $N_2$ cannot be studied by infrared spectroscopy at all but can be studied very successfully by Raman spectroscopies, as this application of CARS shows.

An additional reason for using CARS results from the fact that the light wave mixing occurs in a very small region in the sample. The size of this region is small enough to map the variation of temperature in the flame. Such mapping is called *profiling the flame temperature* and permits a determination of the energetics of combustion.

### Significance of the Study

The chemical reaction we call combustion is a major source of the world's heat energy. The laser, through measurements such as CARS, is helping us better understand this very important process with the promise of even more efficient burning and conservation of fuel and heat energy.

## FOR FURTHER EXPLORATION

One classical application of static light scattering measurements is the determination of molecular weights, polydispersity, and interparticle interactions of polymers and macromolecules. One recent study of the eye lens protein $\alpha$-crystallin[12] measured the intensity of light scattered as a function of scattering angle, solution ionic strength, and protein concentration. An argon ion laser tuned to 488 nm provided the incident light. The experimentalists were able to estimate the molecular mass, hydrodynamic radius, and net charge on the $\alpha$-crystallin particles at the various concentration and ionic strength conditions. Their results provided quantitative evidence for the factors influencing interparticle interaction that may elucidate the origin of age-related malfunctioning of the lens. Light scattering is also well-established as a method for the characterization of the aggregation number, critical micelle concentration, and shape of surfactant micellar systems. A recent report[13] demonstrated that a simple low-power (15 mW) He/Ne laser as a light source provides adequate directionality and intensity to observe light scattering from dilute solutions of a nonionic polyoxyethylene surfactant. The nonionic nature of the surfactant headgroup facilitates studies of the general mechanism of micelle formation but results in a low critical micelle concentration (cmc). The sensitivity of light scattering measurements under the dilute conditions permits determinations of both the cmc and micelle aggregation number as a function of surfactant concentration and temperature. As light scattering permits the direct determination of structural parameters without the need for probe molecules, this technique will continue to be an excellent choice for the characterization of colloidal and polymeric systems.

## DISCUSSION QUESTIONS

1. Why is CARS a reasonable technique to use on samples that easily fluoresce?
2. Why has CARS found significant use studying hostile environments?

## SUGGESTED EXPERIMENTS

Suggested references to experiments that further illustrate the principles of light scattering described in this chapter:

1. Moore, R. J., Trinkle, J. F., Khandhar, A. J., and Lester, M. I., "Experiments in Laser Raman Spectroscopy for the Physical Chemistry Laboratory," *Physical Chemistry: Developing a Dynamic Curriculum* (R. C. Schwenz and R. J. Moore, Eds.), American Chemical Society, Washington, D.C., 1993.
2. Van Hecke, G. R., "Determination of Thermodynamic Excess Functions by Combination of Several Techniques Including Laser Light Scattering," *Physical Chemistry: Developing a Dynamic Curriculum* (R. C. Schwenz and R. J. Moore, Eds.), American Chemical Society, Washington, D.C., 1993.
3. Wirth, F. H., "Dye Laser Experiments for the Undergraduate Laboratory; Experiment 1—Rayleigh Scattering," Laser Science, Inc., Cambridge, MA. An experiment to illustrate the dependence of the intensity of Rayleigh scattering on the incident wavelength.
4. Wirth, F. H., "Dye Laser Experiments for the Undergraduate Laboratory; Experiment 2—Raman Scattering," Laser Science, Inc., Cambridge, MA. An experiment to illustrate the dependence of the intensity and wavelengths of Raman scattering on the concentration and structure, respectively, of the scattering species.

## LITERATURE CITED

1. Cabannes, J., *La Diffusin Moleculaire la Lumiere*, Les Presses Universitaires de France, Paris, 1929, Chapter X.

2. Šegudović, N., and Deželić, G., "Light-scattering in binary liquid mixtures. I. Isotropic scattering." *Croatia Chem. Acta*, 1973, 45, 385–406.

3. Kerker, M., *The Scattering of Light*, Academic Press, New York, 1969, p. 580.

4. Kratohvil, J. P., and Smart, C., "Calibration of light scattering instruments. III. Absolute angular intensity measurements on Mie scatters," *J. Colloid Sci.*, 1965, 20, 875–892.

5. Schmidt, R. L., and Clever, H. L., "Thermodynamics of binary liquid mixtures by Rayleigh light scattering," *J. Phys. Chem.*, 1968, 72, 1529–1536.

6. Myers, R. S., and Clever, H. L., "Excess Gibbs free energy of mixing in some hydrocarbon and alcohol solutions by Rayleigh light scattering, *J. Chem. Thermodynamics*, 1970, 2, 53–61.

7. Redlich, O. and Kister, A. T., "Thermodynamics of nonelectrolytic solutions. Algebraic representation of thermodynamic properties and the classification of solutions," *Ind. Eng. Chem.*, 1948, 40, 345–8.

8. Westervelt, R., and Van Hecke, G. R., previously unpublished results.

9. Verdieck, J. F., Hall, R. F., Shirley, J. A. and Eckbreth, A. C., "Some applications of gas phase CARS spectroscopy," *J. Chem. Educ.*, 1982, 59, 495–503.

10. Hall, R. L., and Eckbreth, A. C., "Coherent anti-Stokes Raman spectroscopy: Applications to combustion diagnostics," *Laser Applications* (R. F. Erf, Ed.), Academic Press, New York, 1981.

11. Hall, R. J., "Pressure-broadened linewidths for nitrogen anti-Stokes Raman spectroscopy thermometry," *Appl. Spectrosc.*, 1980, 34, 700–2.

12. Xia, J., Aerts, T., Donceel, K., and Clauwaert, J., "Light scattering by bovine $\alpha$-crystallin proteins in solution: Hydrodynamic structure and interparticle interaction," *Biophysical J.* 1994, 66, 861–872.

13. Kato, T., Kanada, M., and Seimiya, T., "Measurements of light scattering intensities on extremely dilute solutions of nonionic surfactant," *Langmuir*, 1995, 11, 1867–1869.

## GENERAL REFERENCES

### Light Scattering

Coumou, D. J., and Mackor, E. L., "Isotropic Light Scattering in Binary Liquid Mixtures," *Trans. Faraday*, 1964, 60, 1726–1735.

Deželić, G., "Evaluation of Light-Scattering Data of Liquids from Physical Constants," *J. Chem.*, 1966, 45, 185–191.

Deželić, G., Šegudović, N., "Light Scattering in Binary Liquid Mixtures. II. Anisotropic Scattering," *Croatia Chem. Acta*, 1973, 45, 407–418.

Kerker, M., *The Scattering of Light*, Academic Press, New York, 1969.

Lewis, H. H., Schmidt, R. L., and Clever, H. L., "Thermodynamics of Binary Liquid Mixtures by Total Intensity Rayleigh Light Scattering. II," *J. Phys. Chem.*, 1970, 74, 4377–4382.

Myers, R. S., and Clever, H. L., "Excess Gibbs free energy of mixing in some hydrocarbon + alcohol solutions by Rayleigh light scattering," *J. Chem. Thermodynamics,* 1970, 2, 53–61.

Schmidt, R. L., and Clever, H. L., "Thermodynamics of Binary Liquid Mixtures by Rayleigh Light Scattering," *J. Phys. Chem.*, 1968, 72, 1529–1536.

Schmitz, K. S., *An Introduction to Dynamic Light Scattering by Macromolecules*, Academic Press, New York, 1990.

S˘egudovic´, N., and Dez˘elic´, G., "Light Scattering in Binary Liquid Mixtures. I. Isotropic Scattering," *Croatia Chem. Acta*, 1973, 45, 385–406.

Simonson, J. M., Bradley, D. J., and Busey, R. H., "Excess molar enthalpies and the thermodynamics of (methanol + water) to 573 K and 40 MPa," *J. Chem. Thermodynamics*, 1987, 19, 479–492.

### Raman Spectroscopies

Demtroder, W., *Laser Spectroscopy: Basic Concepts and Instrumentation*, 2nd Ed., Springer-Verlag, Berlin, 1996.

Gardiner, D. J., and Graves, P. R. (Eds.), *Practical Raman Spectroscopy*, Springer-Verlag, Berlin, 1989.

Grasselli, J. G., Snavely, M. K., and Bulkin, B. J., *Chemical Applications of Raman Spectroscopy*, John Wiley, New York, 1981.

Harvey, A. B. (Ed.), *Chemical Applications of Nonlinear Raman Spectroscopy*, Academic Press, New York, 1981.

Kiefer, W., and Long, D. A., *Nonlinear Raman Spectroscopy and Its Chemical Applications*, D. Reidel, Dordrecht, Holland, 1982.

Levenson, M. D., *Introduction to Nonlinear Laser Spectroscopy*, Academic Press, Boston, 1982.

Levenson, M. D., and Kano, S. S., *Introduction to Nonlinear Laser Spectroscopy*, Academic Press, Boston, 1988.

Nibler, J. W., and Knighten, G. V., "Coherent Anti-Stokes Raman Spectroscopy," *Raman Spectroscopy of Gases and Liquids* (A. Weber, Ed.), Springer-Verlag, Berlin, 1979.

Stencil, J. M., *Raman Spectroscopy for Catalysis,* Van Nostrand Reinhold, New York, 1990.

Strommen, D. P., and Nakamoto, K., *Laboratory Raman Spectroscopy,* John Wiley, New York, 1984.

CHAPTER 10

# Laser-Assisted Mass Spectrometry

## Chapter Overview

In this chapter we explore the significant ways in which lasers have advanced the analytical use of mass spectrometry to characterize two distinct types of samples: macromolecules and molecular adsorbates on solid surfaces. In the search for methods to achieve mass spectral analysis of complex, nonvolatile substances, scientists have discovered that large molecules can be volatilized and ionized with little structural damage through a process of laser-induced desorption from a solid surface. The high-power density and short pulse width of the laser excitation beam extend the molecular weight range of mass spectral analysis. With the use of either matrix isolation techniques or methods to achieve a temporally and spatially separated two-step desorption/ionization process, even more extensive analytical gains have been attained.

## The Impetus for Developing Laser-Assisted Mass Spectroscopy

Accurate and direct molecular weight determination is a powerful central feature of mass spectrometry. This technique is also a universal analytical tool for the characterization, structural determination, and quantitative analysis of gases and condensed media. Recall that the fundamental basis of mass spectrometry is the observation that charged particles in motion with the same velocity are deflected by a magnetic field to an extent dependent upon each particle's mass. What types of samples are most effectively analyzed via mass spectrometry? As the principles of the method require a charged gaseous ion for analysis, samples analyzed using commercial instruments are generally restricted to those easily volatilized by direct heating and readily ionized via electron impact. Thus, conventional mass spectrometry precludes the study of molecules with low volatility, as well as those molecules that decompose readily upon ionization with an electron beam. With these restrictions, fragile, high molecular weight biomolecules and polymers represent two classes of molecules typically not amenable to conventional mass spectroscopic analysis.

However, several recent developments involving tunable lasers have dramatically extended the range of samples accessible for characterization by mass spectrometry. As a consequence, high-resolution mass spectrometric analyses of macromolecules as well as of adsorbates on solid surfaces are now feasible through specific applications of laser techniques. These methods first recognize that higher molecular weight substances can be ionized and remain as intact ions through a process of laser-induced desorption from a solid surface. Reproducible analyses without serious sample degradation are possible using this desorption method for substances with molecular weights ≤ 1500 Da. Even more significant advances are possible through modifications of the desorption approach.

Matrix-assisted laser desorption/ionization represents one such revised strategy. By incorporating the target molecule in a solid matrix containing a highly UV-absorbing chromophore, decomposition of the macromolecule is significantly reduced by the energy-absorbing properties of the matrix. The use of a matrix also has the added advantage of isolating macromolecules to prevent intermolecular interaction and subsequent aggregation. Intermolecular hydrogen bonds between macromolecules and between biopolymer chains are the most frequently observed intermolecular forces, but hydrophobic interactions, van der Waals forces, and electrostatic interactions also govern intermolecular associations. Why is it advantageous to limit the likelihood of aggregation? The presence of aggregates complicates the analysis of a mass spectrum by obscuring the detection of the ion corresponding to the isolated macromolecule. Employing lasers for sample desorption and ionization in a matrix environment has produced the highest mass ion to date ($m/z$ = ~450,000).[1]

As an alternative to the matrix isolation technique, a two-step laser mass spectrometry method has been developed for the analysis of adsorbates on solid surfaces. By spatially and temporally separating the desorption step that volatilizes the sample and the ionization step that acquires the charged species, both the retention of the adsorbate's integrity and the sensitivity of the method are enhanced. Subattomole (< $10^{-18}$ mol) detection limits for atoms and subfemtomole (< $10^{-15}$ mol) detection limits for molecules have been achieved.[2]

We illustrate the extraordinary scope currently afforded to mass spectrometry with several examples of laser-assisted investigations. In particular, we focus on the implementation of laser techniques to achieve continued advances in mass range and sensitivity. Developments in such diverse areas as structural biology and heterogeneous catalysis are facilitated by the successes in laser desorption/ionization mass spectroscopy.

## Case Study I: Matrix-Assisted Laser Mass Spectroscopy of Large Biomolecules

### Overview of the Case Study

**Objective.** To use matrix-assisted laser mass spectrometry to detect femtomolar levels of high molecular weight proteins and to analyze mixtures of such proteins.

**Laser System Employed.** A *Q*-switched, frequency-quadrupled Nd:YAG laser emitting at 266 nm in conjunction with a time-of-flight mass spectrometer.

**Role of the Laser System.** To induce matrix vaporization, subsequent desorption of the nonvolatile high molecular weight proteins into the gas phase, and subsequent ionization of the macromolecules.

**Useful Characteristics of the Laser Light for This Application.** Short-duration pulses on the nanosecond timescale, monochromaticity, wavelength tunability.

**Principles Reviewed.** Matrix isolation, desorption, ionization, parent molecular ion.

**Conclusions.** This study demonstrates the feasibility of reproducibly generating and detecting molecular ions from high molecular weight proteins using a matrix-assisted laser-induced desorption and ionization mass spectroscopic technique. Using a liquid matrix on a fibrous substrate with a sufficiently high surface area, femtomolar levels of proteins with molecular weights as high as 97,400 daltons are detected and mixtures of up to five high molecular weight proteins are analyzed.

### Introduction

An ever-increasing demand for the structural characterization of macromolecules challenges chemists daily. Key areas of high-mass interest include the sequencing of proteins and nucleic acids and the determination of both the weights and oligomeric distributions of polymers. However, conventional mass spectrometry of large molecular weight compounds is limited by the low volatility of the sample and the excessive fragmentation of the macromolecule that generally results upon desorption and/or ionization. Modifications in experimental protocol to optimize the production of gaseous molecular ions (the parent molecule minus an electron) from these materials have generally demonstrated little success. For example, heating of such samples to promote volatilization typically results in degradation or pyrolysis of the material. Furthermore, the use of "soft" ionization techniques that limit the internal energy imparted by incident excitation generally minimizes fragmentation but produces low concentrations of ions. Any quantitative assessment of resulting mass spectra is thus plagued by weak signal intensities and poor signal-to-noise ratios.

### The Mechanism of Matrix-Assisted Mass Spectrometry

To overcome these difficulties that hinder an extension of the mass range available for analysis, Hillenkamp and associates introduced a novel matrix-assisted method of laser mass spectroscopy designed for the study of high molecular weight, nonvolatile compounds.[3–5] In this technique the sample to be characterized is dissolved in a liquid or solid matrix with a low boiling point. The matrix initially serves as an efficient chromophore to absorb sufficient laser radiation to induce matrix vaporization and limit the excitation of the macromolecule. The sizable expansion of the volatile matrix molecules upon their vaporization induces the desorption of the macromolecules into the gas phase. Specifically, the volume change disrupts the intermolecular forces between macromolecule and matrix, permitting the transport of the embedded macromolecules into the vapor phase. The rapid cooling of the macromolecules upon expulsion from the matrix stabilizes them against dissociation, although some fragmentation is generally unavoidable. The degree of fragmentation is minimized by increasing the power input rate (therefore increasing the rate of heating) using short laser pulses. The mechanism by which the matrix promotes the ionization of the macromolecule is less clearly understood than the desorption process but is commonly believed to involve either photoionization when ultraviolet photons are used or proton transfer from an electronically excited state of a surface matrix molecule.[6]

Using matrix-assisted ultraviolet laser desorption/ionization techniques, numerous successful mass spectral characterizations of proteins, nucleic acids, carbohydrates, and biopolymers have been reported. Exceptional highlights of these investigations include characterizations of DNA with molecular weights over 410 kilodaltons,[7] detections of femtomole quantities of 66 kD molecular weight,[8] and analyses of mixtures of up to six proteins.[8] Matrix-assisted studies with infrared laser radiation, although only recently attempted,[5] are also feasible and suited to an even wider variety of matrix materials.

## The Laser Systems Employed

For the studies of matrix-assisted ultraviolet laser desorption/ionization (UV-LDI), a *Q*-switched frequency-quadrupled Nd:YAG laser emitting at 266 nm is typically employed in conjunction with a time-of-flight mass spectrometer.[8] To enhance matrix vaporization and macromolecular desorption and ionization, the laser beam is carefully focused on the matrix surface to yield a spot size of 10-50 $\mu$m diameter and irradiances of $10^7$ to $10^8$ W cm$^{-2}$. For matrix-assisted infrared laser desorption/ionization (IR-LDI) investigations, a Q-switched Er: YAG laser is a suitable source.[6] This laser beam is characterized by a 2.94 $\mu$m wavelength and a 200 ns pulse duration. The beam is focused to obtain a 300 $\mu$m spot diameter, and the irradiance on the sample surface is varied between $5 \times 10^4$ and $10^7$ W cm$^{-2}$ (by varying the voltage of the flashlamp which pumps the laser) to achieve the best results depending on the nature of the matrix used.

## Ion Production and Analysis of Conventional Mass Spectra

To analyze the results of these laser-assisted investigations, we first review the process of ion production in a mass spectrometer and the elements of a conventional mass spectrum. In mass spectrometry, interaction of an energetic electron beam with a sample generates positively charged ions of a particular mass ($m$), charge ($z$), and therefore mass-to-charge ratio ($m/z$). Electron bombardment produces singly or multiply charged ions, depending upon the number of electrons released from the gaseous molecules. Negative ions may also result but are discharged from the mass spectrometer. The term *parent molecular ion* describes the ion formed by loss of a single electron with no accompanying fragmentation of the molecule. Fragment ions formed directly from the parent molecular ion are known as *primary fragments*. Many fragment ions may result depending on the molecular weight of the sample and the cleavage pattern. For the hypothetical molecule ABC composed of three atoms with a bonding pattern of *A–B–C*, some singly charged fragment ions that may be observed:

$$ABC + e\text{-} \rightarrow ABC^+ + 2\,e\text{-}$$
$$ABC + e\text{-} \rightarrow A^+ + BC\cdot + 2\,e\text{-}$$
$$ABC + e\text{-} \rightarrow AB^+ + C\cdot + 2\,e\text{-}$$
$$ABC + e\text{-} \rightarrow BC^+ + A\cdot + 2\,e\text{-}$$
$$ABC + e\text{-} \rightarrow AC^+ + B\cdot + 2\,e\text{-}$$

The fifth ion, $AC^+$, requires a rearrangement of atoms during the dissociation process to form bonds not present in the original molecule. Collision of ions with neutral molecules can yield additional ions, *cluster ions*, often with molecular weights greater than the original sample. For example:

$$A^+ + ABC \rightarrow A_2BC^+$$
$$\text{or} \quad ABC^+ + ABC \rightarrow (ABC)_2{}^+$$

Experimental factors that can be varied to control the observed fragmentation patterns and clustering probabilities include sample concentration (pressure) and energy of the impinging electrons.

A conventional mass spectrum is a plot of signal intensity (i.e., relative abundance of each ion) on the ordinate versus $m/z$ value on the abscissa. For a pure sample in the absence of aggregation, the most intense peak will correspond to the parent molecular ion and yield the molecular weight of the molecule. Elemental composition is often revealed from an isotope analysis of the spectrum, and structural information is provided from an analysis of molecular fragments and fragmentation patterns. Sequence determination for proteins, peptides, sugars, and nucleotides, for example, is also possible from the elucidation and analysis of fragmentation patterns. The presence of fragment ions, multiply charged ions, and cluster ions can introduce considerable complexity in the appearance of the mass spectrum and in its interpretation. For example, for the molecule $A_2$, the doubly charged molecular ion $A_2^{2+}$ and the singly charged fragment ion $A^+$ exhibit the same $m/z$ value. Similarly, for the triatomic molecule ABC, identical $m/z$ values are predicted for the parent molecular ion $ABC^+$ and the double-charged cluster ion $(ABC)_2^{2+}$. For higher molecular weight molecules, the need to avoid aggregation, fragmentation, and multiple charge formation is often vital to a successful molecular weight determination.

## Analysis of Matrix-Assisted Mass Spectra

To illustrate the nature of the matrix-assisted mass spectra obtained for varying molecular weight ranges, Figure 10-1 presents the matrix-assisted ultraviolet laser desorption/ionization (UV-LDI) spectra for four proteins: cytochrome c (MW ≈ 12,384), trypsinogen (MW ≈ 24,000), ovalbumin (MW ≈ 45,000), and albumin (MW ≈ 66,000).[8]

These spectra were obtained using a Q-switched frequency-quadrupled Nd:YAG laser emitting at 266 nm in conjunction with a time-of-flight mass spectrometer. The liquid matrix consisted of a mixture of m-nitrobenzyl alcohol, methanol, and water in a ratio of 3:2:1. The hydrogen-bonding capacity of the matrix alcohols facilitates adherence of the protein to the matrix via hydrogen bonding. Fine fibrous paper (1 mm$^2$ surface area) served as a substrate for

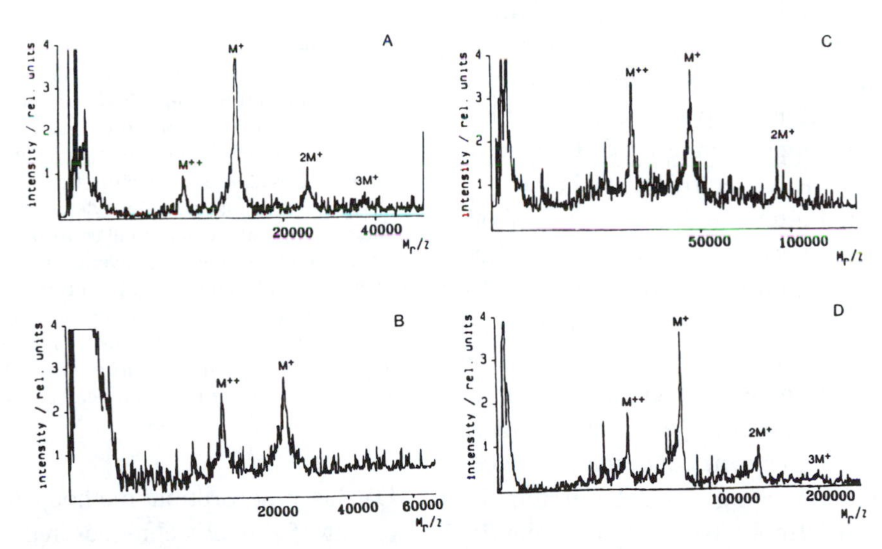


**Figure 10-1**
Mass spectra of (*a*) cytochrome c, (*b*) trypsinogen, (*c*) ovalbumin, and (*d*) albumin, obtained using two-step laser desorption/nonresonant multiphoton ionization. Cytochrome c exhibits a parent molecular ion $M^+$ at an $m/z$ value consistent with its approximate molecular weight of 12,384 as well as signals at $m/z$ values corresponding to the $M^{2+}$, $2M^+$, and $3M^+$ ions. Trypsinogen exhibits signals attributed to $M^+$ ($MW = 24{,}000$) and $M^{2+}$. For ovalbumin, both $M^{2+}$ and $2M^+$ accompany the parent molecular ion $M^+$ at $m/z \approx 45{,}000$. Albumin exhibits the most complex mass spectrum with signals coinciding with $M^{2+}$, $M^+$ ($m/z \approx 66{,}000$), $2M^+$, and $3M^+$. Reprinted with permission from *Anal. Chem.*, 1991, 63, 450–453. Copyright 1991 American Chemical Society.

the matrix and the protein to be analyzed. The spectra illustrate several typical observations:

1. Matrix signals appear in the low mass range, yet generally are similar in intensity to the parent molecular ion signal, despite the fact that matrix molecules are in excess by a factor of more than $10^4$.
2. Only parent molecular ions, cluster ions, and doubly charged ions are observed with no fragment ions, confirming that liquid matrix laser mass spectrometry is a "soft" ionization technique.
3. Cluster ions up to $3M^+$ are always apparent with proteins of MW < 20,000, while generally only $M^{2+}$ and/or $2M^+$ ions accompany the parent molecular ion in spectra of proteins of high molecular weight (> 20,000) (exception: albumin).

These multiply charged and cluster ions can serve as an aid to molecular weight determination rather than as a hindrance. (To see this point, try assigning a $3M^+$ peak in Figure 10-1 to the parent molecular ion. Can you successfully assign the other peaks in the mass spectrum with this incorrect assumption?) The tractability of analyzing high mass compounds by this technique is further enhanced by the reproducibility and simplicity of the types of ions observed. As an upper limit, phosphorylase with a molecular weight of approximately 97,400 daltons has been successfully analyzed by this technique.

The spectra in Figure 10-1 are "single-shot" spectra (i.e., obtained with a single pulse of the Nd:YAG laser) for 5 $\mu$mole quantities. Lower detection limits are possible using this nitrobenzyl alcohol matrix, where a detection limit is defined as the protein amount which gives rise via a single laser pulse to a signal that is twice the noise level. Figure 10-2 exhibits single-shot spectra to define the detection limit for albumin using (*a*) 250 femtomole, (*b*) 25 fmol, and (*c*) 5 fmol quantities of protein deposited on the nitrobenzyl alcohol matrix.[8] As signal averaging of at least several hundred single-shot spectra is possible, samples of attomole ($10^{-18}$) quantities are predicted to be sufficient to generate detectable ion signals.[8] Such extreme sensitivity will facilitate detection of trace and ultratrace quantities of materials in numerous applications.

The spectacular advances in the acquisition of mass spectra of large molecules using matrix-assisted UV-laser desorption methods have prompted the testing of similar techniques in the infrared region. In the first reported study of a matrix-assisted infrared laser desorption/ionization (IR-LDI) investigation,[6] a Q-switched Er:YAG laser was used successfully to ionize large biomolecules incorporated in a wide variety of matrices. An interesting dependence of the mass spectrum on matrix identity was observed. While most matrices with IR desorption produce spectra that closely resemble those obtained with UV-LDI, caffeic acid produces distinct spectra when used as a matrix

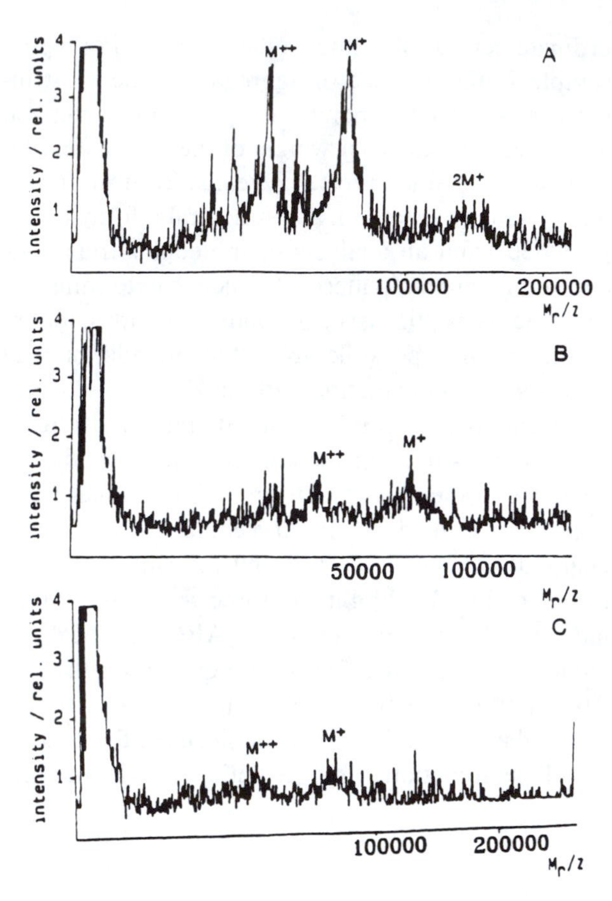


**Figure 10-2**
Single-shot UV-laser desorption mass spectra of varying amounts of albumin deposited on a nitrobenzyl alcohol matrix. The quantities of protein used were (*a*) 250 fmol, (*b*) 25 fmol, and (*c*) 5 fmol. The spectra indicate decreasing signal intensities as albumin concentration is decreased, as well as changes in the number of ions detected. At 250 fmol in (*a*), $M^{2+}$ and $M^+$ dominate with a small contribution from $2M^+$. The ratio of the $M^+$ signal to the average noise level is about 8–10:1. At 25 fmol and 5 fmol only $M^{2+}$ and $M^+$ are detected. In (*b*) the $M^+$ signal-to-noise ratio is on the order of 3:1. For (*c*) the $M^+$ signal is about twice the intensity of the average noise level. Reprinted with permission from *Anal. Chem.*, 1991, 63, 450–453. Copyright 1991 American Chemical Society.

for larger molecular weight proteins.[6] Figure 10-3 illustrates the influence of the matrix on the IR desorption of a monoclonal antibody of MW ≈ 150,000 Da. Extensive multiple charging is observed with caffeic acid (*b*) but not with nicotinic acid (*a*). To date, no model for the formation of such highly charged ions as $M^{10+}$ has been proposed. To add to the difficulty of interpreting these results, multiple charging has not been observed with other derivatives of cinnamic acid as matrices, nor have UV-LDI mass spectra of monoclonal antibodies using a caffeic acid matrix been recorded. The mechanism for ionization of the macromolecule via IR radiation is also relatively obscure. As matrix-assisted IR-LDI studies are relatively new, additional investigations will explore the potentials and limitations of this technique.

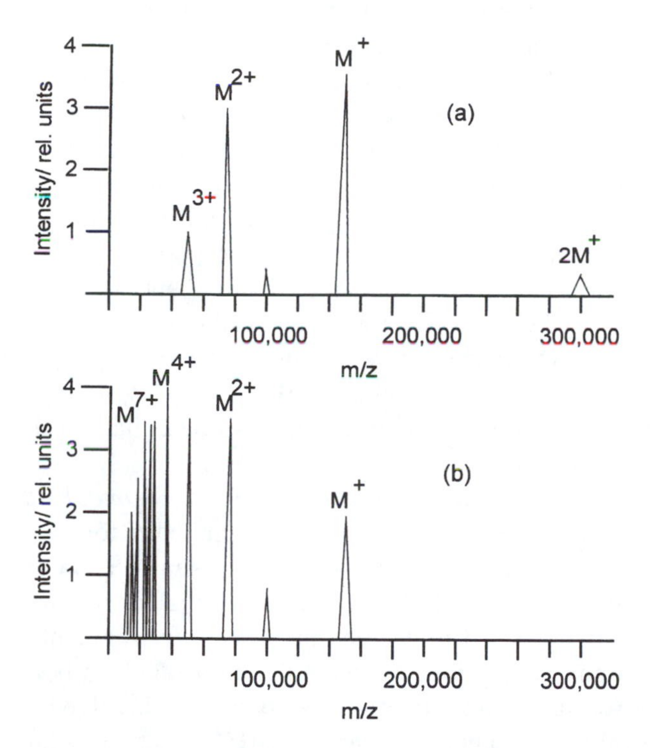


**Figure 10-3**
Schematics of mass spectra of a monoclonal antibody with an approximate molecular weight of 150,000 obtained using the matrix-assisted infrared laser desorption/multiphoton ionization technique. The matrices used correspond to nicotinic acid (*a*) and caffeic acid (*b*). With nicotinic acid, four ions are detected: $M^{3+}$, $M^{2+}$, $M^{+}$, and $2M^{+}$, with the parent molecular ion $M^{+}$ dominant. An extensive number of multiple-charge ions (but no cluster ions) are resolved with caffeic acid as the matrix, with all integral charges from +1 to +10. The signal at an *m/z* ratio corresponding with the $M^{4+}$ species appears to be dominant.

## Future Applications of the Matrix-Assisted Laser MS Technique

Matrix-assisted laser mass spectrometry may be most powerful when used in combination with other analytical techniques. For example, many protein research groups currently couple matrix-assisted laser mass spectroscopy with gel electrophoresis, capillary electrophoresis, or liquid chromatography to determine the molecular weights of peptides and proteins. By using the rapid and efficient separation capabilities of these latter techniques with the sensitivity and molecular specificity of mass spectrometry, exceptionally high molecular weight determinations are possible. Gel or capillary electrophoresis or liquid chromatography provides an initial attempt to separate the components of a protein mixture. Subsequent molecular weight determination using matrix-assisted laser mass spectroscopy, often desorbing the proteins directly from the gel, results in enhanced resolution of the mass spectrum and more accurate molecular weight determinations. As a further development, "ladder sequencing" of proteins is a technique developed by Stephen B. H. Kent of the Scripps Research Institute that combines traditional wet chemistry with laser-assisted mass spectroscopy.[9] By chemically breaking down a protein or peptide into a family of peptide fragments, each differing from the next by one amino acid residue, a sequence ladder can then be rapidly established using matrix-assisted laser mass spectroscopy. Studies to characterize the conformational states of macromolecules, ascertain the stereochemistry of carbohydrates, or monitor the folding of proteins via the matrix-assisted laser mass spectroscopy technique are potential applications, but are at a very early stage of development.

# Case Study II: Laser Mass Spectrometry of Solid Surfaces

## Overview of the Case Study

**Objective.** To use two-step laser-induced mass spectrometry to characterize and quantify molecular adsorbates on solid surfaces, to detect femtomolar levels of high molecular weight compounds, and to analyze mixtures of such macromolecules.

**Laser System Employed.** A high-power pulsed laser for desorption (e.g., a $CO_2$ laser at 10.6 $\mu$m with 10 $\mu$s pulse width and 10 Hz repetition rate) and a high-powered, pulsed, focused laser beam for irradiation to induce ionization (e.g., the 266 nm output of a frequency-quadrupled Nd:YAG laser with ≈ 1 mJ/pulse, 10 ns pulse width, and 10 Hz repetition rate; or the pulsed 248 nm output of a KrF excimer laser).

**Role of the Laser System.** A laser in the far UV or far IR regions with short pulse widths (typically 1–100 ns) provides the rapid heating necessary to avoid pyrolytic decomposition of the adsorbate. A second high-intensity pulsed laser beam in close proximity to the surface facilitates the ionization, selective or nonselective, of neutral molecules of the vaporized ejected sample.

**Useful Characteristics of the Laser Light for This Application.** Short-duration pulses on the nanosecond timescale, short-duration high-intensity pulses, monochromaticity, wavelength tunability.

**Principles Reviewed.** Adsorption and adsorbates, resonance-enhanced multiphoton ionization (REMPI), nonresonant single-photon ionization.

**Conclusions.** The two-step methodology of laser-induced desorption and ionization of adsorbates from a substrate surface effectively reduces the likelihood of thermal decomposition and structural fragmentation of the adsorbed species. Quantitative analytical measurements and sensitive detection are significantly enhanced.

## Introduction

The surface chemist is interested in the characterization of molecular and atomic species adsorbed on solid surfaces, the quantification of these species, and the elucidation of the molecular "architecture" or organization at the solid surface. An understanding at the molecular level is a prerequisite for comprehension of such surface phenomena as adsorption, adhesion, friction, catalysis, lubrication, and corrosion. Laser-assisted mass spectrometry of molecular adsorbates is able to accomplish these analytical tasks with high sensitivity and reproducibility.

## Adsorption at Solid Surfaces

Let's begin with a review of some adsorption terminology. Recall that *adsorption* is distinct from *absorption*—adsorption is the accumulation of particles at a surface, while absorption is the assimilation or incorporation of particles within a bulk substance. The *adsorbate* is the substance that adsorbs or accumulates on the solid surface and is also known as the *substrate.* The process that is the reverse of adsorption is known as *desorption.*

$$\text{Adsorbate(g)} + \text{Adsorbent(s)} \underset{\text{Desorption}}{\overset{\text{Adsorption}}{\leftrightharpoons}} \text{Covered Surface}$$

Two main types of adsorption can be distinguished—*physisorption* and *chemisorption.* In physisorption (the contraction of *physical adsorption*), the forces of adsorption are van der Waals interactions. These long-range and generally weak dispersion or dipolar forces occur over several molecular diameters from the solid surface. As a consequence, multilayer adsorption is possible. Physisorbed molecules retain their identity, although they may be conformationally distorted by the presence of the surface. No appreciable activation energy is involved with physisorption, but a slightly exothermic enthalpy change may be observed ($\Delta H_{physisorption} \approx -20$ kJ/mol). The value of $\Delta H_{physisorption}$ is more dependent on the nature of the adsorbate than on the solid surface.

In contrast, chemisorption (i.e., *chemical adsorption*) is a strong interaction where the forces of adsorption are chemical bonds (usually covalent bonds, particularly coordinate covalent bonds). Chemisorbed molecules may become fragmented upon adsorption. As chemisorption involves bond formation, an activation energy may be involved. A highly exothermic reaction ($\Delta H_{chemisorption} \approx -200$ kJ/mol) is likely, with the magnitude of the enthalpy change dependent on the nature of both the adsorbed gas and the solid surface. One of the most characteristic features of chemisorption is that adsorption leads to, at most, a monolayer of adsorbate on the solid surface to "saturate" the surface. It is important to recognize that chemisorption does not occur indiscriminately on the solid surface but at specific active sites or locations on the surface that can form strong chemical bonds. These characteristics of physisorption and chemisorption are summarized in Table 10-1.

## Laser Desorption of Adsorbates

In the previous case study, we observed that the unique coupling of laser excitation and matrix incorporation of macromolecules enables the direct production of charged high-mass species in the vapor phase for mass spectral analysis. Various additional laser-based techniques have been similarly developed that employ mass spectrometry to characterize materials adsorbed onto solid surfaces. Some of these methods rely on laser-induced sample heating to remove and volatilize the adsorbed layer(s) without fragmentation of the molecular species. Lasers in the far UV and far IR regions with short pulse widths (typically 1–100 ns) provide the rapid heating necessary to avoid pyrolytic decomposition of the adsorbate. Numerous lasers meet this requirement, including: $CO_2$ lasers at 10.6 $\mu$m; frequency-tripled or -quadrupled *Q*-switched Nd:YAG lasers with output at 355 nm and 266 nm, respectively; and excimer lasers with variable gas mixtures that emit at wavelengths ranging from about 190–350 nm. Unlike the matrix-assisted studies, laser desorption of adsorbates produces a small population of ions. In fact, the ion-to-neutral-species ratio upon desorption is typically only $10^{-5} - 10^{-3}$:1[10]. Despite the predominance of neutral species, mass spectral analysis of the ionized fraction of the desorbed material is possible. Clearly, the efficiency of analytical studies would be vastly enhanced if the dominant neutral species were ionized after desorption for subsequent mass spectral analysis. This modification would offer a further advantage, as the more abundant desorbing neutral fraction is likely to be more representative of the surface.

## Two-Step Laser Desorption/Ionization

One characterization technique uses this two-step methodology by decoupling the surface removal and analysis processes through both spatial and temporal separation of the desorption and ionization steps. In this technique of two-step laser mass spectrometry, a pulsed laser probe beam desorbs a small amount of adsorbate from the solid surface under high vacuum. After a time delay, a second high-intensity pulsed laser beam passes in close proximity to the surface to ionize, selectively or nonselectively, neutral molecules of the vaporized ejected sample. The photoions generated are accelerated and focused for time-of-flight mass spectral analysis. A single ionizing laser pulse is sufficient to record the entire mass spectrum. For nonselective (also called nonresonant) ionization of the desorbed species using a high-intensity pulsed ultraviolet laser beam, the mass spectrometer, not the probe laser, performs the chemical differentiation of the desorbed sample. For selective (resonant) ionization using a laser frequency tuned to ionize a particular molecule, the wavelength of the ionizing laser also contributes to the sample characterization.

**Table 10-1**
Distinguishing Characteristics of Physisorption and Chemisorption

| | *Physisorption* | *Chemisorption* |
|---|---|---|
| Forces of adsorption | Long-range and generally weak dispersion or dipolar forces over several molecular diameters | Chemical bonds, usually covalent bonds (particularly coordinate covalent bonds) at specific surface sites |
| Multilayer adsorption? | Possible | No, monolayer only |
| Fragmentation of adsorbate? | No (Conformational distortion possible) | Possible |
| Activation energy? | Not appreciable | Appreciable |
| Enthalpy of adsorption | $\approx -20$ kJ/mol, dependent on the nature of the adsorbate | $\approx -200$ kJ/mol, dependent on the nature of both the adsorbed gas and the solid surface |

The two-step laser technique offers distinct advantages for surface analysis. The operating parameters of the probe laser can be optimized to control the extent of desorption and to minimize molecular fragmentation upon desorption. Separation of the ionization phase from the desorption process increases the efficiency of ionization, provides the flexibility of choosing optically selective ionization or nonselective photoionization, enhances the mass resolution, and permits variable time delays between desorption and ionization. The dramatic accomplishments of this analytical method include subfemtomole (i.e., $< 10^{-15}$ mole or $\approx 10^8$ particles) quantification of molecular adsorbates,[2] picosecond time resolution of the flux of desorbing molecules,[11] and chemical analysis of metal surfaces with a 1 ppm sensitivity.[12] We present here accounts of such studies to demonstrate the powerful contributions of two-step laser mass spectrometry to surface analysis.

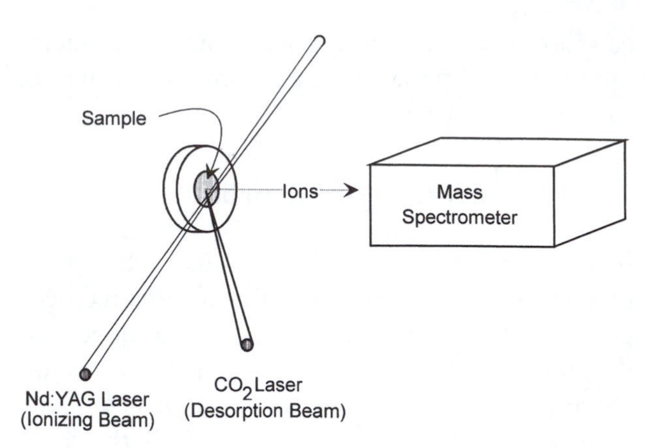


**Figure 10-4**
Schematic of the experimental setup for mass spectral analysis of solid surfaces using a two-step laser desorption/multiphoton ionization technique. A pulsed $CO_2$ laser with emission at 10.6 μm is focused on the sample and serves as the desorption beam. The output of a frequency-quadrupled Nd:YAG laser (266 nm) ionizes the desorbed species.

## Two-Step Laser Desorption/Resonant Multiphoton Ionization: Quantitative Analysis of Molecular Adsorbates on Solid Surfaces

In a recent investigation, R. N. Zare and associates[2] demonstrated the ultrahigh sensitivity and dynamic range of quantification of the two-step laser mass spectrometry technique. A schematic of the experimental setup is presented in Figure 10-4.

In these experiments the adsorbates of interest are deposited as thin films on the inner surface of a glass cup. The glass cup is mounted on a rotating Teflon rod, and these samples are then introduced into a vacuum system (with a pressure of about $10^{-7}$ torr). A pulsed $CO_2$ laser (10.6 μm, 10 μs pulse width, 10 Hz repetition rate) is focused onto the inner surface of the cup at a 45° angle. Laser irradiation induces a rapid heating ($10^8$ K/s) of the glass surface, desorbing molecules before any decomposition occurs. At about 1 cm above the surface, the escaping neutral molecules are irradiated by the 266 nm output of a frequency-quadrupled Nd:YAG laser ($\approx$ 1 mJ/pulse, 10 ns pulse width, 10 Hz repetition rate). The use of this high-powered, pulsed, and focused UV laser beam establishes a condition known as 1 + 1 resonance-enhanced multiphoton ionization (REMPI). The REMPI process involves the absorption of two photons per desorbed molecule—one photon to excite the desorbed species to an excited electronic state and the other to subsequently ionize the excited-state desorbed molecule at the focal point of the laser. The glass cup is positioned in the center of the first of two electrodes that constitute the acceleration region of a simple time-of-flight mass spectrometer. A typical mass spectrum is obtained after 200 shots of the Nd:YAG laser. The selective ionization of a particular desorbed species achieved by the REMPI process further ensures that the observed mass spectrum reflects the adsorbate under investigation.

Figure 10-5 presents mass spectra of three distinct molecules that have been obtained using the laser desorption/resonant multiphoton ionization technique: (*a*) protoporphyrin IX dimethyl ester, a precursor of many plant and blood pigments with a molecular weight of 591 g/mol; (*b*) β-estradiol, an estrogen with a MW of 272 g/mol; and (*c*) adenine, a purine base, MW 135 g/mol. Each spectrum is clearly dominated by the parent molecular ion, confirming the low yield of fragmented species upon ionization. The amount of laser sample desorbed by a laser pulse is calculated to be 5, 50, and 200 femtomole for (*a*)–(*c*), respectively. These calculations assume complete desorption of all $CO_2$-irradiated molecules, an assumption that has been subsequently experimentally demonstrated. With a typical signal-to-noise ratio of 2, these results suggest a cal-

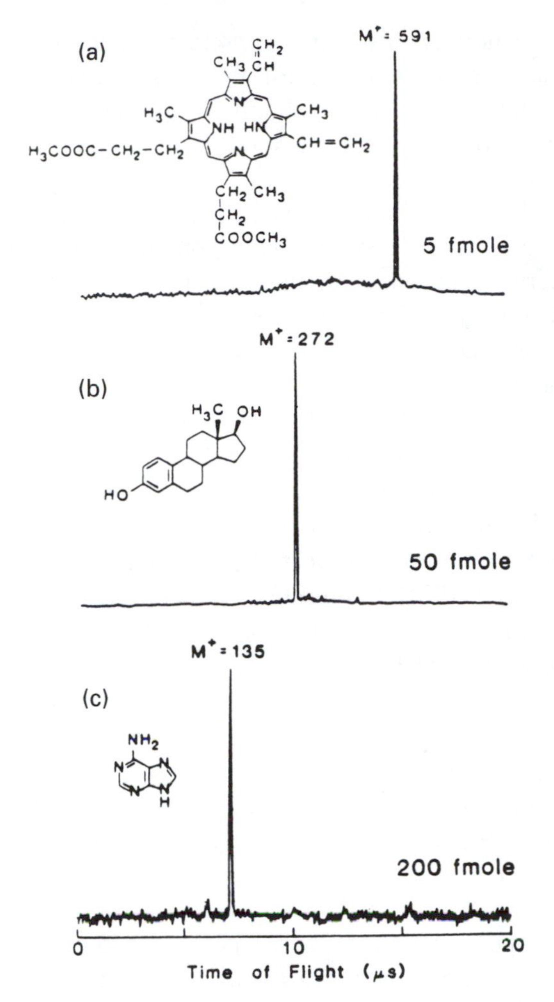


**Figure 10-5**
Mass spectra obtained via two-step laser desorption/ionization of (*a*) 5 fmol protoporphyrin IX dimethyl ester, (*b*) 50 fmol β-estradiol, and (*c*) 200 fmol adenine. The parent molecular ions $M^+$ are detected at *m/z* values of 591, 272, and 135, respectively. Reprinted with permission from *J. Am. Chem. Soc.*, 1987, 109, 2842–2843. Copyright 1987 American Chemical Society.

culated detection limit of $4 \times 10^{-17}$ mole ($\approx 2 \times 10^7$ molecules) for protoporphyrin IX dimethyl ester. Thus, the ultrasensitive and discriminatory powers of this technique make it an extremely powerful probe of solid surfaces.

An additional intriguing application of laser desorption/laser multiphoton ionization mass spectrometry is a recent quantitative analysis of the spatial distribution of organic molecules on the surface of the Allende meteorite.[13] Focusing the study on polycyclic aromatic hydrocarbons (PAHs), those organic molecules proposed to play an important role in interstellar chemistry, the investigators used the same experimental setup for mass spectral analysis as in Figure 10-4. Freshly fractured interior surfaces of the meteorite were irradiated, and desorption of nonvolatile and thermally labile compounds of molecular weights even above 1000 amu was accomplished. The mass spectra of 1 $mm^2$ spots indicate that the meteorite contained PAHs with extensively alkylated rings. Furthermore, the heterogeneous distribution of PAHs is primarily associated with the fine-grained matrix which contains organic polymer and not with the spherical chondrules. Future investigations will explore possible correlations of the organic components with various mineralogical phases and inclusions to characterize the formation and history of meteorites.

## Two-Step Laser Desorption/ Nonresonant Multiphoton Ionization: Elemental Analysis of Impurities on Metal Surfaces

Recall that nonselective (nonresonant) ionization of desorbed molecules achieves a chemically general mass spectral analysis of a surface. A powerful pulsed laser typically delivering focused power densities in excess of $10^9$ to $10^{10}$ W/$cm^2$ [14] photoionizes all types of desorbed species with approximately equal efficiencies. This provides a uniformity of detection essential for an accurate and quantitative representation of the surface composition from a single mass spectrum. As an illustration of the sensitivity and generality of this technique, Becker and Gillen[12] analyzed a sample of a National Bureau of Standards (NBS) copper C1252 standard with bulk impurities in the parts-per-million range. Although an argon ion beam source was used to desorb molecules from the copper C1252 surface, a pulsed laser beam of photons could also be used to accomplish the desorption. Desorption was estimated to remove $\approx 10^{-10}$ g of material (about $10^{-2}$ of a monolayer over $10^{-1}$ $cm^2$). The ionizing laser beam was the pulsed 248 nm output of a KrF excimer laser (focused intensity of $2 \times 10^9$ W/$cm^2$). A portion of the mass spectrum obtained is presented in Figure 10-6. The largest peaks at 126, 128, and 130 are attributed to dimers of $^{63}Cu$ and $^{65}Cu$. The peaks at 107 and 109 are consistent with the NBS impurity analysis of 98 ppm total for $^{107}Ag$

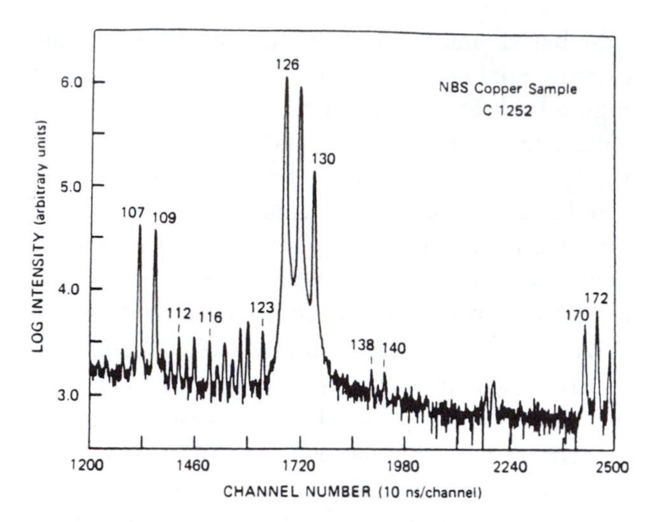


**Figure 10-6**
A segment of the mass spectrum of NBS copper C1252 standard obtained using two-step laser desorption/nonresonant multiphoton ionization. The dominant peaks at 126, 128, and 130 are attributed to dimers of $^{63}Cu$ and $^{65}Cu$; the signals at 107 and 109 are assigned to $^{107}Ag$ and $^{109}Ag$. Thus, the signals at 170, 172, and 174 are ascribed to the corresponding AgCu dimers. Tentative assignments for the weaker signals are as given in the text. Reprinted with permission from *Anal. Chem.*, 1984, 56, 1671–1674. Copyright 1984 American Chemical Society.

and $^{109}Ag$. Thus, AgCu dimers most reasonably account for the appearance of peaks at 170, 172, and 174. Although the NBS could not provide an accurate analysis of additional elements in this region of the mass spectrum, the remaining minor signals are suggestive of trace amounts of such isotopes as $^{112}Cd$, $^{114}Cd$, $^{116}Sn$, $^{118}Sn$, $^{138}Ba$, and $^{140}Ce$. With information on the relative desorption yields of different species, an accurate quantitative assessment of surface concentrations could be attained.

## Two-Step Laser Desorption/ Nonresonant Single-Photon Ionization: Surface Analysis of Bulk Organic Polymers

The two-step laser desorption/ionization technique has also been used to characterize the surface of bulk polymers.[15] With larger molecules, however, a multiphoton ionization process following desorption often leads to at least some (and often complete) nonspecific molecular fragmentation. While the use of lower laser powers in the ionization process usually reduces unwanted bond fragmentation, signal-to-noise ratios are also generally reduced. Alternatively, characteristic mass spectra with less extensive fragmentation are readily obtained with single-photon ionization induced by a vacuum ultraviolet (VUV) source.

Figure 10-7 illustrates the experimental arrangement recently used with single-photon ionization to characterize bulk organic polymer samples.[15] A pulsed argon ion beam from an ion gun is used to desorb surface material; again, a pulsed laser beam could alternatively be used. The desorbed neutral molecules are irradiated with VUV light at about 1 mm above the polymer surface. Frequency tripling of the 355 nm third harmonic of a Nd:YAG laser in a Xe/Ar mixture generates the 118 nm VUV beam with 10 ns pulse duration. After focusing both the 355 nm and 118 nm light beams through a LiF lens, the fundamental beam is dispersed and only the VUV light reaches the sample to nonselectively photoionize the desorbed molecules. An ionization efficiency of approximately 1% is generally achieved.[15] The nonselectivity of photoionization in the VUV region is a consequence of rather uniform and sizable one-photon ionization cross sections for molecules in this region.

Figure 10-8(a) presents the mass spectrum from bulk poly(methyl methacrylate) (PMMA) obtained using single-photon ionization. A number of characteristic peaks dominate the spectrum, including a peak consistent with the monomer of PMMA at $m/z = 100$, a peak indicative of the monomer minus $OCH_3$ at $m/z = 69$, and a peak consistent with $COOCH_3$ at $m/z = 59$. While the multiplicity of peaks exceeds the simplicity

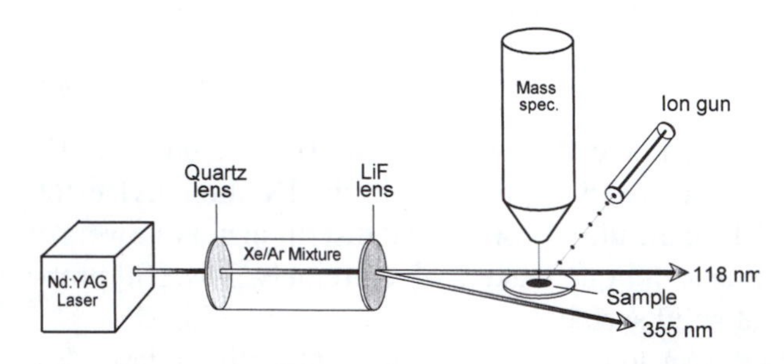


**Figure 10-7**
Schematic of the experimental setup for mass spectral analysis of solid surfaces using a two-step desorption/single-photon ionization technique. A pulsed argon ion gun serves to desorb the adsorbates. The desorbed neutral samples are ionized with the third harmonic of a Nd:YAG laser that has been frequency tripled in a Xe/Ar mixture. The resulting 118 nm emission nonselectively photoionizes the sample.

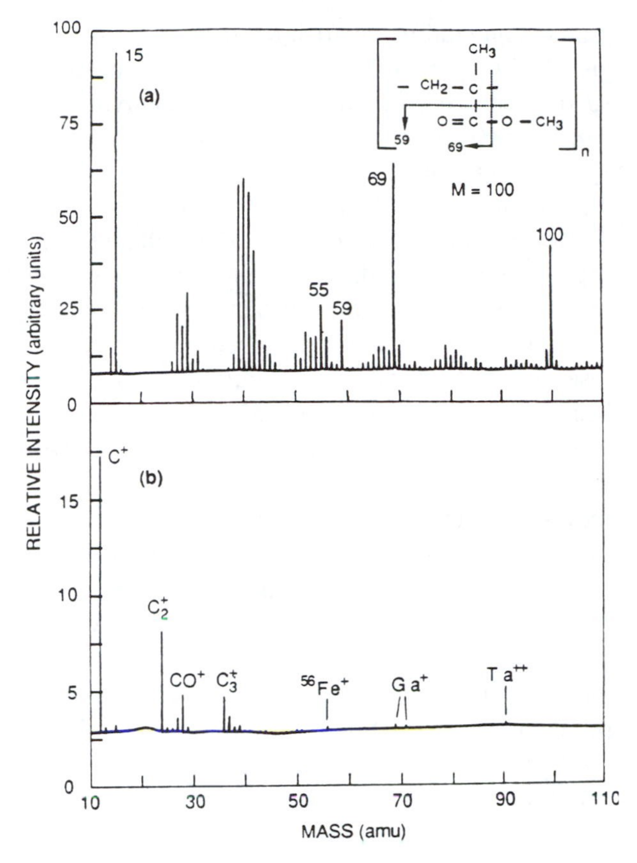


**Figure 10-8**
(*a*) Mass spectrum of poly(methyl methacrylate) obtained via single-photon ionization. Peaks ascribed to the monomer ($M^+$, $m/z = 100$), monomer minus $OCH_3$($m/z = 69$), and the functional grouping $COOCH_3$ ($m/z = 59$) dominate the spectrum. (*b*) Mass spectrum of poly(methyl methacrylate) obtained via multiphoton ionization. The spectrum consists entirely of photofragments, namely $C^+$ ($m/z = 12$), $C_2^+$ ($m/z = 24$), $CO^+$ ($m/z = 28$), and $C_3^+$ ($m/z = 36$). Reprinted with permission from *Anal. Chem.*, 1989, 61, 805–811. Copyright 1989 American Chemical Society.

of the spectra in Figure 10-5, for example, the spectrum is readily interpreted. In fact, the small degree of fragmentation observed with single-photon ionization is often an aid to structural analysis. Figure 10-8(b) illustrates the corresponding mass spectrum of PMMA using multiphoton ionization (MPI). These comparison MPI studies were conducted with 248 nm radiation generated by a KrF excimer laser and attenuated in power using a partially transmitting mirror. Despite these precautions, the MPI process resulted in almost complete fragmentation into atoms, with C and $C_2$ as the primary photofragments. Clearly, no structural information can be resolved from such a spectrum, confirming the need for single-photon ionization methods for such samples.

As a further example, the mass spectrum from bulk polystyrene obtained using single-photon ionization is displayed in Figure 10-9. While several fragmentation patterns are apparent, the peaks are easily identified. The parent monomer styrene peak at $m/z = 104$ ($C_8H_8^+ = M^+$) appears with sizable intensity, with a similarly intense peak at an $m/z$ ratio

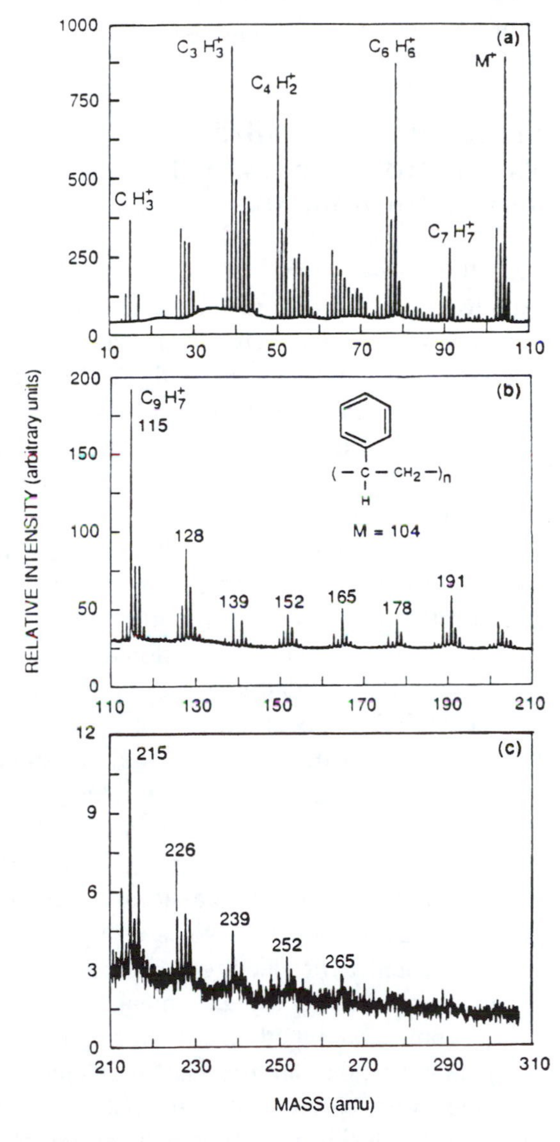


**Figure 10-9**
Mass spectrum of bulk polystyrene obtained via single-photon ionization. The parent monomer styrene ion appears at $m/z = 104$. Other peaks are assigned as specified in the text. Note the change in scale for the three diagrams. Reprinted with permission from *Anal. Chem.*, 1989, 61, 805–811. Copyright 1989 American Chemical Society.

of 78 ($C_6H_6^+$) that is especially characteristic of aromatic polymers. While the peaks at 91 ($C_6H_5CH_2^+$) and 115 ($C_9H_7^+$) are also consistent with aromatic polymers, the reduced intensity of these signals compared with that of $M^+$ eliminates any possible ambiguity in the determination of monomer structure. The progression of signals between $m/z = 115$ and 191 (note the change in relative intensity scales) is attributed to a $[C_9H_7 + (CH)_n]^+$ structure. The 9-carbon fragment is characteristic of a 6-membered ring fused to a 5-membered ring, resulting from an unspecified structural rearrangement.

## The Potential and Realized Contributions of Lasers to Mass Spectrometric Analysis

How will the advent of materials characterization via two-step laser mass spectrometry aid industrial and research chemists? Many chemical reactions and technological applications are critically dependent on the localized chemical composition of solid surfaces. For example, the catalytic activity of metals arises from the adsorption of molecules at specific sites on the metal surface. The type of crystal face, the interatomic spacings on the surface, and the heterogeneity of surface sites play a dramatic role in such processes as the adsorption of $N_2$ and $H_2$ onto iron to facilitate the synthesis of $NH_3$ and the adsorption of $SO_2$ onto platinum to promote oxidation to $SO_3$. In optics research, the varied composition of optical coatings and advances in their deposition on solid substrates afford numerous applications for both industrial and consumer products. Coating formulations are developed for such specific functions as high-resolution coatings for magneto-optical data storage, damage-resistant coatings for optical devices in high-power laser systems, and antireflecting coatings for eyeglasses and ophthalmological equipment.

More sophisticated methods of chemical analysis and quantitative measurements of molecules adsorbed onto solid surfaces would enable further advances in the design of efficient heterogeneous catalysts and optimum surface coatings. These analytical capabilities would also aid in the development of materials in other surface-related capacities such as lubricants, surfactants, and electrodes and in the study of the dynamics and mechanisms of surface reactions including corrosion, adhesion, and catalysis. Clearly, the methodology of laser-assisted mass spectroscopy is a versatile and viable analytical tool. The laser has played several key roles in these mass spectral analyses of macromolecules and surfaces. To accomplish the nondestructive, nonspecific desorption of adsorbed molecules or the volatilization of high molecular weight compounds, the high-power density of the focused laser beam is required for a rapid localized change in temperature. In two-step laser mass spectroscopy, a pulsed ionizing laser beam accomplishes the ionization over a short time interval, minimizing the time for charge redistribution in the molecule, which often leads to fragmentation. For efficient ionization of desorbed surface molecules, both the pulsed nature and the focused power of laser light ensure complete and nondiscriminating ionization. On the other hand, the wavelength tunability of an ionizing laser affords selectivity and quantitative sensitivity for elemental analysis. Temporal analyses of the desorbed molecules and of the distribution of their kinetic energies are also possible using various time delays between desorbing and ionizing laser beams.[11] These distinguishing features of the laser light source make the coupling of laser desorption/ionization to mass spectrometry an extremely powerful means of macromolecular and surface analysis.

## FOR FURTHER EXPLORATION

The mass range available for the mass spectroscopic analysis of biomolecules has been dramatically extended by the matrix-assisted laser desorption/ionization (MALDI) technique. Despite the successes of the method, scientists have been hampered by the limited ability to inject liquid samples directly into the spectrometer to couple liquid separation techniques (e.g., liquid chromatography, capillary electrophoresis) with mass analysis. Only a few liquid matrices have been employed[17,18] with quite restrictive flow rates of less than 5 $\mu$L/min. One promising method of direct liquid introduction is the aerosol method, whereby liquid streams of matrix and analyte flowing at a rate of up to 1 mL/min are sprayed into a mass spectrometer along with a carrier gas (e.g., nitrogen) to form aerosol particles—dispersions of liquid particles in a gas phase. The stream of aerosol particles is heated by a 500°C drying tube to remove solvent, simultaneously dried, and then directed to intersect at 90° with a pulsed laser beam. The continuous aerosol stream and the pulsed laser irradiation to generate ions lead to excessive sample consumption, but increases in the repetition rate of the laser can considerably reduce sample consumption. One study employing a 10 Hz pulsed $Nd^{3+}$:YAG laser frequency-tripled to 355 nm reported a mass resolution of over 300 for a series of peptides.[19] Early aerosol MALDI results show considerable promise for the technique as an interface between liquid separation and mass spectrometry, and thus modifications for enhanced sensitivity and resolution will improve the prospects of this method as a future analytical technique.

## DISCUSSION QUESTIONS

1. By what reasoning can the peaks in the mass spectrum of Figure 10-1 be assigned? Why is the $3M^+$ peak not considered the parent molecular ion $M^+$?
2. What relationship is observed between the molecular weight of a parent ion and the time-of-flight recorded in Figure 10-5?
3. Explain the differences between optically selective ionization and nonselective photoionization. In what circumstances would each technique be preferable?
4. In addition to analysis of surfaces and nonvolatile macromolecules, laser-assisted mass spectroscopy has been employed in the study of combustion reactions. What advantages would this technique offer to the characterization of combustion processes?

## LITERATURE CITED

1. Schilke, D., and Levis, R. J., "A laser vaporization, laser ionization time-of-flight mass spectrometer for the probing of fragile biomolecules," *Rev. Sci. Instrum.*, 1994, 65, 1903–1911.
2. Hahn, J. H., Zenobi, R., and Zare, R. N., "Subfemtomole quantification of molecular adsorbates by two-step laser mass spectrometry," *J. Am. Chem. Soc.*, 1987, 109, 2842–2843.
3. Karas, M., Ihgendon, A., Bahr, U., and Hillenkamp, F., "Ultraviolet-laser desorption/ionization mass spectrometry of femtomolar amounts of large proteins," *Biomed. Environ. Mass Spectrom.*, 1989, 18, 841–843.
4. Karas, M., Bachmann, D., Bahr, U., and Hillenkamp, F., "Matrix-associated ultraviolet laser desorption of non-volatile compounds," *Int. J. Mass Spectrom. Ion Processes*, 1987, 78, 53–68.
5. Karas, M., and Hillenkamp, F., "Laser desorption ionization of proteins with molecular masses exceeding 10000 daltons," *Anal. Chem.*, 1988, 60, 2299–2301.
6. Overberg, A., Karas, M., Bahr, U., Kaufmann, R., and Hillenkamp, F., "Matrix-assisted infrared laser (2.94 m$\mu$) desorption/ionization mass spectrometry of large biomolecules," *Rapid Commun. Mass Spectrom.*, 1990, 4, 293–296.
7. Nelson, R. W., Rainbow, M. J., Lohr, D. E., and Williams, P., "Volatilization of high molecular weight DNA by pulsed laser ablation of frozen aqueous solutions," *Science*, 1989, 246, 1585–1587.
8. Zhao, S., Somayajula, K. V., Sharkey, A. G., Hercules, D. M., Hillenkamp, F., Karas, M., and Ingendoh, A., "Novel method for matrix-assisted laser mass spectrometry of proteins," *Anal. Chem.*, 1991, 63, 450–453.
9. Borman, S., "Biochemical applications of mass spectrometry take flight," *Chem. Eng. News*, 1995, 73(25), 23–35.
10. Williams, P., and Sundqvist, B., "Mechanism of sputtering of large biomolecules by impact of highly ionizing particles," *Phys. Rev. Lett*, 1987, 58, 1031–1034.
11. Eleftheriadis, N., Ludescher, H.-P., Sussbauer, M., and von der Linde, D., "Picosecond time-resolved laser desorption," *Appl. Surf. Sci.*, 1990, 46, 284–287.
12. Becker, C. H., and Gillen, K. T., "Surface analysis by nonresonant multiphoton ionization of desorbed or sputtered species," *Anal. Chem.*, 1984, 56, 1671–1674.
13. Zenobi, R., Philippoz, J. M., Buseck, P R., and Zare, R. N., "Spatially resolved organic analysis of the Allende meteorite," *Science*, 1989, 246, 1026–1029.
14. Daiser, S. M., Welkie, D. G., and Becker, C. H., "Surface analysis by laser ionization," *American Laboratory*, 1990, 22(1), 54–59.
15. Palliz, J. B., Schuhle, U., Becker, C. H., and Huestis, D. L., "Advantages of single-photon ionization over multiphoton ionization for mass spectrometric surface analysis of bulk organic polymers," *Anal. Chem.*, 1989, 61, 805–811.

16. Engelke, F., Hahn, J. H., Henke, W., and Zare, R. N., "Determination of phenylthiohydantoin-amino acids by two-step laser desorption/multiphoton ionization," *Anal. Chem.*, 1987, 59, 909–912.

17. Li, L., Wang, A. P. L., and Coulson, L. D., "Continuous-flow matrix-assisted laser adsorption ionization mass spectrometry," *Anal. Chem.,* 1993, 65, 493–495.

18. Williams, E. R., Jones, Jr., G. C., Fang., L., Nagata, N., and Zare, R. N., "Laser-desorption tandem time-of-flight mass spectrometry with continuous liquid production," in *Applied Spectroscopy in Materials Science II* (Golden, W. G., Ed.), Proc. SPIE 1636, Los Angeles, CA, 1992, pp. 172–181.

19. Fei, X., Wei, G., and Murray, K. K., "Aerosol MALDI with a reflection time-of-flight mass spectrometer," *Anal. Chem.,* 1996, 68, 1143–1147.

CHAPTER 11

# Photoacoustic Thermal Characterization of Liquid Crystals

## Chapter Overview

The absorption of light, particularly the intense monochromatic light of laser excitation sources, will often result in nonradiative heating of a sample. The extent of heating will vary in accordance with the substance's heat capacity and thermal conductivity and the laser's intensity and irradiation time. The heating effect resulting from optical absorption of laser light often complicates a chemical investigation requiring isothermal conditions. In this chapter, however, we investigate a practical application of periodic heating of a sample achieved through the use of pulsed or modulated laser radiation. The phenomenon known as the photoacoustic effect accompanies such periodic heating to generate audible pressure pulses whose frequency matches the pulsing or chopping frequency of the laser. We briefly examine the innovative use of the photoacoustic effect to characterize the thermodynamic properties of liquid crystalline materials.

## Photophones

> . . . Andropov was a bit tired and more nervous than he hoped he showed. From the broad plate-glass window that formed almost one wall of his office he could see the sun pouring down into Dzerzhinskaya Square. . . . The curtains were open only a few inches—the sun flooding through their translucence—but if anyone entered to talk with him, Andropov would certainly have drawn them completely to prevent detection of his conversation by laser-monitoring of the plate-glass vibrations.
>
> H. Salisbury, *Gates of Hell*, Random House, New York, 1975

The observation that objects become warm, even hot, when exposed to light, is certainly not a new discovery. Knowing how to quantify the amount of heat produced and knowing what factors affect the quantity of heat produced are relatively new advances, however. In the 1880s, Alexander Graham Bell showed that sunlight, modulated in time and focused on an absorbing medium, produced audible sounds.[1,2] Crucial to this observation was the time modulation of the light, for by interrupting (in time) the energy falling on the surface of the absorber, the heat effect in the sample became periodic in time. When photons are absorbed by a sample, not all of their energy is reemitted as other photons; instead, some of the energy is converted by so-called nonradiative processes into heat in the sample. The periodic heat waves produced in the sample are responsible for a number of observable effects. Today the operation of Bell's "photophone" (or Andropov's window) is understood by recognizing that periodic heating of a solid sample generates pressure pulses in the gas at the solid sample/atmosphere interface. Those pressure waves are heard as sound. This phenomenon is now known as the photoacoustic or optoacoustic effect and is the basis for a wide variety of spectroscopic techniques and, most recently, for thermal characterizations of samples as well.

## Photoacoustics

How do lasers come into the photoacoustic picture? The heat waves generated in a sample depend on the periodicity of the incident photon flux. Lasers provide intense monochromatic light that is either inherently periodic in time, such as a pulsed nitrogen laser, or that can be modulated in time by a mechanical "chopper" or by other means. Thus, pulsed or modulated lasers are well-suited to photoacoustic investigations. The object of this chapter will be to explore how lasers can aid in the photoacoustic thermal characterization of a substance. As might be imagined, the heat produced in a substance depends critically on the heat capacity ($C_p$), a static (time-independent) thermodynamic property, as well as on the thermal conductivity ($\kappa$) of the sample, a dynamic (time-dependent) thermodynamic property. Thermal characterization will entail the quantitative determination of $C_p$ and $\kappa$ of a sample through the study of laser-generated heat pulses.

Figure 11-1 shows a typical arrangement of a laser, sample cell, and detector for observing the photoacoustic effect in a solid.[2] A brief description of how this apparatus works is as follows. The solid sample is contained within a closed cell containing a gas that has free passage to the face of an acoustic microphone. The microphone acts as a pressure transducer to convert a pressure change to a capacitance

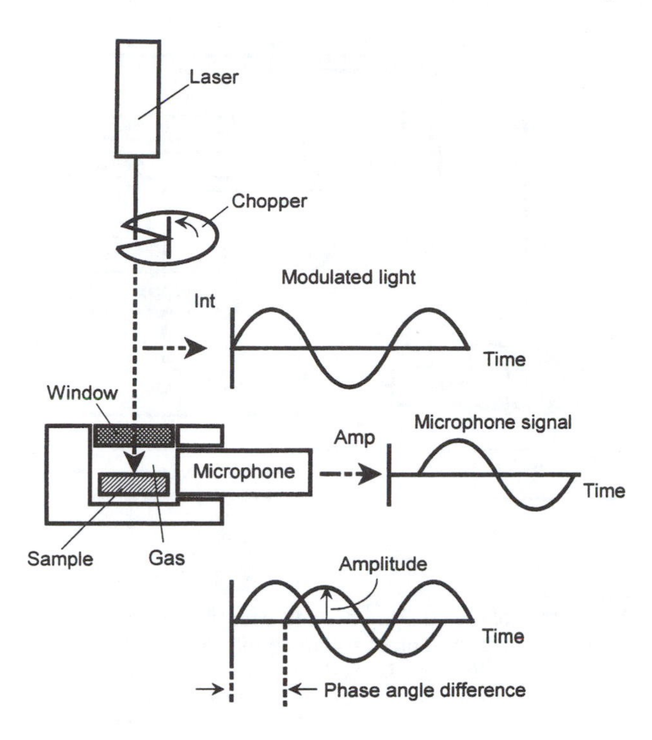


**Figure 11-1**
A typical arrangement of laser, sample cell, and detector for observing the photoacoustic effect in a solid. Time-modulated laser light strikes the sample enclosed in a gas-filled cell. Absorption of light by the sample leads to nonradiative heating of the sample and the surrounding gas. The periodic heating of the gas generates sound wave pressure pulses detected by the microphone. The microphone signal appears with a frequency matching the pulsing or chopping frequency of the laser. The photoacoustic signal generated with sinusoidally modulated incident light is characterized by an amplitude modulation and phase angle shift reflecting the time delay of the output signal relative to the incident time-modulated light.

change. Laser light strikes the surface of the sample and is absorbed by it. As mentioned earlier, a portion of that energy is captured by nonradiative processes to generate heat within the sample. Now recall the critical requirement of time-modulated light. If the laser light of a given initial amplitude (intensity) is chopped (or inherently pulsed), the gas above the sample will subsequently be heated in such a way as to produce pressure pulses whose frequency is that of the laser chopping frequency. These pressure pulses are detected by the acoustic microphone. The output signal will have the same time variation as the laser chopping frequency but will be delayed in time, that is, phase shifted. The fundamental data of this experiment will be the amplitude of the microphone signal and its time delay or phase angle relative to the incident time-modulated laser light, as shown in Figure 11-1. The time delay of the output signal can easily be imagined to arise from the finite-time nonradiative processes that generate heat within the sample. The time for these processes to occur (phase angle) and the extent to which they occur (amplitude) are dependent on bulk thermodynamic properties of the sample absorbing the light: density, heat capacity, thermal conductivity, and optical absorptivity. It is through the dependence of the output signal on heat capacity and thermal conductivity that thermodynamic information can be extracted from the data.

Nevertheless, to extract thermodynamic values from the deceptively simple photoacoustic data requires a rather complex theoretical model. The most commonly used model was formulated by Rosencwaig and Gersho[3] and presented pictorially in Figure 11-2. Of prime consideration are how and where the heat generated by these nonradiative processes can flow in the sample, the gas, and the backing (sample holder). A series of differential equations with boundary conditions is established to describe the heat flows in and across these phases. The equations are solvable, but the details of the solutions will not be discussed here. The interested reader is referred to the original papers and other later expositions of the theory.[4,5] The results give a simple expression for the amplitude of the photoacoustic signal:

$$Q = qe^{-i\Psi} \tag{107}$$

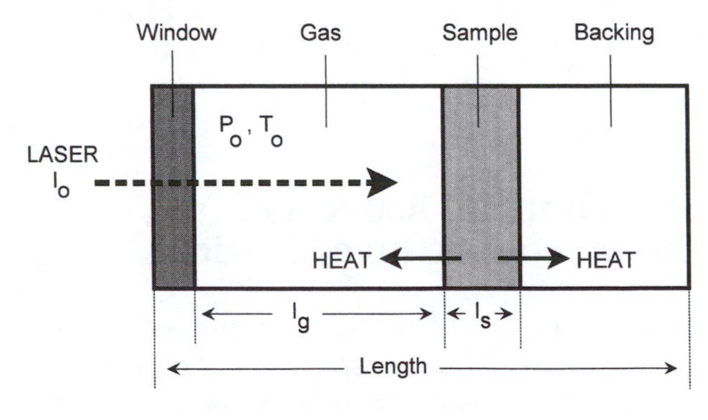


**Figure 11-2**
A model of the heat flow in and across the phases of the photoacoustic sample cell: optical window, surrounding gas medium, sample, and backing (sample holder).

where $q$ is the effective amplitude and $\Psi$ the phase angle. Moreover, $q$ and $\Psi$ are given by:

$$q = \frac{(1 - R)\gamma_g P_o I_o t(2t^2 + 2t + 1)^{-\frac{1}{2}}}{2\sqrt{2}T_o l_g (1 + s)\kappa_g \left(\frac{1}{\mu_g}\right)^2} \tag{108}$$

and

$$\tan\Psi = \left(\frac{1 + t}{t}\right) \tag{109}$$

where $R$ is the sample's optical reflectivity, $\gamma_g$ is $C_p(gas)/C_V(gas)$, $P_o$ is the ambient pressure of the gas, $I_o$ is the incident laser intensity (amplitude), $T_o$ is the ambient gas temperature, $l_g$ is the gas thickness, $\kappa_g$ is the thermal conductivity of the gas, and $\mu_g$ is the thermal thickness of the gas. (One comment on notation: Subscript $g$ refers to the gas phase and no subscript refers to the sample. As it turns out, the equations are independent of the backing material.) The terms $s$ and $t$ have been introduced to simplify the equations and are given by $s = \mu\beta/2$ and $t = [(\mu_g/\mu)(\kappa/\kappa_g)]$, where $\beta$ is the sample's optical absorptivity. Equations 108 and 109 are valid for what are called optically opaque and thermally thick samples. In such samples most of the light that is absorbed is converted into heat in a thickness small compared to the sample thickness, and very little light actually passes through the sample. Just as water waves die out because of the damping action of viscosity, the heat "waves" generated in the sample also diminish from what is called thermal resistance in a distance of travel called the thermal thickness, $\mu$. In this case we assume that the thermal thickness $\mu$ is much less than the sample thickness $l_s$.

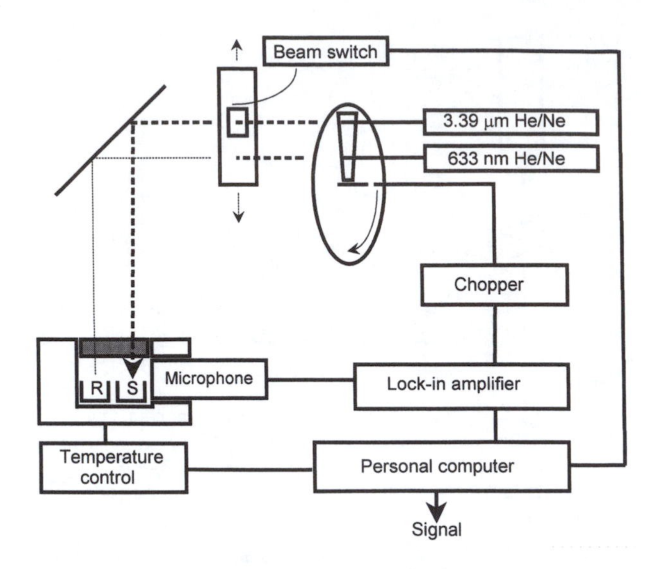


**Figure 11-3**
A schematic of the experimental setup for photoacoustic thermal characterization of a solid sample. Modulated 3.39 $\mu$m light from a He/Ne laser is incident on a sample to heat the sample and surrounding gas and generate a photoacoustic signal. A reference 633 nm He/Ne beam is incident on a reference to characterize the thermal properties of the sample cell and surrounding gas. The time-delayed acoustic microphone output signal is detected by a lock-in amplifier and analyzed by a computer.

## Experimental Aspects of Photoacoustic Investigations

At this point, it is useful for us to discuss some details of a photoacoustic experiment. An overall experimental setup is shown in Figure 11-3. Recall that the goal of a photoacoustic investigation is to determine the heat capacity and thermal conductivity of the sample, each as a function of temperature, from the photoacoustic signal. In practice, a convenient modulation frequency is selected, and then the photoacoustic amplitude and phase angle are measured at various temperatures. A lock-in amplifier or phase-sensitive detector is critical to the experimental measurements, as a consequence of this device's almost incredible ability to detect signals buried inside very noisy signals (almost no pun intended). In fact, signals can often be detected when the amplitude of the noise may be from $10^3$ to $10^7$ times greater than the signal itself. In practice, a dual channel lock-in amplifier is used, one channel to detect the amplitude, and the other to detect the phase angle (time delay) of the signal. Look again at Figure 11-1 to appreciate the relationship between the incident modulated laser light signal and the time-delayed acoustic microphone output signal.

Although it is indeed possible to make absolute measurements of material constants in Equations 108 and 109 ($R$, $\gamma_g$, etc.), usually the use of a reference substance is the preferred means to obtain the sample photoacoustic amplitude $q$ and the sample phase angle $\Psi$. The sample values can be related to reference and measured values by:

$$q(T) = q_{\text{ref}}\left(\frac{q^m(T)}{q^m_{\text{ref}}(T)}\right) \qquad (110)$$

$$\Psi(T) = \Psi_{\text{ref}} + [\Psi^m(T) - \Psi_{\text{ref}}(T)]$$

where the superscript $m$ refers to an actual experimentally measured value.[6,7] Here $q_{\text{ref}}$ and $\Psi_{\text{ref}}$ are usually obtained from known properties of the reference material using Equations 108 and 109 or from the sample itself under special conditions.[7]

To analyze the data we need to observe that Equations 108 and 109 are really just two equations expressed in two unknowns, $s$ and $t$, assuming the other constants are specified by using the reference compound described above. Solving the equations for $s$ and $t$ at each value of temperature for which data have been collected yields $s(T)$ and $t(T)$ and consequently $C_p(T)$ and $\kappa(T)$ because of the following relationships:

$$C_p(T) = \left(\frac{2}{\omega}\right)\left(\frac{1}{\rho(T)}\right)\left[s(T)\left(\frac{\kappa_g}{\mu_g}\right)\right]\left(\frac{\beta(T)}{2t(T)}\right) \qquad (111)$$

$$\kappa(T) = \left(\frac{2s(T)t(T)}{\beta(T)}\right)\left(\frac{\kappa_g}{\mu_g}\right) \qquad (112)$$

The optical absorptivity $\beta$ should be considered as analogous to the molar absorptivity parameter $a$ in Beer's Law: $A = abC$. The parameter $\rho$ is the sample density, and $\omega$ is the modulation frequency of the incident light.[8]

## Photoacoustic Thermal Characterization of Liquid Crystals

Liquid crystals are intriguing substances for a variety of reasons. In particular, these substances are unusual as liquids with some degree of microscopic ordering, that is, individual molecules having some orientational and/or positional relationship to each other that extends in space over many molecular lengths. This ordered structure and the ability to control the extent of order by the application of pressure, temperature, or external electric or magnetic fields have given liquid crystals great commercial value. A second intriguing aspect of liquid crystals is the observation that a given substance may exhibit many phases as a function of temperature. Thus, numerous calorimetric studies of such materials are conducted to ascertain the heat capacities of the phases and the enthalpy changes associated with phase transitions between the liquid crystalline phases.

One of the useful aspects of the photoacoustic technique is that in one experiment both heat capacity, a static thermal property, and thermal conductivity, a dynamic thermal property, are easily measured. While many measurements of their heat capacities exist for liquid crystals, few measurements of their thermal conductivities are available. The case study below describes a photoacoustic thermal characterization of a binary liquid crystalline mixture to obtain both thermodynamic parameters.

## Case Study: Photoacoustic Thermal Characterization of Liquid Crystals

### Overview of the Case Study

**Objective.** To measure the temperature dependence of both the heat capacity and the thermal conductivity of two homologous liquid crystalline compounds and to determine the binary phase diagram of these substances.

**Laser Systems Employed.** An infrared 3.39 $\mu$m He/Ne laser and a normal 633 nm red He/Ne laser are suitable for heating the sample and providing a reference signal, respectively.

**Role of the Laser Systems.** To provide intense monochromatic light that is either pulsed in nature or that can be modulated in time to induce a photoacoustic signal.

**Useful Characteristics of the Laser Light for this Application.** Intensity, monochromaticity, pulsed output.

**Principles Reviewed.** Photoacoustic effect, heat capacity, binary phase diagrams.

**Conclusions.** The photoacoustic approach to the determination of binary phase diagrams is a sensitive and rapid technique for the acquisition of thermodynamic data.

## Liquid Crystal Phase Transitions

Calorimetric studies are often conducted on pure liquid crystal materials to determine the temperatures and enthalpies of phase transitions. The results of one photoacoustic thermal characterization study[6,7,9] are presented in Figure 11-4 which shows $C_p(T)$—Figure 11-4(a)—and $\kappa(T)$—Figure 11-4(b) for the compound 8CB, or 4–octyl-4′–cyanobiphenyl. Most liquid crystal displays are comprised of homologous compounds of this "classic" liquid crystal. The two observed heat capacity phase anomalies reflect the phase transitions involving the liquid crystalline smectic A and nematic phases of this compound. A nematic liquid crystal phase is a liquid phase characterized by orientational order where the molecules are ordered relative to each other so that they more or less point in the same direction. The smectic A phase is characterized by both orientational order and positional order. In the smectic A phase, molecules form layers such that all molecules point in the same direction (orientational order) with all of their centers of mass in the same plane (positional order), but within the layers the positions of the molecules are random. The transition detected around 32°C is the phase transition between the smectic A and nematic phases. The transition observed around 40°C is the phase transition between the nematic and isotropic phases. The isotropic phase is the normal, random liquid. One striking result is immediately evident when the $C_p$ and $\kappa$ results are compared. While the $C_p$ data exhibit two anomalies indicative of two phase transitions (smectic A $\rightarrow$ nematic and nematic $\rightarrow$ isotropic), the $\kappa$ data exhibits only one such anomaly. No discontinuity occurs between the thermal conductivity of the liquid crystalline phases, while a discontinuity in $\kappa$ is observed for the transition from the nematic liquid crystalline phase to the isotropic phase. One practical conclusion of these results is not to rely on thermal conductivity measurements for the detection of phase transitions in liquid crystals. Why does the thermal conductivity of the smectic A phase pass smoothly into that of the nematic and vice versa? This observation still awaits an explanation.

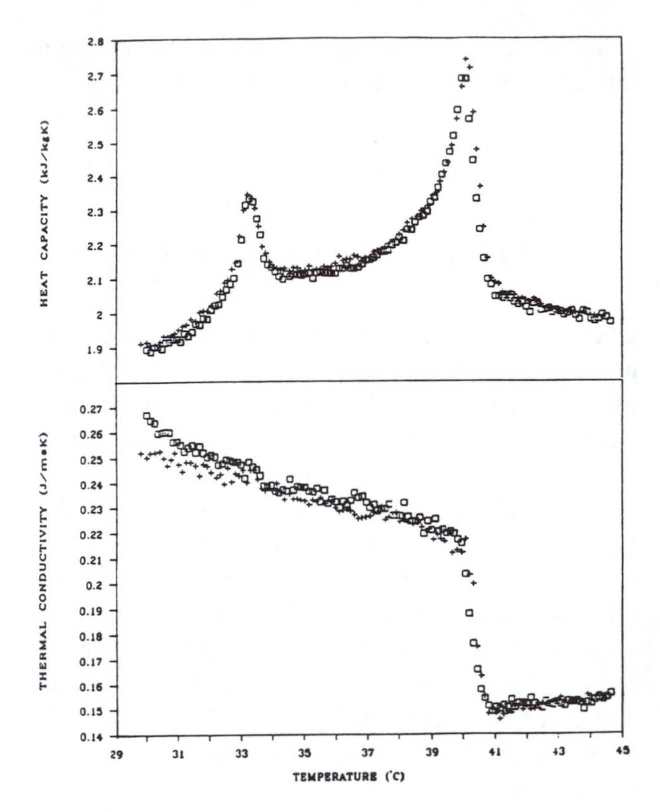


**Figure 11-4**
Temperature dependences of the heat capacity $C_p$ (upper diagram) and thermal conductivity $\kappa$ (lower diagram) of the liquid crystal 8CB derived from photoacoustic measurements. Each □ denotes data in the absence of a magnetic field; each + denotes data in the presence of a magnetic field perpendicular to the sample surface. The transition in the heat capacity at 32°C denotes the transition between the smectic A and nematic liquid crystalline phases. An analogous transition is absent in the thermal conductivity data. Both $C_p$ and $\kappa$ measurements reflect the transition between the nematic liquid crystalline phase and the isotropic liquid phase at 40°C. Reprinted with permission from Thoen, J., Schoubs, E., and Fagard, V., "Photoacoustics Applied to Liquid Crystals," in *Physical Acoustics*, O. Leroy & M. A. Breazeale (eds.), © 1991 by Plenum Publishing Corporation, New York.

## Binary Mixtures: Phase Diagrams

In addition to studies of pure compounds, calorimetric studies are also directed toward binary mixtures of liquid crystals, particularly in light of the great interest in binary isobaric temperature-composition phase diagrams (i.e., two-component $T$–$x$ phase diagrams at constant pressure). Figure 11-5 shows "raw" ampli-

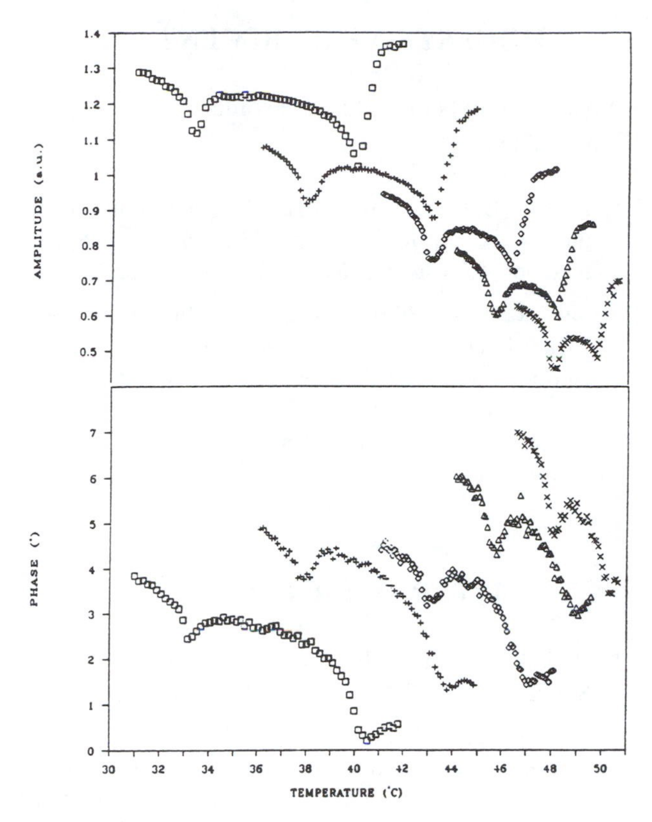


**Figure 11-5**
The photoacoustic amplitude and phase angle data as a function of temperature for various binary mixtures of the liquid crystals 8CB and 9CB. The crosses and the squares are, respectively, pure 8CB and 9CB. For the mixtures, the percentages of 8CB are: 70.0 for the pluses, 33.8 for the diamonds, 16.6 for the triangles. Reprinted with permission from Thoen, J., Schoubs, E., and Fagard, V., "Photoacoustics Applied to Liquid Crystals," in *Physical Acoustics*, O. Leroy & M. A. Breazeale (eds.), © 1991 by Plenum Publishing Corporation, New York.

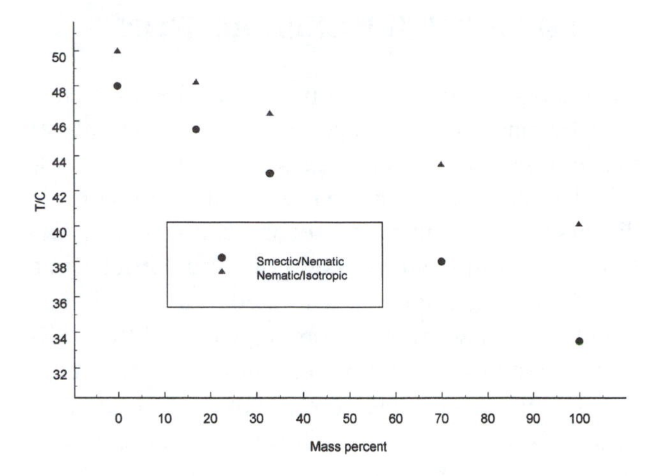


**Figure 11-6**
The binary isobaric phase diagram of 8CB and 9CB derived from the photoacoustic measurements presented in Figure 11-5. Transitions between the smectic A and nematic liquid crystalline phases are denoted by circles and between the nematic liquid crystalline phase and the isotropic liquid phase by triangles.

tude and phase angle data as functions of temperature for various fixed composition mixtures of two cyanobiphenyl-based liquid crystals: 9CB, or 4–nonyl–4′–cyanobiphenyl, and 8CB. The $C_p$ and $\kappa$ data for 8CB are given in Figure 11-4 as discussed. Even without converting the amplitude and phase angle data to heat capacity and thermal conductivity data, it is quite easy to construct the phase diagram shown in Figure 11-6 by essentially reading the phase transition temperatures from the temperature locations of the local minima of the signals shown in Figure 11-5. While the determination of binary phase diagrams is often a tedious and difficult experimental project, here the laser in the photoacoustic experiment has made light of a difficult task.

The information in the phase diagram has very practical significance to those interested in making mixtures suitable for liquid crystal displays. For example, the phase diagram indicates that the smectic A, nematic, and isotropic phases are completely miscible in all proportions within certain temperature ranges. Thus, if certain physical properties of the smectic A phase were desirable for an application, the phase diagram suggests that any mixture of 8CB and 9CB in the temperature range 32–47°C could be used. Since physical properties such as heat capacity and thermal conductivity vary with temperature and composition, the designer of the mixture has a wide range of choices for the mixture best suited to the application. Potentially severe restrictions exist on mixture design when phases are not completely miscible.

## Summary of Results

The use of the laser in the photoacoustic thermal characterization experiment provides calorimetric results whose sensitivity compares favorably with alternating current scanning calorimetric measurements. The ability to obtain dynamic thermodynamic information (the thermal conductivity) in combination with the static calorimetric information (the heat capacity) makes the experiment unique. The small samples and short measurement times represent additional experimental advantages not to be overlooked.

## FOR FURTHER EXPLORATION

An interesting phenomenon that can be characterized by photoacoustic spectroscopy is the sol-gel phase transition. Sols and gels are examples of colloidal systems in which particles of one phase are dispersed in a medium of a second phase. A sol is defined as a colloidal dispersion of solid particles in a liquid. A gel is often characterized as a cross-linked structure formed by the interconnection of the originally isolated solid colloidal particles to form a continuous dispersed phase throughout the continuous liquid dispersing medium. During the sol-gel transition, or gelling process, a material forms with the mechanical properties of solids, yet with a flexible and elastic nature as a consequence of the liquid matrix. As you might surmise, substantial interest exists in the use of sol-gel systems as alternatives to polymers. Scientists have demonstrated that photoacoustic spectroscopy provides direct and sensitive evidence of the gelation process of two classic sol-gel systems—a tetramethoxysilate/methanol system and a titanium(IV) butoxide/butanol-based mixture.[10] By frequency doubling and tripling the 1064 nm fundamental emission of a pulsed $Nd^{3+}$:YAG laser, wavelengths that are transparent to the gels under study are obtained (532 nm and 354 nm, respectively). Unlike other approaches to the study of sol-gels, the photoacoustic signal produced contains both species-specific information (e.g., number and size of siloxane polymers during gelation) as well as evidence of macroscopic gel properties (e.g., viscosity, average molecular weight, transmittance of light). This simultaneous characterization of both microstructure and global material properties will likely be a capability exploited in future applications.

## DISCUSSION QUESTIONS

**1.** While the effect was not discussed, what was the purpose of measuring the photoacoustic signal in the presence and absence of a magnetic field?

**2.** The laser used in these experiments is one whose output is in the infrared energy region equivalent to about 3350 $cm^{-1}$. Why is this a particularly good energy to use on organic materials?

## SUGGESTED EXPERIMENTS

Two experiments that illustrate some of the concepts discussed here are:

1. Erskine, S. R., and Bobbitt, D. R., "Determination of the pK of an Indicator by Thermal Lens Spectroscopy," *J. Chem. Educ.*, 1989, 66, 354–357.
2. Salcido, J. E., Pilgrim, J. S., and Duncan, M. A., "Time-Resolved Thermal Lens Calorimetry with a Helium-Neon Laser," *Physical Chemistry: Developing a Dynamic Curriculum* (R. C. Schwenz and R. J. Moore, Eds.), American Chemical Society, Washington, D.C., 1993.

## LITERATURE CITED

1. Bell, A. G., "On the Production and Reproduction of Sound by Light," *Am. J. Sci.*, 1880, 120, 305–324.
2. Luscher, E., "Photoacoustic effect in condensed matter—historical development," in *Photoacoustic Effect Principles and Applications* (E. Luscher, et al, Eds.), Fried. Vieweg & Sonn, Braunschweig, 1984, 1–20.
3. Rosencwaig, A., and Gersho, A., "Theory of the photoacoustic effect with solids," *J. Applied Phys.*, 1976, 47, 64–69.
4. Rosencwaig, A., "Solid state photoacoustic spectroscopy," in *Optoacoustic Spectroscopy and Detection* (Y. Pao, Ed.), Academic Press, New York, 1977, 193–238.
5. Korpium, P., "Thermodynamic models of the photoacoustic effect," in *Photoacoustic Effect: Principles and Applications* (E. Luscher, et al., Eds.), Fried, Vieweg & Sonn: Braunschweig, 1984, 40–51.
6. Thoen, J., Glorieux, C., Schoubs, E., and Lauriks, W., "Photoacoustic thermal characterisation of liquid crystals," *Mol. Cryst. Liq. Cryst.*, 1990, 191, 29–35.
7. Thoen, J., Schoubs, E., and Fagard, V., "Photoacoustics applied to liquid crystals," *Symposium on Physical Acoustics*, Leuven, June 1990.
8. Strictly $\beta$, $\rho$, $\kappa_g$, and $\mu_g$, are all functions of temperature but are assumed to be independent of temperature here.
9. Glorieux, C., Schoubs, E., and Thoen, J., "Photoacoustic characterization of liquid crystal phase transitions," *Mat. Sci and Eng.*, 1989, A122, 87–91.

10. Puccetti, G., and Leblanc, R. M., "Direct evidence of the liquid-solid transition of a sol-gel material using photoacoustic spectroscopy," *J. Phys. Chem.,* 1996, 100, 1731–1737.

## GENERAL REFERENCES

Bell, A. G., "On the Production and Reproduction of Sound by Light," *Am. J. Sci.,* 1880, 120, 305–324.

Glorieux, C., Schoubs, E., and Thoen, J., "Photoacoustic Characterization of Liquid Crystal Phase Transitions," *Mat. Sci. and Eng.,* 1989, A122, 87–91.

Korpium, P., "Thermodynamic Models of the Photoacoustic Effect," *Photoacoustic Effect: Principles and Applications* (E. Lüscher, P. Korpiun, H. J. Coufal, R. Tilgner, Eds.), Fried, Vieweg & Sonn, Braunschweig, 1984, 40–51.

Luscher, E., "Photoacoustic Effect in Condensed Matter—Historical Development," *Photoacoustic Effect: Principles and Applications* (E. Lüscher, P. Korpiun, H. J. Coufal, R. Tilgner, Eds.), Fried, Vieweg & Son, Braunschweig, 1984, 1–20.

Rosencwaig, A., and Gersho, A., "Theory of the Photoacoustic Effect with Solids," *J. Applied Phys.,* 1976, 47, 64–69.

Rosencwaig, A., "Solid State Photoacoustic Spectroscopy," *Optoacoustic Spectroscopy and Detection* (Y. Pao, Ed.), Academic Press, New York, 1977, 193–238.

Thoen, J., Glorieux, C., Schoubs, E., and Lauriks, W., "Photoacoustic Thermal Characterization of Liquid Crystals," *Mol. Cryst. Liq. Cryst.,* 1990, 191, 29–35.

Thoen, J., Schoubs, E., and Fagard, V., "Photoacoustics Applied to Liquid Crystals," *Symposium on Physical Acoustics,* Leuven, June 1990.

PART III

# The Laser as a Reagent

From the point of view of a chemist, perhaps one of the most exciting applications of lasers is the use of laser photons as chemical reagents. The potential to effect chemical change with laser light offers possibilities to selectively control the direction of a reaction and the nature of the products obtained. While the role of the laser as an initiator of chemical reactions is still in the developmental stage, several applications have emerged as promising avenues for further exploration. The following chapters detail successful demonstrations of laser-induced chemical transformations. Chapters 12 and 13 describe the research and industrial uses of photochemical reaction pathways to synthesize organic molecules and fine chemicals, to catalyze polymer production, and to remove impurities from bulk chemicals. Chapter 14 focuses on the role of laser-induced photochemical and photobiological reactions as the basis for the medical treatment known as photodynamic therapy. Chapter 15 concludes our presentation with perhaps the most exciting and potentially powerful use of lasers in synthetic chemistry today—the selective excitation of specific chemical bonds to effect an express chemical reaction. From these promising advances, we can only anticipate a growing future demand for laser photons as reagents in the synthesis and processing of chemical substances.

CHAPTER 12

# Laser-Induced Chemical Reactions

## Chapter Overview

Laser excitation sources offer the synthetic chemist exceptional control over the outcome of photochemical reactions. Precise selections of the reactant(s) to be photoexcited, the excited state(s) to be generated, and the product(s) to be observed are governed by such laser parameters as wavelength tunability, monochromaticity, intensity, and mode of operation (CW vs. pulsed). In addition to prescribing the yield and distribution of possible products, the use of lasers in synthetic applications can reveal microscopic details of reaction pathways. This chapter illustrates these advantages of laser initiation of photochemical reactions through a consideration of the photoinitiated intramolecular cycloaddition of the ketone carvone.

## Organic Chemical Syntheses

An organic chemical synthesis focuses on the preparation of a desired organic structure, a "target molecule," using readily available compounds in a planned sequence of steps. Each step in the reaction pathway from starting materials to end product involves the formation or breaking of chemical bonds and possibly the introduction, alteration, or removal of a functional group. The ultimate objective of an organic synthesis may be the preparation of a naturally occurring or physiologically active compound, the manufacture of an industrially important material, the development of a new process, or the generation of a novel structure of theoretical interest or use in reactivity and structural studies. The feasibility of a particular synthetic approach depends on the reactivity of organic molecules under set conditions to produce the target species in high yield and at an acceptable rate with minimum interference from competing reactions.

The organic chemist can employ both conventional thermal techniques as well as photochemical methods as synthetic avenues. Ordinary chemical reactions rely on the thermal energy of colliding reactant particles to provide the necessary energy of activation. However, the feasibility of the reaction is restricted by the requirement of a thermodynamically favorable transition from reactants to products. Kinetic considerations further govern the distribution of possible thermodynamically favored products. Thus, syntheses involving thermally activated steps are controlled by the activation enthalpies, entropies, and kinetic rate constants of the reactions involving the initial reactants and subsequent intermediates.

## Organic Photochemistry

Organic photochemistry has rapidly emerged as an alternative synthetic approach. The principal basis for this preparative scheme is the initiation of chemical reactions involving organic molecules via the absorption of visible or ultraviolet light. The interaction of light with molecules generates an excited electronic state—a species distinct from the ground-state molecules with an electron distribution that is not the lowest energy configuration. This excited state is not easily generated by thermal excitation of the ground-state molecule and possesses a different chemical reactivity than the ground-state. The inherent instability of the excited state dictates its conversion to another state of lower energy, enhancing the probability of formation of a new chemical species. Thus, photochemical reactions generate energetic, albeit short-lived, excited states which on thermodynamic grounds provide for a greater range of potential final products. The differing electron configurations of ground and excited states may alter the observed product distribution, widening synthetic possibilities. What are the possible "paths" for an excited-state species? An excited-state molecule might undergo isomerization, internal rearrangement, or fragmentation (unimolecular processes), or reaction with solvent or solute to produce new chemical species (multimolecular processes). The limited excited-state lifetime, however, is a critical determinant of whether a chemical transformation will ensue. Radiative processes (e.g., fluorescence and phosphorescence) and nonradiative processes (e.g., internal conversion, intersystem crossing, thermal deactivation) are competing photophysical reactions which dissipate the energy of the excited state and may limit the observation of photochemical reaction. Thus, both thermodynamic and kinetic considerations of the excited state species are integral factors that dictate photochemical reaction pathways.

## The Laser as a Photochemical Tool

What specific advantages can lasers contribute as initiators of photochemical processes? As we have repeatedly demonstrated in other applications, the unique properties of laser radiation—monochromaticity, wavelength control, high intensity, and spatial coherence—provide the most significant advantages for the use of lasers in photochemical studies.

Clearly, the tunability of lasers enables precise wavelengths to be used for irradiation. Even in the case of organic molecules which have broad absorption spectra, wavelength selectivity may be essential to avoid the simultaneous irradiation of reactants and products or to irradiate a single component in a mixture of reactants. One rapidly developing area of photochemistry in which wavelength selectivity is critical is the area of laser isotope separation techniques. Small changes in the electronic, vibrational, and rotational energy levels of molecules are induced upon replacement of an atom in a molecule by its isotope. Selective excitation of one isotopic component is thus possible with the judicious choice of irradiation wavelength. The narrow bandwidth and the wavelength tunability of laser sources make such selective excitation experimentally feasible.

The intensity of laser radiation is an important advantage for initiating multiphoton reactions. In conventional photochemical experiments, the Law of Photochemical Equivalence dictates a one-to-one relationship between the number of photons absorbed and the number of reactant molecules converted to an excited state. The underlying conditions governing this law are the very short lifetimes of excited-state species and the relatively small number of photons in conventional beams of electromagnetic radiation. The high intensity of laser beams, however, increases the likelihood that a molecule may successively absorb additional photons before the initial excited state is deactivated. Thus, reactions typically requiring significant amounts of thermal energy can be conducted at room temperature via photochemical initiation. The opportunities for the observation of multiphoton phenomena are important outcomes of the application of lasers to photochemical reactions.

Finally, the low divergence of lasers permits the light from laser sources to be propagated for long distances while maintaining the cross-sectional area of the beam. This property of lasers is an important design advantage for photochemical experiments, as the light source can be physically removed from the locale of the chemical experiment. Photochemistry in hostile environments, such as flames and plasmas, as well as in systems at significant distances from the experimenter, is readily feasible with laser sources.

The following case study illustrates the use of lasers to selectively excite a reactant in order to control the yield and distribution of products from a photochemical reaction.

## Case Study: Photoisomerization of Carvone

### Overview of the Case Study

**Objective.** To illustrate the effectiveness of laser monochromaticity, tunability, intensity, and mode of operation (CW vs. pulsed) in governing the efficiency, yield, and product composition of a photochemical reaction.

**Laser Systems Employed.** A pulsed XeF excimer laser with 350 nm output, a pulsed third-harmonic YAG laser with 354 nm output, a pulsed XeCl excimer laser with 308 nm output, and a continuous wave krypton ion laser with both 350.7 and 356.5 nm output.

**Role of the Laser Systems.** To enable selective excitation of a single reactant in a photochemical reaction and sufficient generation of a particular excited state to control the product composition of the photochemical reaction.

**Useful Characteristics of the Laser Light for this Application.** Monochromaticity, wavelength tunability, intensity, mode of operation (CW vs. pulsed).

**Principles Reviewed.** Photolysis, multiphoton processes, selective excitation, singlet and triplet states, photoisomerization, reaction yield.

**Conclusions.** This study demonstrates that by using lasers of appropriate wavelength, intensity, and mode of operation (i.e., pulsed or continuous), careful control is possible of the yields of the two products resulting from the photoisomerization of carvone.

## Photocycloaddition of Carvone—Early Studies

One of the earliest photochemical reactions studied involves an intramolecular cycloaddition of the ketone carvone (I) to carvone-camphor (II), as shown in Figure 12-1. This reaction was first observed by Ciamician and Silber in 1908 [1] to occur via the irradiation of alcoholic solutions of I with sunlight for periods of up to one year. The structure of the photoproduct II was confirmed fifty years later by Buchi and Goldman.[2] The cyclization represents the first synthesis of a four-membered ring by the intramolecular [2+2] photoaddition of two double bonds. A second product (III in Figure 12-1) results from photoaddition of the solvent (ethanol), 1–exo,5–dimethyl–syn–2 [(ethoxycarbonyl)–methyl] bicyclo[2.1.1]hexane.

While the low yield (9%) of II is attributable to the reversibility of the reaction in the presence of ultraviolet light, a higher yield and a faster reaction time are possible with the use of high-pressure mercury vapor lamps or black-light fluorescence lamps as light sources.[3] Furthermore, the rate of cyclization of camphor greatly increases as the polarity of the solvent increases, with especially convenient reaction times in the presence of methanol, ethanol, or aqueous dioxane.[3] Use of these optimized conditions can result in a 44% yield of II after irradiation of carvone for 4 days. During the course of experiments, the concentration of carvone-camphor reaches a maximum and then declines as photolysis of II leads to the

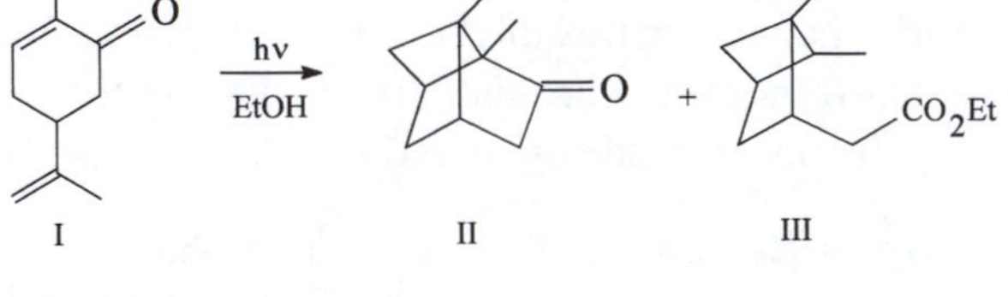

**Figure 12-1**
The cycloaddition of carvone (I) to carvone-camphor (II) via an intramolecular [2+2] photoaddition of two double bonds to form a four-membered ring. Photoaddition of the ethanol solvent produces a second product (denoted as III): 1-exo, 5–dimethyl–syn–2–[(ethoxycarbonyl)–methyl]bicyclo[2.1.1] hexane.

formation of III. For example, continued irradiation of carvone in ethanol for 7 days leads to a 97% yield of III. A higher intermediate concentration of II is observed with black-light irradiation ($\lambda_{max} \approx 355$ nm) than with irradiation via the high-pressure mercury-vapor lamp ($\lambda_{max} \approx 310$ nm). Meinwald and Schneider concluded that these observations supported the notion that the effective wavelength for the cyclization of carvone to produce II was longer than the wavelength required for the photolysis of II to III. In other words, the absorption spectra of II and III must be partially overlapping, with the spectrum for II shifted to longer wavelengths compared to the spectrum for III.

## Laser-Induced Investigations

### *Impact of Laser Monochromaticity and Tunability*

The broad absorption spectra of organic molecules often pose a challenge to photochemists in situations demanding the selective excitation of a single reactant within a mixture of reagents. Advantageous use of the particular characteristics of laser radiation, however, can eliminate this difficulty. With the objective of enhancing the photocycloaddition yield of II relative to III, Zandomeneghi et al. in 1980 [4] used a powerful continuous wave krypton ion laser to irradiate carvone. The choice of laser was dictated by their primary aim to irradiate the long wavelength tail of the absorption band of I and avoid the absorption band of the cycloadduct II. Both the 350.7 and 356.5 nm emission lines of the krypton laser were used jointly to accomplish such an irradiation of an ethanolic solution of carvone. Within 1.5 hours, 88% of carvone was converted to II, with only an 8% yield of III. The remaining species in solution consisted of trace amounts of carvone as well as polymers. The enhanced yield demonstrates that the selective irradiation of I in a region removed from the absorption band of II suppresses the photolysis of II to III. This study strikingly illustrates the effectiveness of laser monochromaticity and tunability in controlling the product composition of a photochemical reaction.

### *Impact of Tunability and Intensity*

Additional mechanistic information is often revealed when high-intensity pulsed lasers are selected as excitation sources for photochemical reactions. For example, to establish the mechanism of the photoisomerization reaction of carvone, two independent studies[5,6] investigated the photolysis using four different laser sources: a pulsed XeF excimer laser with 350 nm output, a pulsed third-harmonic YAG laser with 354 nm output, a pulsed XeCl excimer laser with 308 nm output, and a continuous wave krypton ion laser with both 350.7 and 356.5 nm output. Both studies observed differences in the product yields with the various excitation sources. These differences are illustrated by the results presented in Table 12-1 from the study of Malatesta et al.[5] Most notably, the two pulsed sources at ≈ 350 nm (XeF and third-harmonic YAG) yield negligible amounts of III, while the CW krypton laser at essentially the same wavelength produces predominantly II, but also four times as much of III as present with pulsed sources. In contrast, the pulsed XeCl source at the shorter wavelength of 308 nm affords III in higher yield or even exclusively. Malatesta et al. observed the latter case, with a 98% yield of III and only traces of carvone-camphor (II), while Brackmann and Schafer [6] noted the former result with an intensity-dependent ratio of II to III ranging from 1.8 at low intensities to a maximum of 6.9 at high intensities.

**Table 12-1**
Percent Composition of the Reaction Mixture upon Photolysis of Carvone

| *Laser/Wavelength (nm)* | *Mode* | *II* | *III* | *I* |
|---|---|---|---|---|
| XeF/350 | Pulsed | 94.4 | 2.1 | 3.5 |
| YAG(3v)/354 | Pulsed | 96.1 | 2.3 | 1.6 |
| Kr ion/350.7, 356.5 | CW | 81.7 | 9.3 | 9.0 |
| XeCl/308 | Pulsed | | 98 | 2.0 |

These laser-based studies substantiate previous results on tunability effects as well as introduce new findings on intensity effects. The observation of II as the main photoproduct with 350 nm irradiation and of III as the primary or enhanced photoproduct with 308 nm excitation is consistent with the earlier results of Zandomeneghi et al.[4] The small extinction coefficient of II at 350 nm accounts for the inefficiency of the ring-opening process to form III that is observed when carvone is irradiated with 350 nm excitation. Similarly, the appreciable absorption of II at 308 nm results in the formation of the ethyl ester III with XeCl laser radiation.

To account for the differences observed with pulsed vs. CW laser excitation at 350 nm, similar mechanisms for the overall system involving multiphoton processes were proposed by both research groups,[5,6] as illustrated in Figure 12-2. Irradiation of I produces the excited-state singlet of $^1$[I], which undergoes intersystem crossing with high efficiency to form the triplet of carvone $^3$[I] with a relatively long solvent-dependent lifetime (≈ 50–75 ns). From here, several alternative thermal and photochemical routes are available.

1. A first pathway involves photocycloaddition leading to II, with further ring opening to III with 308 nm irradiation. The decreasing yield of conversion of II to III with higher excitation intensities is proposed to arise from a photon loss mechanism whereby higher XeCl laser intensi-

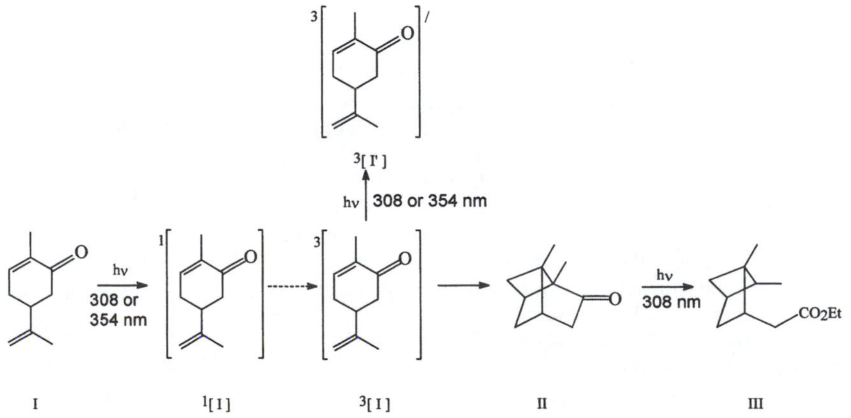

**Figure 12-2**
Overall proposed reaction mechanism[5,6] for the photoinduced cycloaddition of carvone (I) to camphor (II). Irradiation with either 308 or 354 nm excitation produces the excited-state singlet of I, $^1$[I]. Intersystem crossing generates the triplet state of carvone, $^3$[I]. Two alternative pathways are depicted for subsequent reaction of $^3$[I]. Photocycloaddition to form II with additional ring opening to III is induced by 308 nm light. Alternatively, pulsed irradiation at either 308 or 354 nm can form a second excited triplet, $^3$[I′].

ties promote II to an ionization and dissociation continuum.[6]

2. A second pathway for triplet carvone arises from the relatively long lifetime of this species, which enables the absorption of a second photon with pulsed irradiation to form a second excited triplet, $^3$[I′], of undefined structure. This upper triplet state must correlate more closely with II than with III, enhancing the formation of II upon relaxation.
3. A third and final reaction for triplet carvone involves the direct formation of III, as supported by the results of Malatesta et al.[5]

The energy diagram of Figure 12-3[6] further illustrates the proposed reaction scheme. Absorption of either 308 nm or 354 nm light elevates carvone in the ground state ′[I] to a long-lived excited state, $^3$[I]. Presumably this excited state is the lowest triplet state of carvone, reached via an excited singlet state. Reaction of $^3$[I] with rate constant $k_1$ can produce the ground state of carvone-camphor, II. Alternatively, $^3$[I] could absorb a second photon ($\lambda = 308$ nm or 354 nm) to produce a higher energy state $^3$[I′] with a shorter lifetime than $^3$[I] (i.e., $\tau_2 < \tau_1$). A fast reaction of $^3$[I′] to II (rate constant $k_2$) is possible, although absorption of an additional photon by $^3$[I′] could promote carvone to an ionization and dissociation continuum and hence decrease the quantum yield of II. Carvone-camphor can absorb a photon of 308 nm light to reach a long-lived ($\tau_3 \approx \tau_1$) state II$^*$, producing the ester product III with a rate constant $k_3$. Higher intensity irradiation is presumed to promote II$^*$ to the ionization and dissociation continuum, decreasing the quantum yield of III. This figure does not depict the direct transformation of $^3$[I] to III (i.e., triplet carvone → III), as proposed by Malatesta et al.[5]

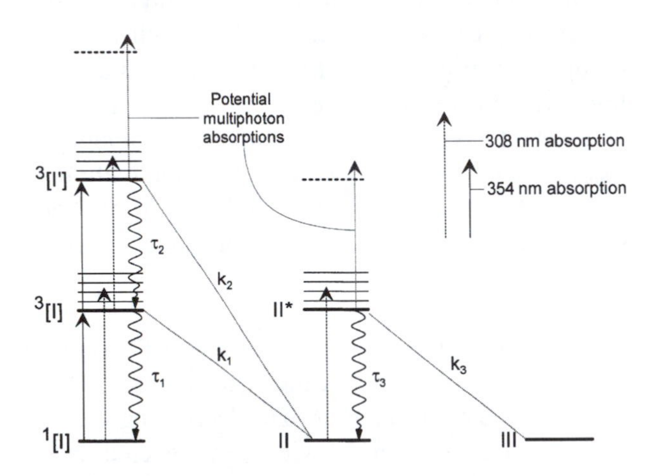


**Figure 12-3**
Energetics of the proposed reaction pathway for the photoisomerization of carvone. Ground-state carvone ′[I] is promoted to a long-lived excited state ($^3$[I] with lifetime $\tau_1$) with either 308 or 354 nm excitation. Here $^3$[I] represents the triplet state of carvone, and thus the absorption process reflects both the initial excited singlet state formation and the subsequent intersystem crossing to the triplet state. The ground state of carvone-camphor (II) results when $^3$[I] reacts with rate constant $k_1$. Alternatively, absorption of a second photon ($\lambda = 308$ or 354 nm) by $^3$[I] produces a higher energy state $^3$[I′] with a shorter lifetime ($\tau_2$) than $^3$[I]. $^3$[I′] can either absorb an additional photon to dissociate carvone by promoting ′[I] to an ionization and dissociation continuum, or $^3$[I′] can undergo a fast reaction (rate constant $k_2$) to generate carvone-camphor II. II can absorb a 308 nm photon to yield II$^*$; II$^*$ can combine with solvent to form the ester product III via a reaction characterized by a rate constant $k_3$. II$^*$ can also be promoted to the ionization and dissociation continuum with absorption of additional radiation.

## Significance of the Study

Thus, by using lasers of appropriate wavelength and with either pulsed or continuous operation, careful control is possible of the yields of the two products resulting from the photoisomerization of carvone. The selectivity of excitation provided by a laser source and the possibility for multiphoton processes afford the chemist alternative, convenient routes to the desired products. The opportunity for efficient, high-yield syntheses with laser photochemical methods may ultimately have an impact on the production of fine materials throughout the chemical industry. This promise may not be realized, however, because current laser costs make such syntheses economically unfeasible. Reductions in the cost per photon are possible with the continued development of laser technology. Nevertheless, the opportunities for laser syntheses are most promising for those synthetic routes which demand either selective excitation, reaction specificity, efficiency, or rapid product formation. Thus, the most auspicious applications of photochemistry to industrial chemical syntheses are those that take full advantage of the particular properties of laser radiation.

## FOR FURTHER EXPLORATION

Chemists have developed recent laser photochemical techniques to modify chemically the inert or hydrophobic surfaces of many synthetic materials to make them more amenable to applications in aqueous or biological systems. As an example, the inert, high molecular weight fluorocarbon polymer known as Teflon has numerous mechanical and physical properties that make it an ideal material for applications requiring strength, flexibility, and resistance to chemical attack. Medical scientists would like to take advantage of these attributes and use Teflon for medical implants, but cell growth over the implant is prohibited by the inert surface. Japanese researchers[7] have developed a photochemical solution, however, using 193 nm ArF laser light to irradiate a Teflon surface coated with a thin film of water.[7] Photodissociation of the water molecules by the ultraviolet light into hydrogen and hydroxyl radicals, H• and •OH, leads to extraction of fluorine from the surface by H• and replacement on the surface by the hydrophilic OH groups. The resultant surface is no longer inert but can attract and bond other hydrophilic substances. As an illustration, this chemical modification enables surgeons to assist the opening of weak blood vessels with Teflon scaffolding. Similar aqueous-based chemical modifications of polymethylmethacrylate surfaces, the constituent of hard contact lenses, are being tested for their ability to improve the comfort level of these lenses. The dyeing and adhesion characteristics of polypropylene, a widely used plastic, have also been improved by ArF laser-induced photochemical modification in water.[8] For applications in an oil medium, the surface of Teflon and other fluororesins can be made more hydrophobic by irradiating the fluororesin in a reaction chamber filled with gaseous $B(CH_3)_3$.[9] The C-F bonds (with a bond energy of approximately 536 kJ $mol^{-1}$) can be broken directly by the ArF laser radiation (193 nm equivalent to a photon energy of 615 kJ $mol^{-1}$). Fluorine atoms at the surface are replaced by methyl ($CH_3$) functional groups, enhancing the affinity of the material for oil. The chemical and physical treatment methods employed by industry to accomplish surface modification of a variety of materials will likely be revolutionized by the success of these photochemical surface modifications.

## DISCUSSION QUESTIONS

1. What experimental variables might a laser photochemist alter in attempting a laser-induced organic synthesis?
2. Referring to Figure 12-3, what advantage might result from a multiphoton absorption ($\lambda$ = 354 nm) of I to produce II rather than a single-photon absorption ($\lambda$ = 354 nm) pathway? What disadvantage might be observed from the multiphoton route over the single-photon mechanism?
3. A chemical firm might consider a laser synthesis for a new pharmaceutical. In general, what cost-saving advantages would have to be realized to make the selection of such a method worthwhile?

## SUGGESTED EXPERIMENTS

Suggested references to experiments that illustrate the principles described in this chapter:

1. Wirth, F. H., "Dye Laser Experiments for the Undergraduate Laboratory; Experiment 4—Actinometry," Laser Science, Inc., Cambridge, MA. An experiment to illustrate the promotion of a chemical reaction with laser light.
2. McMillan, G. R., Calvert, J. G., and Pitts, Jr., J. N., "Detection and Lifetime of Enol-Acetone in the Photolysis of 2-Pentanone Vapor," *J. Am. Chem. Soc.*, 1964, 86, 3602–3605. An example of a photochemical reaction whose progress is easily monitored with infrared spectroscopy. Although this account describes a reaction initiated with a mercury arc as the light source, an experiment could be devised to use a nitrogen laser for photoinitiation.

## LITERATURE CITED


1. Ciamician, G., and Silber, P. "The chemical action of light," *Ann. Chim. Phys.*, 1908, 16, 474–520.
2. Buchi, G. and Goldman, I. M., "Photochemical reactions. VII. The intramolecular cyclization of carvone to carvone camphor," *J. Am. Chem. Soc.*, 1957, 79, 4741–4748.

3. Meinwald, J., and Schneider, R. A., "Photochemical synthesis and reactions of carvone camphor," *J. Am. Chem. Soc.*, 1965, 87, 5218–5229.
4. Zandomeneghi, M., Cavazza, M., Moi, L., and Pietra, F., "Laser photochemistry: The intramolecular cyclization of carvone to carvone camphor," *Tetr. Lett.*, 1980, 21, 213–214.
5. Malatesta, V., Willis, C., and Hackett, P. A., "Laser-induced cycloadditions: The carvone photoisomerization," *J. Org. Chem.*, 1982, 47, 3117–3121.
6. Brackmann, U., and Schafer, F. P., "Photocyclization of carvone to carvone camphor using rare gas halide lasers," *Chem. Phys. Lett.*, 1982, 87, 579–581.
7. Leggett, K., "Laser renders Teflon sticky," *Photonics Spectra*, 1996, 30,*(3)*, 28–30.
8. Murahara, M., and Okoshi, M., "Photochemical surface modification of polypropylene for adhesion enhancement by using an excimer laser," *J. Adhesion Sci. Technol.*, 1995, 12, 1593–1599.
9. Murahara, M., and Toyoda, K., "Excimer laser-induced photochemical modification and adhesion improvement of a fluororesin surface," *J. Adhesion Sci. Technol.*, 1995, 12, 1601–1609.

CHAPTER 13

# Lasers in Industrial Chemical Processes

## Chapter Overview

Numerous nonchemical applications of lasers are commonplace in industrial settings, particularly for high-power uses such as drilling, welding, cutting, and heating. Large-scale chemical processing involving lasers, however, is less prevalent. Indeed, as we explored in the previous chapter, laser-driven chemical syntheses have unique advantages for controlled initiation of reactions and selectivity of product distribution. But the choice of synthetic method in the chemical industry requires the added consideration of economic viability. In this chapter we briefly survey both the potential and real uses of lasers in processing chemicals in industry.

## Useful Laser Characteristics for Industrial Chemical Processing

From *Star Wars* to parts-per-million purification techniques of bulk chemicals, lasers cannot be ignored for their potential and real impact on today's world. In industrial settings, incredibly useful roles for lasers have been and are being developed in the area of materials processing. Thermal or "nonreactive" processing with lasers is commonplace—the application of heat can be used to weld and cut samples, anneal defects on surfaces, recrystallize and harden materials, and induce the local formation of alloys. In the area of "chemical processing," that is, the separation of a specific chemical from a mixture of others or the synthesis of product chemicals from reactants, laser applications are not as routine. Nevertheless, developments in laser-driven chemical processing are especially facilitated by three of the laser's unique set of properties. Specifically, these properties are pure color (i.e., monochromaticity), high power, and spatial coherence. How do these properties assist applications in chemical processing? The first of these properties provides finely tuned energy to effect either physical or chemical transformations. The second attribute provides the potential for sufficient energy to carry out the finely tuned energy process. The third characteristic allows the highly selected power to be transmitted to or applied at convenient distances from the laser light power source. Thus, lasers furnish the intensity, selectivity, and control of energy needed in chemical processing and synthesis.

## Laser-Based Applications of Chemical Processing

Current potential and real applications for lasers in chemical processing are classified in Table 13-1. We will look in detail at examples from the manufacture of fine chemicals, the initiation of chain reactions, the separation of impurities, and the synthesis of ultrafine particles.

Any industrial process must ultimately pass the bottom-line test: The process that provides the required need at the lowest overall cost will be the process of choice. Today the overall cost must also include the cost of unreacted chemicals, waste products, and waste heat. Often a laser-assisted process can reduce "waste" chemicals and/or waste thermal energy and provide the economic incentive for laser use. However, in general, photons as reagents for photoinitiated processes are not inexpensive, and photons from lasers are even less inexpensive. Laser costs in industrial chemical processing can be broken into two major categories [1]. One category is the cap-

**Table 13-1**
Areas of Applications for Lasers in Chemical Processing

| *Laser Photochemistry* |
|---|
| Stoichiometric reactions: translational, rotational, vibrational, electronic energy input |
| fine chemicals |
| isotope separation |
| impurity removal |
| Nonstoichiometric reactions: |
| chain reaction initiation |
| *Surface Chemistry* |
| Surface modifications—laser depositions |
| Surface ablation |
| *Solid Preparations* |
| Powder syntheses |
| Catalyst syntheses |
| *Polymer, Semiconductor Processing* |

ital cost of the equipment, an expenditure that can be depreciated. A second expense category is the operating cost, particularly as related to electricity, gas supplies, and cooling requirements. It can be anticipated that the initial and operating costs of lasers will decrease with time with a concomitant increase in reliability. But, for the near term, laser costs will very much dictate that only a few select applications will be economically viable.

What are some specific examples of laser developments in chemical processing? Suitable laser-based applications are those processes where the cost of the laser is but a fraction of the total cost. One application meeting this criterion is the photoinitiated synthesis of pharmaceutical materials. Vitamin D is one such high-value-added reaction product for which the cost of laser radiation is justified (see Case Study I). On the other hand, for large-scale bulk syntheses involving photochemical reactions, laser photons are perhaps too expensive a source of reagent photons on a stoichiometric basis. Alternatively, laser separation of small amounts of reagent impurities or side products may indeed be economical. For example, hydrogen sulfide separation from various synthesis gas streams using laser techniques is a cost-effective method to purify gas-phase reagents for subsequent reactions (see Case Study III). Another generally practical and economic use of lasers today is in processes where the photoinitiation is catalytic or nonstoichiometric. A potentially huge application of lasers is in photoinduced chain reactions where a small amount of laser light, even though expensive, goes a long way to yield valuable product. The production of vinyl chloride is an excellent example of a photoinduced chain reaction that is economically viable with laser initiation (see Case Study II). Finally, the cost of using laser excitation can be economically warranted when laser-initiated synthesis generates products that either cannot be attained from conventional thermal syntheses or have properties or purities that surpass those that result from traditional synthetic routes. Laser-based syntheses of ultrafine clusters and particles using organometallic reactants illustrate a promising technique for the production of highly pure transition metal-containing species, often with unique chemical and physical properties (see Case Study IV). The four case studies described below highlight these application areas.

## Case Study I: Lasers in the Synthesis of Fine Chemicals

### Overview of the Case Study

**Objective.** To illustrate the dependence of the product distribution of a photochemical reaction on the photolyzing wavelength.

**Laser Systems Employed.** Lasers with varying output wavelengths, including a KrF excimer laser at 248 nm, a XeCl excimer laser at 308 nm, a nitrogen laser at 337 nm, and a YAG laser with a third harmonic at 353 nm.

**Role of the Laser Systems.** To provide excitation of specific wavelength and narrow spectral width to effect specific chemical reactions.

**Useful Characteristics of the Laser Light for this Application.** Monochromaticity, wavelength tunability.

**Principles Reviewed.** Photolysis, photoisomerization.

**Conclusions.** With the proper choice of excitation wavelength, a specific photoreaction can be favored to optimize the yield of the desired product and eliminate or reduce interferences from competing reactions.

### Synthetic Pathway

The industrial production of vitamin D is an example of a fine chemical synthesis. While vitamin D regulates bone growth in humans, this vitamin is also used commercially as a dietary supplement in feed for raising poultry. The synthesis of vitamin D is outlined in Figure 13-1.[1,2] The reagent 7-dehydrocholesterol (7-DHC) is the starting point for the photochemical synthesis of the vitamin. In short, the vitamin is obtained from 7-DHC via a photolytic ring opening. The transformation of 7-DHC to the final vitamin D product involves the formation of an intermediate, or precursor, known as previtamin D or $P_3$, and a subsequent thermal process to convert $P_3$ to vitamin D. The complex distribution of products that

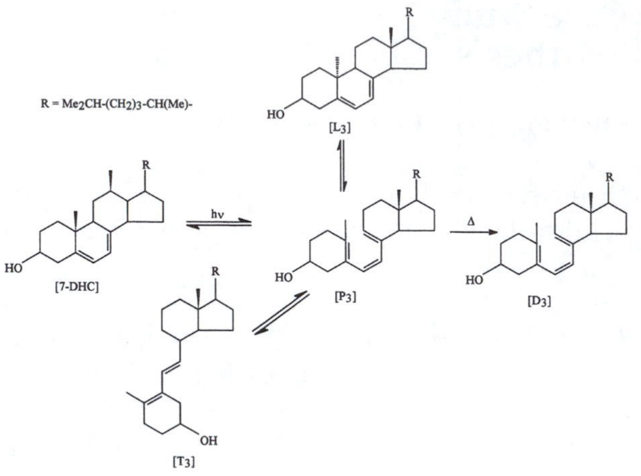

**Figure 13-1**
Reactive pathways for the photoinduced synthesis of vitamin D ($D_3$). Photolysis of 7-dehydrocholesterol (7-DHC) yields previtamin ($P_3$). Three parallel reactions are possible to form the three structural isomers of $P_3$: (1) a thermal reaction to yield vitamin D ($D_3$), (2) a UV-photoactivated reaction to form lumisterol ($L_3$), and (3) a second UV-photoactivated process to generate tachysterol ($T_3$). Short-wavelength excitation of 7-DHC (e.g., 248 nm emission of a KrF excimer laser or 254 nm output from a low-pressure mercury lamp) can yield $T_3$ directly via high-energy, high-intensity multiphoton processes. Longer wavelength radiation (near 350 nm) transforms $T_3$ to $P_3$.

is observed from the photoconversion of 7-DHC arises because the intermediate $P_3$ can form three structural isomers: vitamin D, lumisterol ($L_3$), and tachysterol ($T_3$). In essence, three parallel reactions compete for the available previtamin D as a reactant. The transformations of $P_3$ to lumisterol and tachysterol are photoactivated by UV light, while the formation of vitamin D from $P_3$ is a thermal process. Typical syntheses yield a variety of products, adding separation steps (and further costs) into the process. To obtain the maximum yield of vitamin D from the thermal reaction of $P_3$, the isomerizations to $L_3$ and $T_3$ must be eliminated or minimized. By tailoring the light source and mode of irradiation, more optimal yields of vitamin D are possible.

## Single-Wavelength Photolysis Products

How does the wavelength of the photolyzing light affect the composition of the final product mixture? Table 13-2 presents the yields of products by various photoinitiated pathways.[3] The typical medium-pressure mercury lamp process (excitation $\lambda = 302$ nm) yields primarily $P_3$ (previtamin D), yet the yield is only 53% and some purification is still required. Excitation with XeCl excimer laser light (308 nm) also generates a significant amount of previtamin D, but an even higher percentage of lumisterol, $L_3$, is present. However, when 7-DHC is subjected to the shorter excitation wavelength of either KrF excimer laser light (248 nm) or low-pressure Hg lamp excitation (254 nm), $T_3$ (tachysterol) is the primary result. This high yield of $T_3$ is most likely attributable to high-intensity, high-energy multiphoton processes transforming 7-DHC to $T_3$. The low yield of $T_3$ with longer wavelength radiation arises from a red-shifted absorption band in $T_3$ that converts most $T_3$ to precursor $P_3$.

**Table 13-2**
Composition of the Product Mixture following 7-DHC Photolysis[a]

| *Photolysis λ/nm* | *% 7-DHC* | *% $P_3$* | *% $T_3$* | *% $L_3$* | *% $D_3$* |
|---|---|---|---|---|---|
| 248 | 2.9 | 25.8 | 71.3 | — | — |
| 308 | 13.3 | 35.5 | 3.41 | 42.3 | 4.5 |
| (248 + 337) | 8.8 | 79.8 | 1.5 | 9.8 | — |
| (248 + 353) | 0.1 | 80.1 | 11.0 | 8.7 | — |
| 254[b] | 1.5 | 20.0 | 75.0 | 2.5 | — |
| 302[c] | 3.4 | 53.0 | 26.0 | 17.0 | — |

[a] Results from laser-induced photolyses given in reference 3; results from photolytic studies using mercury lamps given in reference 1.
[b] Low-pressure mercury lamp for excitation source.
[c] Medium-pressure mercury lamp for excitation source.

## Two-Stage Laser Photolysis

A two-stage laser photolysis procedure results in a significant improvement in the yield of $P_3$ with a sizable decrease in the extent of side-reaction contamination. Table 13-2 shows the results of two such two-stage photolyses. In each example, 248 nm KrF excimer laser excitation is initially used to generate a product mix of $T_3$ (major component) and $P_3$ (minor component). The second photolysis step takes advantage of the existence of the appreciable absorption at long wavelength (i.e., near 350 nm) in $T_3$ that converts $T_3$ back to $P_3$ via $E \rightarrow Z$ isomerization. Such long-wavelength absorbance is absent in either $P_3$ or $L_3$. The absorbance of $T_3$ is red-shifted enough so that either $N_2$ or YAG laser light is readily absorbed to induce the photoisomerization, yielding a very high percentage of $P_3$. A weak absorbance of $P_3$ at 337 nm may account for the subsequent cycloaddition reactions that occur to generate some $L_3$ and 7-DHC.

While these improvements in the yield of $P_3$ were established using lasers, the results were also useful in the design of modifications to the conventional photochemical process to enhance yields. In particular, a two-stage photolysis is employed with a low-pressure Hg lamp (254 nm) to maximize $T_3$ production and a suitably filtered medium-pressure Hg lamp (302 nm) to isomerize $T_3$ to $P_3$. Such lamps provide the correct energies and enough intensity to accomplish the desired transformations. Lasers continue to be used in the production of special vitamin $D_3$ metabolites for medical diagnostics.[1]

## Case Study II: Lasers as Photocatalysts

### Overview of the Case Study

**Objective.** To use lasers for photochemical initiation of free-radical chain reactions as an economic application of laser-driven synthesis.

**Laser System Employed.** An excimer XeCl laser with 308 nm output.

**Role of the Laser System.** To initiate a polymerization reaction at low temperatures by decoupling the reaction initiation step from subsequent chain propagation steps. As a consequence, side reactions with higher activation energies are eliminated, thus achieving a higher yield of the desired product.

**Useful Characteristics of the Laser Light for this Application.** High energy and spatial coherence of the laser beam ensure sufficient energy to initiate the desired reactions in gas streams to avoid heterogeneous catalytic effects of the system's walls.

**Principles Reviewed.** Chain reactions, free radicals, nonstoichiometric reactions.

**Conclusions.** Photoinitiation of a chain reaction via laser excitation enhances the yield of product formation and offers an example of cost-effective laser-assisted chemical synthesis.

As an example of a nonstoichiometric, catalytic use of laser light, consider a photoinitiated reaction for the production of vinyl chloride from 1,2-dichloroethane, or DCE. The production of vinyl chloride starting from DCE has been extensively studied at the University of Heidelberg and Dow Chemical.[2–4] The overall reaction given below can be described as an elimination of HCl from DCE, typically conducted via a thermal radical reaction involving HCl:

$$ClH_2CCH_2Cl \rightarrow ClHC = CH_2 + HCl \quad (113)$$

The reaction is of major industrial importance largely for the production of vinyl chloride monomer, the precursor of polyvinyl chloride polymers. Recent rates of production were estimated to be almost 12 billion pounds in the U.S. alone.[5] The photoinitiated conversion begins with the production of photoactivated DCE and the subsequent unimolecular decay to yield the Cl free radical, that is, the chain carrier. These two steps are represented by:

$$\begin{gathered} ClH_2CCH_2Cl + h\nu \rightarrow ClH_2CCH_2Cl^* \\ ClH_2CCH_2Cl^* \rightarrow ClH_2CCH_2\bullet + Cl\bullet \end{gathered} \quad (114)$$

The actual overall production of vinyl chloride involves two additional steps:

$$\begin{gathered} ClH_2CCH_2Cl + Cl\bullet \rightarrow ClHCCH_2Cl\bullet + HCl \\ ClHCCH_2Cl\bullet \rightarrow ClHC = CH_2 + Cl\bullet \\ \hline ClH_2CCH_2Cl \rightarrow ClHC = CH_2 + HCl \end{gathered} \quad (115)$$

This overall process is schematically illustrated in Figure 13-2.[3] In actuality, the total picture of vinyl chloride production is anything but simple. The direct loss of Cl from the free radical $ClHCCH_2Cl\bullet$ yields the desired vinyl chloride product and regenerates the chain carrier Cl radical. However, alternative reactions of this free radical with another Cl• free radical or with another DCE reactant molecule, for example, generate undesired side products such as ethylene, acetylene, ethylene chloride (1-chloroethane = $C_2H_5Cl$), and chloroprene (2-chlorobutadiene = 1,3-$C_4H_5Cl$).

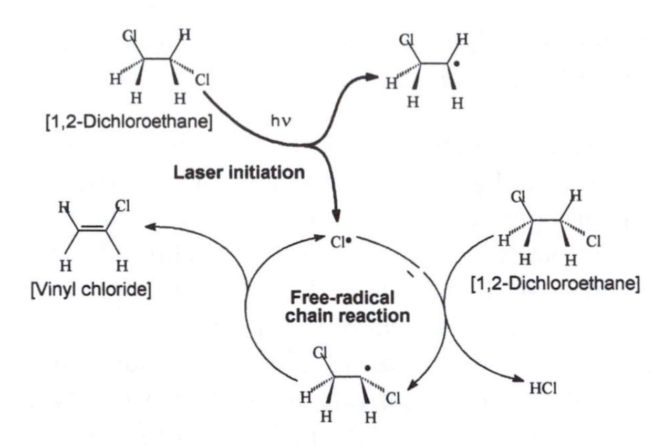


**Figure 13-2**
Reaction pathway for laser-initiated production of vinyl chloride from 1,2-dichloroethane (DCE). Laser initiation yields the chlorine free radical (Cl•) that serves as a chain carrier in a free-radical chain reaction. Reaction of Cl• with additional DCE generates HCl and a second free radical that dissociates to yield vinyl chloride and re-generate Cl•.

How can laser photoinitiation be effective at enhancing the selectivity toward the desired product and/or increasing the product yield? The experimental conditions under which the reaction is conducted play a dramatic role in determining product yield and distribution. When the chain reaction is initiated thermally, the initiation step (the step that generates the chain carrier) is the rate-determining step. With photoinitiation of the chain reaction, the rate-determining step becomes the unimolecular decay of excited DCE to yield the Cl• free radical. This unimolecular decay has a low energy of activation, and thus laser initiation allows the reaction to be conducted at lower temperatures. In turn, higher conversions to vinyl chloride and fewer undesirable side products are observed. Furthermore, in practice the laser-induced reaction is conducted in gas streams where the high energy and spatial coherence of the laser beam can initiate the reactions away from the heterogeneous catalytic effects provided by the walls of the reactor. This situation is also advantageous for high product yield and low side-product synthesis. For example, in one study,[6] the conventional process when conducted at 500°C exhibits a product yield of 75% and a selectivity toward the desired product of 85%. With the use of a KrF excimer laser to generate the chain carriers, the operating temperature can be lowered to 300°C to maintain the product yield at 75% and increase the selectivity toward vinyl chloride to 99%. The energy to generate the KrF laser photons represents only 6% of the energy used to heat the 1,2-dichloroethane from 300 to 500°C. In a second study using a XeCl excimer laser, reduction in undesired side products was afforded by laser initiation as illustrated in Figure 13-3.[3] Relative to the existing thermal process at 480°C, use of the excimer XeCl laser at 308 nm, 80 Hz, and 350°C reduces the by-products—chloroprene, acetylene, and ethylene - by factors of 2 to 3. Nevertheless, laser initiation requires that additional factors be taken into consideration. For example, practical concerns about the density and concentration variations in the reactor that dephase and defocus the laser beam are being addressed in pilot plant testing.[2]

In this application, the laser is a nonstoichiometric device because a few photons cause the formation of many vinyl chloride molecules. In other words, the quantum efficiency (moles of product/moles of photons) is very high. This increase in quantum efficiency comes about by the fact that both $C_2H_2Cl$• and Cl• contribute to maintaining the chain reaction. In fact, $10^4$ to $10^5$ vinyl chloride molecules are produced per photon. Even costly photons can be used to produce huge quantities of the inexpensive vinyl chloride because the laser cost is but a fraction of the total production costs. From this successful application of laser-driven synthesis, progress in the area of

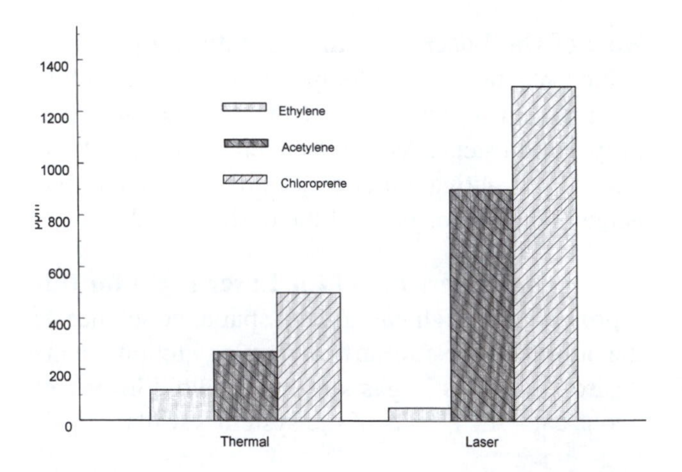


**Figure 13-3**
Comparison of by-product yields for vinyl chloride synthesis using thermal (■) or laser (□) initiation. Use of a XeCl excimer laser with 308 nm emission reduces the yields of the undesired side products of chloroprene, acetylene, and ethylene by factors of 2 to 3 compared with thermal initiation.

laser-induced radical chain reactions is likely to continue.

## Case Study III: Lasers in Impurity Removal

### Overview of the Case Study

**Objective.** To use laser-induced photodecomposition reactions to reduce or eliminate impurities from reagents for chemical syntheses.

**Laser System Employed.** A Raman-shifted argon fluoride excimer laser with 210 nm output.

**Role of the Laser System.** To induce a specific photodecomposition reaction.

**Useful Characteristics of the Laser Light for this Application.** Wavelength tunability.

**Principles Reviewed.** Absorption cross section, quantum yield.

**Conclusions.** Laser photodecomposition of trace impurities in reagents for chemical synthesis is an effective and economically competitive ultrapurification technique.

We have previously emphasized that the cost of laser photons for synthesis of chemicals in bulk is likely to continue to be too great for some time to come. While lasers may not be cost-effective in the production of the major products of a desired chemical transformation, their application may be economically competitive for the removal of undesired side products or impurities. The application of lasers for impurity removal has been suggested for the production of highly pure silane, $SiH_4$,[7,8] the starting point for the ultrapure silicon necessary for the semiconductor industry, and for the removal of sulfur from syngas mixtures.[3,9] We will illustrate this promising application by discussing syngas purification.

Synthesis gas, syngas for short, is mainly a mixture of $H_2$ and CO which can be generated from coal combustion or by combination of off-gas products from many chemical processes. Syngas is an increasingly important chemical feedstock for the synthesis of many very important bulk chemicals, especially methanol and glycol. For example, the components in syngas are reactants in an important pathway to methanol production.

Since the source chemicals used to yield syngas almost always contain sulfur, $H_2S$ is invariably a component of raw syngas:

$$CO_{(g)} + 2H_{2(g)} = CH_3OH_{(g)} \qquad (116)$$

Depending on the syngas application, the presence of $H_2S$ may or may not present difficulties. Most applications of syngas for chemical synthesis require transition metal catalysts which are almost universally "poisoned," that is, rendered ineffective, by reaction with $H_2S$. The activity of a metal catalyst relies on the availability of surface sites for adsorption of reactant gases. Impurities of $H_2S$ either adsorb on the metal surface to reduce the effectiveness of the heterogeneous catalyst or consume the metal through the formation of metal sulfides. The removal of $H_2S$ from syngas and other chemical product streams is thus a task of some importance to maintain catalytic reactivity and lifetime. While effective low-cost removal techniques include activated solid adsorbents such as charcoal or chemical scrubbing with absorbing solvents, the removal of $H_2S$ by these means still leaves sizable impurity concentrations on the ppm level. For metal catalysts, the $H_2S$ impurity level must reach 0.5 ppm levels or lower in order to provide adequate catalyst lifetimes for viable processes.[9]

To reduce $H_2S$ levels in syngas, Chen and Borzileri [10] have suggested a scheme based on the laser-induced photodecomposition of $H_2S$ and the subsequent collection of the reduced sulfur onto sacrificial metal plates. The process is briefly summarized as:

$$\begin{aligned} H_2S + h\nu(\lambda<260\text{ nm}) &= H + HS\bullet \\ HS\bullet + M &= S - M + H\bullet \\ HS\bullet + S - M &= S_2 - M + H\bullet \end{aligned} \qquad (117)$$

The photodissociation of $H_2S$ yields an H–S• fragment that undergoes further fragmentation as a consequence of the subsequent adsorption of S atoms on the sacrificial metal surfaces. Further reaction of the adsorbed sulfur species with an additional gaseous H-S• fragment generates bound polymeric sulfur. Thus, depending on the extent of reaction, either the metal sulfide is formed or polymeric sulfur is deposited on the metal surface.

The key to the success of the scheme to reduce $H_2S$ levels is absorption of the high-power UV laser light either only by, or preferentially by, $H_2S$ and not the components of the syngas. What wavelengths of laser radiation would most effectively accomplish this selective absorption? While both molecular CO and $H_2S$ absorb in the 180–260 nm region, the wavelength-dependent cross sections ($\sigma^2$) presented in Table 13-3 suggest that a spectroscopic selectivity is possible. For example, at 210 nm (the wavelength of light available from a $H_2$ Raman-shifted ArF excimer laser), the absorption cross section of $H_2S$ is $2 \times 10^{-18}$ cm$^2$ while the absorption for syngas (made up of pure reagent gases to approximate typical compositions) is $3 \times 10^{-25}$ cm$^2$. Thus the cross-section ratio that represents the factor by which absorption of $H_2S$ exceeds that of syngas is $2 \times 10^{-18}$ cm$^2$/$3 \times 10^{-25}$ cm$^2 \approx 10^7$.[9] Hence, the use of 210 nm output from a Raman-shifted ArF excimer laser could effect the selectivity desired. Note that the 210 nm output is the Stokes Raman line that results from excitation of a sample of hydrogen gas by the 195 nm ArF excimer emission, a practical use of the Raman effect. The use of 195 nm light from an ArF excimer laser (i.e., not Raman-shifted) would take advantage of a greater absorption cross section for $H_2S$ ($\approx 5 \times 10^{-18}$ cm$^2$) but suffer from the fact that the cross section of syngas has increased two orders of magnitude ($\approx 7 \times 10^{-22}$ cm$^2$), more than negating the advantage.

**Table 13-3**
UV Absorption Cross Sections of $H_2S$ and Synthesis Gas[a]

| | $\sigma^2$ / cm$^2$ | |
|---|---|---|
| *λ/nm* | *$H_2S$* | *Synthesis Gas* |
| 195[b] | $5 \times 10^{-18}$ | $7 \times 10^{-22}$ |
| 210[c] | $2 \times 10^{-18}$ | $3 \times 10^{-25}$ |
| 248[d] | $3 \times 10^{-19}$ | $7 \times 10^{-24}$ |

[a] Results from reference 10.
[b] Corresponds to the wavelength of light available from an ArF excimer laser.
[c] Corresponds to the wavelength of light available from a $H_2$ Raman-shifted ArF excimer laser.
[d] Corresponds to the wavelength of light available from a KrF excimer laser.

The quantum yield for this photoinitiated process is defined as:

$$\text{quantum yield} = \frac{\#\ H_2S \text{ molecules removed from syngas}}{\#\ \text{photons absorbed by } H_2S \text{ molecules}} \qquad (118)$$

and has been measured to be about 1 at 10 torr and about 0.4 at 1 atm. A complicating feature, however, relates to the chemistry of the HS• radical. At low pressure, atomic hydrogen could initiate the formation of HS• via the sequence of reactions:

$$H_2S + H\bullet = H_2 + HS\bullet$$
$$HS\bullet + M = S - M + H\bullet \text{ (chain repeats)} \qquad (119)$$

Thus, the observed quantum yield would reflect a quantum yield enhanced by these additional reactions of HS• radicals. At higher pressures, recombination reactions of HS• radicals might occur to regenerate $H_2S$ and thereby lower the measured quantum yield of $H_2S$ photodissociation:

$$HS\bullet + HS\bullet = H_2S + S \qquad (120)$$

A chemical process for $H_2S$ removal to 0.1 ppm levels is available using ZnO and is based on the reversal of roasting zinc blend ZnS to obtain Zn:

$$ZnO + H_2S = ZnS + H_2O \qquad (121)$$

The authors present an economic comparison of the laser and ZnO processes.[10] At 1980 prices, the technologies are cost comparable. The majority of the laser cost is initial outlay and maintenance. For the ZnO method, the major costs are for the initial reactor bed and heating costs to maintain the reactor at the necessary temperature for rapid conversion. At this writing, the capital and maintenance costs for the laser technology must be much reduced, but it is not known whether any commercial methanol plant using syngas feedstock has installed the technology. Since it often takes on the order of ten years to bring a new process onstream, it may be premature to expect the technology to actually be in place.

# Case Study IV: Laser-Based Synthesis of Ultrafine Materials

## Overview of the Case Study

**Objective.** Gas-phase synthesis of ultrafine clusters and powders via pyrolysis of gaseous reactants to yield high surface area materials with unique chemical and physical properties.

**Laser System Employed.** A tunable $CO_2$, YAG, or excimer laser with single-photon or multiphoton excitation.

**Role of the Laser System.** To rapidly heat gaseous reactants to induce pyrolysis or cleavage of ligands.

**Useful Characteristics of the Laser Light for this Application.** Wavelength tunability, high power for rapid heating and for selective pyrolysis.

**Principles Reviewed.** Heterogeneous catalysis.

**Conclusions.** The ultrafine particle size, high purity, variable particle size distribution and composition, and extensive surface area of laser-synthesized particles offer both superior and unique characteristics for heterogeneous catalysis studies and for the production of ceramic materials.

Solid catalysts perform a vital role in the production of many leading industrial chemicals. Most heterogeneous catalysts are metals, particularly transition metals (e.g., Fe, Co, Ni, Pd, Pt, Cr, Mn, Cu, Ag, and W), and metal oxides (e.g., $Al_2O_3$, $Cr_2O_3$, $V_2O_5$, $Fe_2O_3$, ZnO, and NiO). For example, the oxidation of $SO_2$ to $SO_3$, a step in sulfuric acid synthesis, relies on Pt metal or solid $V_2O_5$ for efficient conversion of reactants. The production of ammonia from the gas-phase reactants $N_2$ and $H_2$ involves a solid catalyst composed primarily of iron metal with small amounts of added potassium and aluminum oxides. The cracking of high molecular weight hydrocarbons to yield high-octane gasoline is facilitated by $SiO_2/Al_2O_3$ catalytic surfaces. The hydrogenation and dehydrogenation of hydrocarbons is aided by oxides of nickel, aluminum, and chromium.

Recent studies [11–13] have demonstrated that ultrafine particles of some catalysts can be produced with laser excitation. For example, ultrafine (10 to $10^2$ nm) powders of a wide range of metal carbides ($M_xC_y$), nitrides ($M_xN_y$), sulfides ($M_xS_y$), and oxycarbides ($M_xC_yO_z$) have been produced using $CO_2$, excimer, and YAG lasers. While these catalysts can be prepared using standard synthetic methods, can lasers offer any specific advantages in the production of heterogeneous catalysts? There is considerable evidence that laser-induced synthesis of catalytic materials yields physical properties of the heterogeneous catalyst that surpass those of materials synthesized by conventional procedures. What physical and chemical properties of a solid substrate influence its activity as a heterogeneous catalyst? The reactivity and selectivity of a solid surface in catalyzed reactions are functions of the chemical composition, particle size, surface area, and surface purity of the catalyst. In laser-driven syntheses, laser excitation of volatile metals and metal-containing compounds in the presence of gaseous reactants or inert gas species can produce monodisperse, unaggregated particles with greater surface area and reduced surface contamination. All of these properties are crucial for superior catalytic activity. The features of ultrafine particle size, nonagglomerated particles, and greater particle surface area arise from the rapid heating rates that can be achieved by laser excitation—often exceeding $10^6$ degrees per second. The confined space in which a reactant interacts with the spatially coherent laser beam reduces the contact of the products with the surface of the reactor and enhances catalyst purity. In effect, reactions are conducted in a wall-less environment where the reaction zone consists of only the region where the reactant gas stream and laser beam interact. In addition to improved physical characteristics, laser-driven syntheses can provide new or altered chemical compositions of solid catalysts. For example, selective dissociation of ligands on organometallic compounds using tunable lasers can be used to generate free metal atoms and small metal clusters as well as new metallic and even bimetallic compounds.

Laser-based production of heterogeneous catalysts also offers economic advantages for the chemical industry. In addition to higher product purity, greater catalytic yields, and reduced raw material costs, laser synthesis may be more economical than a conventional synthetic route as a consequence of lower energy demands. The rapid heating provided by lasers eliminates the need for high-temperature furnaces, thus reducing energy costs as well as providing more control over product distributions and yields. The industrial environment is also well-suited for the lasers typically employed in these syntheses. For example, $CO_2$, YAG, and excimer lasers are already used for welding, cutting, and drilling applications and are rugged and stable enough for industrial operation.

## SUMMARY

This chapter has highlighted a few real and potential uses of lasers in the area of chemical processing in industrial settings. It seems clear that the unique capabilities of lasers will continue to impact chemical processing, particularly in those cases where a small amount of laser light input can make a substantial difference in the yield and characteristics of the final product.

## FOR FURTHER EXPLORATION

Laser-based synthesis of superconducting materials is a promising application on the horizon. Traditionally, solid-state synthesis requires high temperatures to facilitate diffusion and reaction in this rather immobile phase. The observed products of high-temperature reaction chemistry are those materials thermodynamically stable at the elevated temperatures. The prospect of kinetically driven, low-temperature, and even low-pressure, solid-state synthesis intrigues chemists for the possibility of yielding new materials with novel physical properties. A methodology known as pulsed laser ablation and deposition (PVD) employs high-energy pulsed lasers to generate new solids under nontraditional reaction conditions. One research group has demonstrated that PVD affords the synthetic chemist significant control in the design of layered structures that represent ideal model systems for copper oxide superconductors.[14,15] Ablation of $(Sr_{1-x}Ca_x)_{1-\delta}CuO_2$, $Sr_{2-x}Ca_xCuO_{3+\delta}$, and $Sr_{2-x}Nd_x$-$CuO_{3+\delta}$ targets using a frequency-doubled pulsed Nd:YAG laser or a KrF excimer laser subsequently deposits the ablated material onto single-crystal $SrTiO_3$ and MgO substrates. Novel layered crystalline structures result from the low-temperature conditions, with the product structure and physical properties exhibiting a dependence on both temperature and the nature of the substrate. The systematic exploration of the dependence of such properties as superconductivity on the structural parameters of copper oxide materials is an exciting prospect of controlled solid-state synthesis via PVD.

## DISCUSSION QUESTIONS

**1.** In the two-stage photolysis of 7-dehydrocholesterol (7-DHC), the composition of the product mixtures is approximately 80% $P_3$ and 9% $D_3$, whether obtained using a nitrogen laser or a YAG laser. However, the relative ratios of $T_3$ and $L_3$ vary with the wavelength of the second photolyzing laser. How can you account for the 9% contribution of $L_3$ and the minor yield of $T_3$ with the nitrogen laser, and the 11% contribution of $T_3$ and the virtual absence of $L_3$ using the third-harmonic YAG laser?

**2.** Following the argument that lasers might be well used in impurity removal, can you suggest other processes where removal of impurities by laser-induced photocomposition might be useful?

**3.** Using the data provided in Table 13-3, what would be the absorption cross-section ratio that represents the factor by which absorption of $H_2S$ exceeds that of syngas using excitation at 248 nm?

## LITERATURE CITED

1. Kleinermanns, K., and Wolfrum, J., "Laser chemistry-What is its current status?" *Angew. Chem. Int. Ed. Eng.*, 1987, 26, 38–58.

2. Woodin, R. L., Bomse, D. S., and Rice, G. W., "Lasers in chemical processing," *Chemical and Engineering News*, 1990, 68, 20–31.

3. Malatesta, V., Hackett, Peter A., and Willis, C., "Laser photochemical production of vitamin D," *J. Am. Chem. Soc.*, 1981, 103, 6781–6783.

4. Wolfrum, J., "Laser-induced chemical reactions in combustion and industrial processes," *Laser Chem.*, 1986, 6, 125–147.

5. Reisch, M. S., "Top 50 chemicals production stagnated last year," *Chemical and Engineering News*, 1992, 70, 16–22.

6. Hackett, P. A., "Making light work—Applications of lasers to chemical production," *Laser Chem.*, 1988, 9, 75–106.

7. Clark, J. H., and Anderson, R. G., "Silane purification via laser-induced chemistry," *Appl. Phys. Lett.*, 1978, 32, 46–49.

8. Hartford Jr., A., Huber, E. J., Lyman, J. L., and Clark, J. H., "Laser purification of silane: Impurity reduction to the sub-part-per-million level," *J. Appl. Phys.*, 1980, 51, 4471–4474.

9. Chemists are more familiar with the concept of molar absorptivity measuring the amount of light a given species will absorb at a specific wavelength. Absorption and molar absorptivity are related through Beer's Law, $A = abC$, where $a$ is absorptivity, $b$ is path length, and $C$ is concentration. The absorptivity has units of area per particle, that is, a cross-section for absorbance, whenever $C$ is measured in units of number of absorbers per unit volume and $b$ has units of length.

10. Chen, H. L., and Borzileri, C., "Laser cleanup of $H_2S$ from synthesis gas," *IEEE J. Quant. Elec.*, 1980, QE-16, 1229–1232.

11. Musci, M., Notaro, M., Curcio, F., Casale, C., and De Michele, G., "Laser synthesis of vanadium-titanium oxide catalysts," *J. Mater. Res.*, 1992, 7, 2846–2852.

12. Chaiken, J., "Laser chemical synthesis of clusters and ultrafine particles using organometallics," *Appl. Organomet. Chem.*, 1993, 7, 163–172.

13. Rice, G. W., "Laser-driven synthesis of transition-metal carbides, sulfides, and oxynitrides," *Laser Chemistry of Organometallics*, ACS Symposium Series, 1993, Chapter 19, pp. 273–278.

14. Niu, C., and Lieber, C. M., "Exploiting laser based methods for low-temperature solid-state synthesis: Growth of a series of metastable $(Sr_{1-x}M_x)_{1-\delta}CuO_2$ materials," *J. Am. Chem. Soc.*, 1993, 115, 137–144.

15. Morales, A. M., Yang, P., and Lieber, C. M., "Preparation of layered $Sr_2CuO_{3+\delta}$ by pulsed laser deposition: Rational synthesis and doping of a metastable copper oxide material," *J. Am. Chem. Soc.*, 1994, 116, 8360–8361.

CHAPTER 14

# Laser Photons as Medicine

## Chapter Overview

Lasers offer several powerful and unique advantages for medical diagnoses, therapeutic treatments, and internal surgeries. The ability to direct and focus energy of requisite wavelength and power density (often via optical fiber) ensures selective absorption by the desired biological target. Such directionality also minimizes traumatization of surrounding tissue. As a consequence, laser-assisted treatments are often more effective with less pain for the patient, faster healing, and shorter hospital stays than conventional techniques. Absorption of laser radiation may facilitate a range of procedures, including surgical incisions, soft tissue welding, ablation of hard tissues, photocoagulation of blood vessels, and bleaching of pigmented tissue. After a survey of the merits and limitations of laser uses in medicine, this chapter focuses on current efforts to detect and treat malignant tumors using the technique of photodynamic therapy. This procedure uses laser-activated photosensitizers as exogenous chromophores that preferentially localize in cancerous tissues and eradicate such tissue through a light-induced chemical reaction.

## Interaction of Laser Light with Biological Tissue

Since the introduction of the first laser device in 1960, lasers have rapidly achieved a prominent position in medical applications. Laser systems serve as routine tools for surgical and clinical practice, particularly in the areas of ophthalmology, dermatology, dentistry, arthroscopy, angioplasty, and orthopedics. The fundamental basis for medical uses of lasers is the specific, often noninvasive, absorption of laser light by soft tissues and bones. This absorption process is often a complex interaction, requiring an intricate delivery of light to the desired target to initiate a reaction that is dependent on the wavelength range of the absorbed photons. For example, infrared radiation (e.g., from a $CO_2$ laser) is typically absorbed by the water in tissues, leading to the rapid heating and evaporation of water and the breakdown of tissue structure analogous to the surgical cutting of tissue. Visible light (e.g., the output of an argon ion or ruby laser) is transmitted by water but may be absorbed by pigmented tissues to coagulate tissues, treat blood clots, and remove skin disorders and lesions. High-energy ultraviolet wavelengths (e.g., emitted by excimer lasers) lead to the photoablation of tissue as needed in angioplasty to open blood vessels. Thus, the absorption characteristics of tissue constituents span the range of wavelengths available as laser radiation.

An initial survey of the multitude of medical laser applications currently established in practice or under development reveals an almost overwhelming array of uses. Nevertheless, the nature of the interaction of laser light with human tissue can be categorized according to four general types of interactions and four basic subsequent effects. When laser light is directed to a specific tissue, what interactions are possible? Incident light may:

1. be *reflected* by the surface,
2. enter the tissue and be *absorbed* by a chromophore,
3. permeate the tissue, be *scattered* internally, and then absorbed at a location removed from the site at which light entered the tissue, or
4. be *transmitted* all the way through the tissue.

While light may interact with a tissue in these manners, one basic premise for therapeutic and surgical uses must be remembered. Only *absorbed* light can produce a subsequent effect for a potential medical application.

What factors determine the role a laser will play in medicine? The wavelength of light absorbed, the identity of the absorbing chromophore, and the impact on the absorbing chromophore and the surrounding tissue all determine the medical procedure or treatment possible with laser light. Four principal effects of laser light are generally classified:

1. Photothermal effects involve the local heating of tissue as a consequence of the absorption of light. This heating leads to such temperature-induced phenomena as protein denaturation at 40°C, protein coagulation above 50°C, and the boiling of cellular fluid and the vaporization of tissue above 100°C.
2. Ablative effects are defined as the removal of cellular material by breaking molecular bonds

with negligible heating to damage nearby tissue. Ablation is the basis for such techniques as ophthalmological corneal sculpting and atherosclerotic plaque removal.

3. Photoreaction is the general term for absorption by a chromophore that leads to fluorescence emission, triplet formation, and/or chemical reaction. Photoreactions are the essence of the photodynamic therapy treatment of cancers and procedures for blood purification and virus inactivation.
4. Acoustical fragmentation consists of using pulsed laser light to generate an acoustical wave to disintegrate a target tissue, such as kidney stones and gallstones.

## Benefits and Limitations of Lasers in Medical Applications

The effective utilization of lasers in medical applications arises, as we have seen in so many other situations, from the unique aspects of laser light. What characteristics of laser systems would be advantageous in medicine? One obvious gain is the ability to deliver high energy at precise wavelengths to minimize invasive damage. The technology of optical fibers facilitates such precise delivery of very high energy to small tissue regions. Absorption by the target tissue and the depth of penetration are both wavelength dependent, and thus several kinds of lasers have developed important roles in medicinal and therapeutic applications. Lasers are also capable of irradiating a small cross-sectional area with low divergence and often with high power density. These features are requisite criteria for many medical procedures to reduce damage to surrounding tissue and provide effective treatment. Often the capacity to choose either continuous or pulsed illumination is an added benefit in the design of laser applications. Table 14-1 provides an overview of a vast range of laser medical treatments that are currently in practice.

The use of lasers in medicine is not without complications. The propagation of light to the targeted tissue is hampered by scattering and differential absorption through multiple layers in tissue. While optical fibers often serve as delivery systems to transmit the laser beam directly to the tissue or chromophore of interest, some lasers are limited by the lack of suitable optical fibers. For example, the 10.6 $\mu$m radiation of a $CO_2$ laser is absorbed by all currently available fibers. Suitable optical materials still need to be developed for this purpose. For those applications where laser techniques have not yet yielded significant medical advantages over conventional surgery or treatment, functional considerations, such as the cost, sheer size, and durability of the laser system, may slow the establishment of laser practices. Economic competition in the rapidly changing medical industry will ensure that lasers continue to play a unique and effective role in medical and clinical practice.

**Table 14-1**
Lasers in Medical Applications

*Argon Ion Lasers—Absorption of the blue-green lines (476.5–514.5 nm) by hemoglobin and pigmented tissue to cause photocoagulation and local constriction of blood vessels*

| Application | Laser Effect |
|---|---|
| Ophthalmologic surgery for diabetic retinopathy | Sealing of broken blood vessels in the retina damaged from elevated blood sugar levels from diabetes in order to minimize fluid buildup and scarring. |
| Vascular lesions | Triggering of photocoagulation of skin marks caused by overgrowth of blood vessels. |
| Glaucoma treatment—laser trabeculoplasty | Reduction of the intraocular pressure in the eye associated with glaucoma and reduced peripheral vision by the introduction of "burns" in the trabecular network to allow drainage of the aqueous humor. |
| Cosmetic and restorative dentistry | Polymerization of sealants in pits and fissures via the simultaneous action of cutting and coagulation. |

(continued)

**Table 14-1**
Continued

| *$CO_2$ Lasers—10.6 μm output to heat water for tissue vaporization* | |
|---|---|
| **Application** | **Laser Effect** |
| Arthroscopic surgery | Resection of soft-tissue abnormalities in the knee joint. |
| Dermatological surgery | Removal of skin lesions such as warts, skin cancers, ingrown toenails, and benign facial growths. |
| Burn surgery | Arresting of bleeding and removal of dead tissue to stimulate healing. |
| Soft-tissue treatment in dentistry | Removal and reshaping of gums. |
| *Nd:YAG Lasers—1.064 nm output to dissect tissue by directing energy via heat and light to a sapphire scalpel* | |
| **Application** | **Laser Effect** |
| Laparoscopic cholecystectomy | Removal of gallstones. |
| Laser therapeutic thoracoscopy | Removal of lung tumors. |
| Radial thermal keratoplasty | Selective vaporization of cells within the cornea to induce a collapse of the surface in order to change corneal shape to treat nearsightedness (myopia), farsightedness (hyperopia), and moderate astigmatisms (misshaped corneal surfaces). |
| *Ho:YAG Lasers—2.1 μm output to heat water for tissue vaporization* | |
| **Application** | **Laser Effect** |
| Diskectomy | Treatment of herniated disks through vaporization of a small portion of the disk. |
| Orthopedic surgery | Repair of damaged cartilage in the knee and other joints by cutting and sculpting of a torn meniscus. |
| Radial thermal keratoplasty | Thermocoagulation of a series of spots (6–9 μm in size, generally 16 in number) within the cornea to induce shrinkage of corneal collagen and steepen the curvature of the corneal surface to treat farsightedness (hyperopia). |
| *Er:YAG Lasers—2.94 μm output* | |
| **Application** | **Laser Effect** |
| Hard tissue applications in dentistry | Ablation of hard tissue (tooth enamel). |
| *Diode Lasers—Variable-power 810 nm output to heat tissue for cutting, coagulation, or vaporization* | |
| **Application** | **Laser Effect** |
| Tissue welding | Replacement and/or supplement for conventional surgical closures (sutures) using the laser as a thermal energy source at low-power levels (300 mW to 5 W). |
| Ophthalmologic surgery for retinopexy | Formation of adhesions surrounding a retinal tear to correct retinal detachment. |
| Ophthalmologic surgery for transcleral photocoagulation | Thermocoagulation of the ciliary body within the eye via the passage of light through the sclera (white of the eye). |
| *Excimer (ArF) Lasers—193 nm output with uniform beam profile for even cutting and removal of cellular material by breaking molecular bonds in one cell at a time* | |
| **Application** | **Laser Effect** |
| Radial keratotomy or photorefractive keratectomy | Ablation of tissue layers from the corneal surface to reduce the curvature of the eye to treat nearsightedness (myopia). |

# Case Study: Photodynamic Therapy

## Overview of the Case Study

**Objective.** Photodynamic therapy is a light-activated therapy for cancer involving the injection into the body of photosensitive dyes or drugs that are selectively retained in malignant tumors. Laser illumination of the localized photosensitizer initiates a series of photochemical reactions that destroy the neoplastic (cancerous) tissue with limited damage to the surrounding healthy tissue.

**Laser System Employed.** Dependent on the photosensitizer, the typical laser system is a continuous wave, argon ion pumped dye laser tuned in the 600–700 nm range.

**Role of the Laser System.** To selectively activate a photosensitizer at the required depth of penetration within a cancerous tissue with minimal invasion of normal tissue.

**Useful Characteristics of the Laser Light for this Application.** Wavelength tunability, high energy, high power density.

**Principles Reviewed.** Chromophores, photosensitizers, tetrapyrroles.

**Conclusions.** The combination of a tuned and high-powered laser source and a photoactive dye or drug is the basis for a promising therapeutic procedure to treat malignant tumors.

## Introduction

### *Endogenous and Exogenous Chromophores in Human Tissues*

Initial investigations in laser medicine focused on the optical properties of two *endogenous* (naturally occurring) chromophores in human tissues: the melanin pigment in skin and the hemoglobin of red blood cells.[1] Various skin lesions such as malignant melanoma can be selectively targeted with the ruby laser at 694.3 nm to destroy the malignant cells. Cell destruction occurs via thermophotolysis—the absorption of light to generate heat and raise tissue temperature. Vascular lesions arising from abnormal vascular vessels such as port-wine stain birthmarks and facial spider vessels can be treated with laser excitation absorbed by hemoglobin. The blue-green lines of the argon laser (488.0 and 514.5 nm) are effective in destroying these blood clots that form in the skin as a result of the perforation of capillary vessels.

Another promising prospect for lasers in medicine, however, involves the interaction of light with tissue to induce photochemical and photobiological reactions involving *exogenous* chromophores. One technique receiving considerable attention is that of photodynamic therapy (PDT) whereby a photosensitizing molecule, injected into the body and preferentially deposited in malignant tumors, interacts with visible laser light. This tumor-selective drug can absorb light of the appropriate wavelength to fluoresce, thus permitting an estimate of the extent of the tumor as well as a determination of its precise location. Furthermore, the photosensitizing drug can be activated by another wavelength of visible light to generate cytotoxic agents that destroy the tumor. The overall process is thus a form of photochemotherapy—eradication of a cancerous tissue through a light-induced chemical reaction. The future of this modality for cancer treatment is especially promising, and here we review the chemical nature of current photosensitizers, the mechanism of action of the photosensitizing drug, and the role of the laser in initiating the photodynamic effect.

### *Photosensitizers*

What properties of a photosensitizer are important for its clinical application in photodynamic therapy treatments? A compound of low toxicity that preferentially localizes in a tumor and that exhibits photodynamic activity is essential. Studies of light penetration in human tissue reveal that absorption in the red region of the electromagnetic spectrum is required for high tissue penetration and adequate delivery of light to the malignant tissue. The two most common photosensitizing drugs administered to patients are hematoporphyrin-derivative (HpD) and Photofrin® II. The chemical nature of these photosensitizers is complex.[2] Each is composed of variable proportions of an inactive component and an active component responsible for tumor localization, photosensitizing activity, and fluorescence. In HpD the inactive component generally comprises 50% by weight of the overall content, while in Photofrin® II the inactive component amounts to only 15–20%.[2] The inactive component in both HpD and Photofrin® is composed of multiple monomeric porphyrins: Hp, hematoporphyrin; HVD, hydroxyethylvinyldeuteroporphyrin; and PP, protoporphyrin. The basic monomeric hematoporphyrin structure is illustrated in Figure 14-1, with one of the $-CHOH-CH_3$ substituents replaced with $-CH{=}CH_2$ in HVD and both $-CHOH-CH_3$ moieties replaced with $-CH{=}CH_2$ in PP. The active material consists of ether-linked and ester-linked porphyrin units (designated DHE, dihematoporphyrin ether or ester) ranging in size from the dimer to oligomers with six porphyrin units. In Photofrin® II, the mix of ester- and ether-linked units is approximately 1:1. The covalent linkages occur at the hydroxyethylvinyl groups. Dimerization via an ester bond occurs through the reaction of the alcohol chains of HP or HVD and the acid chains of HP, HVD, or PP. The method of synthesis of the photosensitizer dictates the ratio of ester- to ether-linked units in oligomers. Aggregation of DHE through noncovalent forces has been observed to occur quite readily.[2]

Other photosensitizers have been developed to take advantage of the greater penetration in tissue of longer wavelength laser radiation. In addition to the enhanced absorption in the red or near-infrared region, the need for a nontoxic, photochemically active compound with selectivity for malignant tissues must still be met. Tetrapyrroles are one class of compounds that fits these criteria.[3] As the name implies, tetrapyrroles are macrocyclic compounds consisting of four pyrrole groups (heterocyclic rings of four carbons and one nitrogen) linked by methene bridges. Examples of tetrapyrrole photosensitizers include hydroxy-phenyl substituted porphyrins,[4,5] purpurins,[6,7] and metallo and nonmetallo phthalocyanine derivatives.[8–16] The porphyrin and phthalocyanine parent structures are illustrated in Figure 14-2.

An additional important consideration in the design of new photosensitizers is to include structural features that prevent or limit the formation of intermolecular aggregates of photosensitizer molecules. Why is aggregation a concern? Dimeric or oligomeric photosensitizers exhibit a greatly reduced activity toward potential substrates as a consequence of significant changes in the excited-state properties of the aggregated system.[17] Thus, bulky substituents or charged moieties are added to the ring systems to reduce aggregation. The insertion of central metal ions is also effective if one or two axial ligands are coordi-

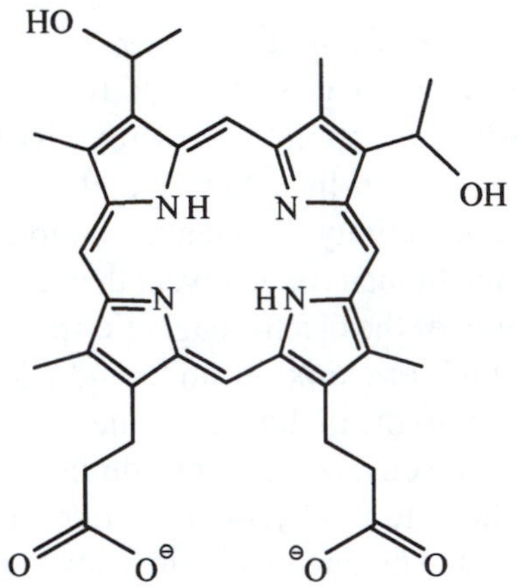

**Figure 14-1**
The chemical structure of monomeric hematoporphyrin, the parent structure for the photosensitizing drugs HpD and Photofrin® II.

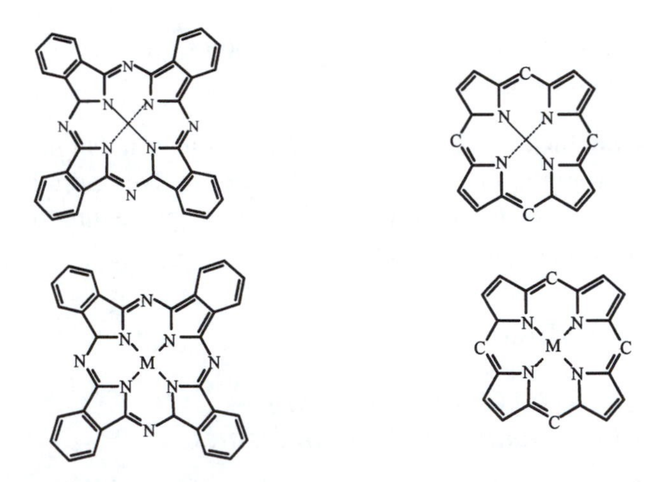

**Figure 14-2**
The parent chemical structures of two classes of tetrapyrrole photosensitizers: porphyrins (on the right) and phthalocyanines. Both metallo and nonmetallo structures are illustrated. In a metal-containing derivative, the metal atom is chelated to the four nitrogen atoms in the center of the heterocyclic ring. In metal-free structures, hydrogen atoms are bonded to two of the nitrogen atoms in the center of the heterocyclic ring. Purpurins constitute a third class of tetrapyrrole photosensitizers (not shown) in which the double C=C bond of one pyrrole ring is reduced to a single C–C bond.

nated perpendicular to the molecular plane to add considerable steric hindrance. While no single compound has achieved clinical significance for all applications, each of these new classes of photosensitizers has advantageous characteristics. For example, phthalocyanines generally exhibit extremely efficient photochemical processes, purpurins have the advantage of extinction coefficients 15–20 times those of porphyrins, and hydroxy-phenyl substituted porphyrins often exhibit 25–30 times greater potency than HpD. Advances are likely to continue in the development of potentially useful photosensitizers.

## Photochemical Mechanisms of Action

There are two primary mechanisms for the action of photosensitizers in biological systems.[18–21] Photosensitizers act either via a free-radical mechanism (type I) or an energy transfer mechanism (type II). Both of these mechanisms rely on the absorption of light to produce an excited singlet that generates a triplet state of the sensitizer via intersystem crossing:

$$^{1}P + h\nu \rightarrow {}^{1}P^{*}$$
$$^{1}P^{*} \rightarrow {}^{3}P^{*}$$

A major distinction between the two mechanisms centers on whether the triplet state of the photosensitizer reacts directly or indirectly with the substrate or tissue of the biological system. In a type I mechanism, the photogenerated triplet state of the sensitizer reacts *directly* with biological molecules in the vicinity via an electron (or H atom) transfer process. The efficiency of the electron donation or abstraction is greater with the triplet state as the reactant than with the singlet state of the photosensitizer as reagent. These reactions generate a free-radical form of both the substrate and the sensitizer. The radical species can undergo further reaction, often with oxygen, to yield additional products. The highly reactive superoxide and hydrogen peroxide species may also be produced, leading to possible cell damage.

The second mechanism (type II) involves an *indirect* reaction of the photogenerated triplet state of the sensitizer with the biological target. Instead, the photosensitizer in its triplet state reacts directly with ground-state oxygen (also a triplet) dissolved in tissue. This energy transfer process produces a ground-state sensitizer and an electronically excited singlet state of molecular oxygen. The reactive and long-lived $^{1}\Delta_{g}$ form of $O_2$ injures the biological system by oxidizing a variety of biological substrates, including certain amino acid residues in proteins (cysteine, histidine, methionine, tryptophan, and tyrosine residues, in particular), guanine residues in nucleic acids, unsaturated lipids and phospholipids, cholesterol, quinones, and metal coordination compounds.[15] Such damage to essential cellular components precedes cumulative deleterious alterations of the tumor tissue that lead to necrosis. The efficiency of a photosensitizing reaction by either a type I or type II mechanism is significantly dependent on the photosensitizer triplet state lifetime.[22]

A reaction scheme consistent with the above discussion would include the following steps:

| | |
|---|---|
| 1. Excitation | $^{1}P \rightarrow {}^{1}P^{*}$ |
| 2. Intersystem crossing | $^{1}P^{*} \rightarrow {}^{3}P^{*}$ |
| 3. Fluorescence and nonradiative decay modes | $^{1}P^{*} \rightarrow {}^{1}P$ |
| 4. Phosphorescence and nonradiative decay modes | $^{3}P^{*} \rightarrow {}^{1}P$ |
| 5. Type I reaction | $^{3}P^{*} + M \rightarrow X_1\cdot + X_2\cdot$ |
| 6. Singlet oxygen generation | $^{3}P^{*} + {}^{3}O_2 \rightarrow {}^{1}P + {}^{1}O_2^{*}$ |
| 7. Type II reaction | $^{1}O_2^{*} + M \rightarrow X_3$ |
| 8. Singlet oxygen deactivation | $^{1}O_2^{*} \rightarrow {}^{3}O_2$ |

Steps 5, 6, and 7 are bimolecular reactions requiring a local concentration of an additional molecule in the vicinity of the photogenerated triplet state of the photosensitizer. Type II reactions are clearly limited by the availability of ground-state molecular oxygen, which is not likely to be transported quickly to the site of the photosensitizer to replenish oxygen consumed in step 6.

## Laser Initiation of the Photodynamic Effect

The major excitation peak for HpD and Photofrin® II, as is typical of porphyrin compounds, occurs at about 400 nm. However, the optimal excitation wavelength to observe the photodynamic effect with these compounds is about 630 nm.[23] For hydroxy-phenyl substituted porphyrins, the optimal wavelength occurs at 648 nm.[2] The drug excitation efficiency, the transmission characteristics of the tissue, and the optical penetration distance play a role in determining this longer optical excitation wavelength. The usual laser source for photodynamic therapy treatments with HpD and Photofrin® II is a continuous wave, argon ion pumped dye laser tuned to emit at 630 nm. A recent alternative laser source is the pulsed gold vapor laser with emission at 628 nm. This system has the advantage of simpler operation with a single laser device as well as significantly higher peak power, on the order of 50–100 kW. Interferences from high-power induced nonlinear effects might be anticipated with the gold vapor laser, but none have been observed to date.

Phthalocyanines (PCs) are structurally analogous to porphyrins, containing a ring system comprised of four isoindole units linked by aza nitrogen atoms (compared with four pyrrole units linked by methine carbon atoms in porphyrins). The isoindole units provide a longer conjugated pathway in PCs than the pyrroles in porphyrins, resulting in a longer absorption wavelength for PCs. Figure 14-3 presents the structure of a substituted metal-free phthalocyanine. The visible absorption region of a phthalocyanine may have one or two bands in the 600–700 nm range. Dimeric metal-free PC in aqueous solution exhibits one band at approximately 674 nm, while monomeric metal-free PC in ethanol has two peaks near 655 and 690 nm.[15] Monomeric and dimeric metallo-PCs exhibit a single absorption band in the red region.[15] Tunable argon ion pumped dye laser systems are ideal laser sources for photodynamic therapy applications with these photosensitizers. Light penetration by both skin and human tissue is enhanced at these longer wavelengths. These advantages are realized even more readily using purpurins as photosensitizers, with strong absorption bands near 700 nm.

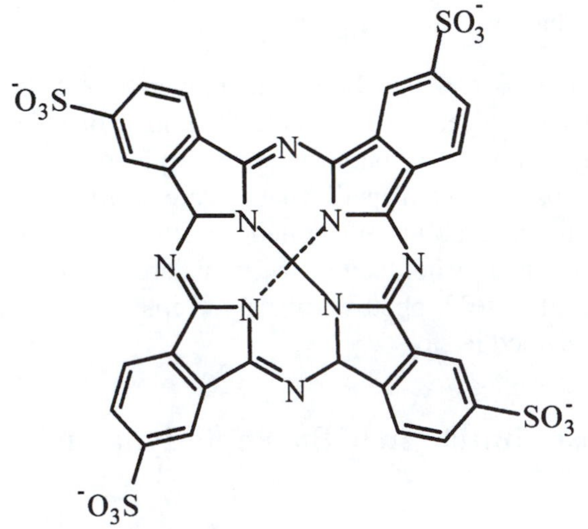

**Figure 14-3**
Chemical structure for a substituted metal-free phthalocyanine photosensitizer, tetrasulfophthalocyanine.

Several technical difficulties involving the light source and light delivery systems are common to all PDT treatments. A sizable initial cost of a complete dye laser system must first be overcome. The general lack of movability of the system can be accommodated by locating the unit adjacent to surgical suites and using optical fibers for light delivery. As the power and wavelength of radiation are critical for clinical treatments, careful monitoring of these variables must be maintained. Knowledge of the ideal power density of the light dosage and the optimal irradiation time are also critical elements for effective photodynamic treatment.

## Tumor Retention of Photosensitizers

The effectiveness of photodynamic therapy is clearly dependent on the selective retention of a photosensitizer by malignant tissue. For most organs, the photosensitizer distribution between tumor and normal tissue is between 2-5:1, although an exceedingly high ratio of 28:1 has been observed for brain tissue.[24] Details on the mechanism of transport of photosensitizer to tumor tissue are not well understood.[25] The accumulation of photosensitizers in tumors may be hypothesized to be a response to differences in oxygenation, thermal profiles, cell proliferation, blood flow, and lymphatic drainage. The generally low concentration ratio of photosensitizer between tumor and normal tissue can place the healthy tissue at risk and demands the accurate delivery of laser light to the damaged tissue. The use of multiple fiber optics to administer activating light enables a clinician to more uniformly distribute the input light within the tumor as well as to tailor the treatment to irregularly shaped tumors. In addition, multiple fiber light delivery with varying light dosages allows irradiation of the central as well as peripheral malignant tissue at optimal light levels to reduce the damage to surrounding normal tissue.

Clearly, the chemical structure of the photosensitizer affects its affinity for normal or malignant tissues. To cross cell membranes, amphiphilic or hydrophobic substances are needed. As a consequence, these photosensitizers will partition into nonpolar compartments, such as mitochondrial, lysosomal, and cytoplasmic membranes. Investigations of the subcellular distribution patterns of substituted porphyrins show several emerging trends that will aid in the design of photosensitizers for photodynamic therapy.[26] Localization in mitochondria was observed for porphyrins with cationic side chains, while uptake into lysosomes was associated with porphyrins with anionic character. This differentiation may be useful in the formulation of a targeting strategy, as there is strong evidence for considerable difference in the mitochondria of normal and tumor tissue.[26]

## Future Developments in Photodynamic Therapy

The promising results in all aspects of photodynamic therapy research over the last 15 years ensure that this modality will continue to grow as an effective therapeutic treatment for malignant tumors. Research in the future will focus on the development of new photosensitizers with more selective retention in tumor tissue and with enhanced photoactivity. Parallel developments in the laser industry will likely play an important role in the design of new photosensitizing drugs. For example, the availability of inexpensive diode lasers with emission in the red and near infrared region will dramatically lower the cost and increase the flexibility of PDT treatments. New and improved optical fibers for the delivery of both a variety and multiplicity of laser wavelengths will also enhance the number of compounds that are useful for photodynamic therapy. Even the design of new lasers specifically suited for particular photosensitizers or applications may be anticipated in the future.

## Summary of Results

### *Properties of Potent Photosensitizers for Photodynamic Therapy*

- Chemical stability to retain photophysical properties
- Low toxicity
- Affinity for and retention in malignant cells—"selectivity"
- High absorptivity in the red and near-infrared (600–700 nm) regions
- High quantum yield of triplet formation (intersystem crossing)—"efficiency"
- Small quantum yield of fluorescence to enable determination of location in tissue

### *Photochemical Mechanisms of Action of Photosensitizers*

#### *Type I Mechanism*

- Direct interaction of triplet state of photosensitizer with biological target
- Electron transfer or hydrogen abstraction process to yield free-radical form of photosensitizer and substrate
- Cell damage resulting from subsequent reactions of free-radical species

#### *Type II Mechanism*

- Indirect interaction of triplet state of photosensitizer with biological target
- Energy transfer by triplet state of photosensitizer to oxygen to yield ground state of photosensitizer and excited state of oxygen
- Malignant tissue damaged via oxidation by reactive form of oxygen

## FOR FURTHER EXPLORATION

Are you interested in reading about other applications of lasers in medicine? One well-documented and extremely successful use of excimer lasers is in the surgical procedure known as corneal sculpting or photorefractive keratectomy (PRK). Myopia or nearsightedness occurs when the cornea is highly curved or elongated, causing light rays to be focused in front of the retina. Selective removal of corneal tissue via photoablation with the 193 nm emission of an ArF excimer laser is used to correct myopia. Using a fiber optical delivery system, the ultraviolet light is delivered to the cornea to ablate and vaporize successive layers of tissue of the prescribed area and depth. Water has limited absorption in the far UV region, but the peptide linkages of proteins within the cornea, particularly the primary structural component collagen, strongly absorb these short wavelengths. The high photon energy achieves bond breaking and the significant abundance of these chromophores leads to ablation of the tissue. The ejection of both debris and low molecular weight gases from the corneal surface has been documented.[27] Laser intensity and pulse rate affect both the efficiency of ablation and the extent of heating of the surrounding tissue.[27] PRK procedures have been explored with other UV excimer and solid-state lasers, including the fifth harmonic of a Nd:YAG laser at 213 nm, diode-pumped Nd:YAG and Nd:YLF emission at 213 and 209 nm, respectively, and the tunable fourth-harmonic Ti:sapphire emission from 205 to 220 nm.[28] While these laser systems and other alternative laser technologies are only in the developmental stage, the success of ArF laser corneal sculpting will continue to generate interest in laser refractive surgery.

## DISCUSSION QUESTIONS

1. Are useful photosensitizers for photodynamic therapy highly fluorescent species? Why or why not?
2. What kinds of species will act as efficient photosensitizers according to a type I reaction?
3. What aspects must be considered in the development of photosensitizers?
4. Distinguish the four possible interactions of light with human tissue.
5. Why can't the emission of $CO_2$ lasers be delivered via optical fibers to tissues or chromophores within the human body?
6. Describe the two primary mechanisms for the action of photosensitizers in biological systems.

## LITERATURE CITED

1. Goldman, L. "Chromophores in tissue for laser medicine and laser surgery," *Lasers in Medical Science,* 1990, 5, 289–292.
2. Dougherty, T. J., "Photosensitizers: Therapy and detection of malignant tumors," *Photochem. Photobiol.*, 1987, 45, 879–889.
3. Roeder, B., "Tetrapyrroles: A chemical class of potent photosensitizers for the photodynamic treatment of tumours," *Lasers in Medical Science,* 1990, 5, 99–106.
4. Berenbaum, M. D., Akande, S. L., Bonnett, R., Kaur, H., Ivannow, S., White, R. D., and Winfield, U. J., "Meso-tetra(hydroxyphenyl) porphyrins, a new class of potent tumour photosensitizeres with favourable sensitivity," *Br. J. Cancer,* 1986, 54, 717–725.
5. Bonnett, R., McCarvey, D. J., and Harriman, A., "Photophysical properties of meso-tetraphenylporphyrin and some meso-tetra(hydroxyphenyl) porphyrins," *Photochem. Photobiol.,* 1988, 48, 271–276.
6. Morgan, A. R., Garbo, G. M., Kreimer-Birnbaum, M., Keck, R. W., Chaudhuri, K., and Selman, S. H., "Morphological study of the combined effect of purpurin derivatives and light on transplantable rat bladder tumors," *Cancer Res.,* 1987, 47, 496–498.
7. Morgan, A. R., Garbo, G. M., Keck, R. W., and Selman, S. H. "New photosensitizers for photodynamic therapy: Combined effect of metallopurpurin derivatives and light on transplantable bladder tumours," *Cancer Res.,* 1988, 48, 194–198.
8. Ben-Hur, E., and Rosenthal, I., "The phthalocyanines: A new class of mammalian cell photosensitizers with a potential for cancer phototherapy," *Int. J. Radiat. Biol.,* 1985, 47, 145–147.
9. Ben-Hur, E., and Rosenthal, I., "Photosensitized inactivation of Chinese hamster cells by phthalocyanines," *Photochem. Photobiol.,* 1985, 42, 129–133.
10. Ben-Hur, E., and Rosenthal, I., "Factors affecting the photokilling of cultured Chinese hamster cells by phthalocyanines," *Radiat. Res.,* 1985, 103, 403–409.
11. Ben-Hur, E., and Rosenthal, I., "Photosensitization of Chinese hamster cells by water-soluble phthalocyanines," *Photochem. Photobiol.,* 1986, 43, 615–619.

12. Bown, S. G., Tralau, C. J., Coleridge-Smith, P. D., Akdemir, D., and Wieman, T. J., "Photodynamic therapy with porphyrin and phthalocyanine sensitization: Quantitative studies in normal rat liver," *Br. J. Cancer,* 1986, 54, 43–52.

13. Selman, S. H., Kreimer-Birnbaum, M., Chadhuri, K., Garbo, G., Seanar, D., Keck, R., Ben-Hur, E., and Rosenthal, I., "Photodynamic treatment of transplantable bladder tumors in rodents after pretreatment with chloroaluminum tetrasulfophthalocyanine," *J. Urol.,* 1986, 136, 141–145.

14. Brasseur, N., Hasrat, A., Langlois, R., Wagner, J. R., Rousseau, J., and van Lier, J. E., "Biological activities of phthalocyanines. V. Photodynamic therapy of EMT-6 mammary tumors in mice with sulfonated phthalocyanines," *Photochem. Photobiol.,* 1987, 45, 581–586.

15. Spikes, J. D., "Phthalocyanines as photosensitizers in biological systems and for the photodynamic therapy of tumors," *Photochem. Photobiol.,* 1986, 43, 691–699.

16. Ben-Hur, E., Green, M., and Prager, A., "Phthalocyanine photosensitization of mammalian cells: biochemical and ultrastructural effects," *Photochem. Photobiol.,* 1987, 46, 651–656.

17. Jori, G., "Factors controlling the selectivity and efficiency of tumour damage in photodynamic therapy," *Lasers in Medical Science,* 1990, 5, 115–120.

18. Dougherty, T. J., Kaufman, J. E., Goldfarb, A., Weishaupt, K. R., Boyle, D., and Mittleman, A., "Photoradiation therapy for the treatment of malignant tumors," *Cancer Res.,* 1978, 38, 2628–2635.

19. Hayata, Y., Kato, H., Konaka, C., Ohno, J., and Takizawa, N., "Hematoporphyrin derivatives and laser photoradiation in the treatment of lung cancer," *Chest,* 1982, 81, 269–277.

20. Kessek, D., "Hematoporphyrin and Hpd: photophysics, photochemistry, and phototherapy," *Photochem. Photobiol.,* 1984, 39, 851–859.

21. Nakajima, A., Hayashi, H., Ohshima, K., Yamazaki, K., Kubo, Y., Samejima, N., Kakiuchi, Y., Shindoh, H., Koshimizu, I., Sakata, I., and Yamauchi, N., "Tumor imaging with [$^{111}$In]mono-DTPA-ethyleneglycol-Gadeuteroporphyrin," *Photochem. Photobiol.,* 1987, 46, 783–788.

22. Takemura, T., Ohta, N., Nakajima, S., and Sakata, I., "Critical importance of the triplet lifetime of photosensitizer in photodynamic therapy of tumor," *Photochem. Photobiol.*, 1989, 50, 339–344.

23. Bottiroli, G., Croce, A. C., Ramponi, R., and Vaghi, P., "Distribution of di-sulfonated aluminum phthalocyanine and photofrin II in living cells: A comparative fluorometric study," *Photochem. Photobiol.,* 1982, 55, 575–585.

24. Tralau, C. J., Barr, H., Sandeman, D. R., Barton, T., Lewis, M. R., and Bown, S. G., "Aluminum sulfonated phthalocyanine distribution in rodent tumours of the colon, brain and pancreas," *Photochem. Photobiol.,* 1987, 46, 777–781.

25. Nelson, J. S., Liaw, L.-H., and Berns, M. W., "Tumor destruction in photodynamic therapy," *Photochem. Photobiol.,* 1987, 46, 829–835.

26. Woodburn, K. W., Vardaxis, N. J., Hill, J. S., Kaye, A. H., and Phillips, D. R., "Subcellular localization of porphyrins using confocal laser scanning microscopy," *Photochem. Photobiol.,* 1991, 54, 725–732.

27. Pettit, G. H., Ediger, M. N., and Weiblinger, R. P., "Excimer laser ablation of the cornea," *Optical Engineering,* 1995, 34, 661–667.

28. Lin, J. T., "Critical review on refractive surgical lasers," *Optical Engineering,* 1995, 34, 668–675.

CHAPTER 15

# Laser-Induced Selective Bond Chemistry

## Chapter Overview

To be able to selectively energize a chemical bond for further chemical reaction is the dream of virtually every chemist interested in synthesis of new compounds. The advent of the laser brought immediate visions of bond selective chemistry revolutionizing synthetic chemistry. The practice has proven more difficult than first thought. Significant milestones have recently been achieved, however, by finding molecular systems with bonds that can be considered localized. Localized bonds can indeed be selectively excited by laser light, and the direction of chemical reaction involving such bonds is subsequently largely controllable.

## The Great Quest

If there is a "great quest" among chemists who use lasers to facilitate chemical reactions, it would be the discovery of how to tune the laser to a selected bond's vibrational frequency and activate that specific bond for chemical reaction. In essence, chemists seek to conduct *bond selective chemistry*. The quest has not yet been fulfilled, but several significant mileposts have been passed. In particular, chemists seeking to selectively excite a particular molecular motion have learned that a molecule rapidly distributes the energy absorbed by a single vibration among several or all of its other molecular motions. Indeed, early attempts at bond selective chemistry did cast a pall on the prospects of success. To meet the challenge, molecules are required where at least some, if not all, of the molecular vibrations can be considered localized. What do we mean by a localized vibration? Recall that a polyatomic molecule with $N$ atoms should have $3N$-6 independent vibrational motions, each describable by a normal coordinate. (We discuss normal coordinates in Chapter 9 in connection with Raman spectroscopy.) If the motions were truly independent, the possibility of energy flow from one motion to another would be unlikely. Unfortunately in the real world, these motions are not independent, and the absorption of energy by one supposedly independent molecular vibration can quickly "leak" throughout the molecule by the coupling, in essence the extent of interdependence, of the molecular motions. Certain molecules, however, seem to have particular motions that can be well-modelled as effectively independent of other motions. Such independent vibrations are often referred to as *localized*.

Water is a substance whose OH stretching motions can be considered localized, and it should be no surprise that the earliest demonstrations of bond selective chemistry involved water. A bond which is localized can be excited to higher vibrational states and remains in those states long enough to react before the molecular motion once again "leaks" energy to other motions. The examples that will be highlighted in the following case study use different excitations of the localized OH stretch in water. Other essentially localized motions in different molecules are being utilized for bond selective chemistry as well. For example, the umbrella motion of the planar ion $NH_3^+$ has also been shown to undergo bond selective reactions, as we will soon see below.

## Case Study: Bond Selective Chemistry of Light Atom Molecules

### Overview of the Case Study

**Objective.** To demonstrate that the direction of a bimolecular chemical reaction can be controlled by initial selective excitation of a specific chemical bond with laser light.

**Laser Systems Employed.** Three laser systems are used. For bond excitations a Nd:YAG frequency-doubled laser is used to pump either a dye laser or an optical parametric oscillator. The photodissociation energy necessary to fragment the second reactant is provided by a frequency-tripled Nd:YAG laser, again pumping a dye laser. The laser-induced fluorescence (LIF) detection is accomplished with either a frequency-quadrupled Nd:YAG laser or an excimer laser pumping another dye laser.

**Role of the Laser Systems.** The first laser system selectively excites the chemical bond of interest. The second system photodissociates selected molecules to yield the atoms necessary for the intended atom-molecule reaction. The third laser system is used to induce fluorescence in product molecules for subsequent detection by a photomultiplier.

**Useful Characteristics of the Laser Light for this Application.** The Nd:YAG laser provides high-energy pulses which are capable of generating high-energy frequency harmonics to drive a variety of tunable dye lasers or an OPO. Tunability is necessary to provide the specific energy control for bond excitation or induced fluorescence.

**Principles Reviewed.** Vibrational energies and spectroscopy, chemical reaction rates, activated complexes.

**Conclusions.** Deuterated water HOD shows bond selective chemistry when either the OH or OD stretch of HOD is excited. The planar ion $NH_3^+$ also shows selectivity toward reaction with $ND_3$ when the umbrella motion of $NH_3^+$ is excited.

## Chemical Reactions of HOH and HOD

Some of the earliest hints that bond selective chemistry was feasible involved reactions of atoms (represented below by *L*) with water, following one of the two possible reactions:

$$L + \mathrm{HOH} \rightarrow \mathrm{H_2} + L\mathrm{OH} \qquad (1)$$
$$L + \mathrm{HOH} \rightarrow L\mathrm{H} + \mathrm{OH} \qquad (2)$$

Selective excitation of the water OH stretching motion was observed to increase the rate of the reaction, especially for reaction 2. Specifically, Crim and his co-workers discovered that the rates of the reactions:

$$\mathrm{H} + \mathrm{HOH} \rightarrow \mathrm{H_2} + \mathrm{OH} \qquad (3)$$
$$\mathrm{Cl} + \mathrm{HOH} \rightarrow \mathrm{HCl} + \mathrm{OH} \qquad (4)$$

were accelerated by selective excitation of the OH stretching vibration but not influenced by selectively exciting the bending vibration of water.[1,2] The degree by which the rate of reaction 3 was increased showed a dependence on the exact vibrational state for each of the two vibrations in water. For example, a state called |03 > has one OH bond excited with three vibrational quanta while the other OH bond is still in its ground state. A state close-by in terms of total vibrational energy in the water molecule is the |12 > state, which is characterized by one quantum of vibrational energy in one OH bond and two quanta in the other. The rate of reaction 3 starting from the |03 > state was found to be significantly faster than the rate starting from the |12 > state. This type of discrimination in reaction rates that depends on bond vibrational excitation is suggestive of bond selective chemistry but is not conclusive evidence since each bond in HOH is the same: OH. Reactions of atoms with HOD would better demonstrate bond selectivity because the two bonds would at least differ in vibrational frequency.

Crim and his co-workers[1,2] and Zare and his co-workers[3] both demonstrated that HOD can be made to selectively react, breaking either the OD or the OH bond, depending on which bond was vibrationally excited. The basic idea is represented by the reactions:

$$(\mathrm{HOD} + h\nu_{\mathrm{OH}}) + L \rightarrow \mathrm{OD} + \mathrm{H}L \qquad (5)$$
$$(\mathrm{HOD} + h\nu_{\mathrm{OD}}) + L \rightarrow \mathrm{OH} + \mathrm{D}L \qquad (6)$$

The differences in the experimental approaches and results of the two groups are slight but important. Crim and co-workers studied reaction 5 with $L$ = H or Cl and a water molecule with four $h\nu_{OH}$ quanta of vibrational energy. Zare and his co-workers studied reactions 5 and 6 using $L$ = H and water with one $h\nu_{OH}$ quantum of energy (reaction 5) and with one $h\nu_{OD}$ quantum of energy (reaction 6). The results of these studies are summarized in Table 15-1. Clearly, exciting the OH bond in HOD favorably activates that bond toward reaction with H or Cl. It is envisioned that the activation promotes enhanced OH stretching, and the direction of stretching becomes the reaction coordinate of the activated complex [O–H——L]. Moreover, the OD bond can be viewed as a spectator bond, with little of its energy transferred to the OH bond or vice versa.

What is also clear from the results is that the bond selective chemistry can be accomplished with only one $h\nu$ quantum of energy, either $1h\nu_{OH}$ or $1h\nu_{OD}$. Exciting the OD bond yields HD while exciting the OH bond yields $H_2$. The bond not excited takes on a spectator role. Zare and his co-workers note that, with their experimental uncertainties, it is not clear whether the OD excitation yields less product than the OH excitation.

**Table 15-1**
Bond Selective Chemistry of HOD and $NH_3^+$

| *Reactants* | *Products (Relative Yields x)* |
|---|---|
| $[HOD + 4\nu_{OH}] + H$ | $100x \rightarrow OD + H_2$<br>$1x \rightarrow OH + HD$ |
| $[HOD + 1\nu_{OH}] + H$ | $25x \rightarrow OD + H_2$<br>$1x \rightarrow OH + HD$ |
| $[HOD + 1\nu_{OD}] + H$ | $1x \rightarrow OD + H_2$<br>$8x \rightarrow OH + HD$ |
| $[HOD + 4\nu_{OH}] + Cl$ | $8x \rightarrow OD + HCl$<br>$1x \rightarrow OH + DCl$ |
| $[NH_3^+ + nh\nu_{umb} + mh\nu_{breth}] + ND_3$ | $\rightarrow NH_3D^+ + ND_2$ deuterium abstraction<br>$\rightarrow NH_3 + ND_3^+$ charge transfer<br>$\rightarrow\rightarrow\rightarrow NH_2 + ND_3H^+$ proton transfer |

## Reactions of $NH_3^+$

In another study with a more complex molecule, $NH_3^+$, further evidence of bond selective chemistry was observed by Zare and co-workers.[4] Allowing $NH_3^+$ and $ND_3$ to react yields three possible outcomes, as shown in Table 15-1: deuterium abstraction, charge transfer, or proton transfer. Both the so-called umbrella motion and the so-called breathing motion of $NH_3^+$ can be considered localized vibrations for independent excitation. When either of two excited states of $NH_3^+$, one with $n = 2$ and $m = 2$ (see Table 15-1 for definition of $n$ and $m$) and the other with $n = 5$ and $m = 0$, are allowed to react with $ND_3$, proton transfer is the preferred result. The extent of the reaction is apparently independent of the excitation of the breathing vibration because $m = 1$ or $m = 0$ makes little difference in the products found. Moreover, even though there is some dependence on the relative kinetic energies of the reactants, the vibrational activation seems to be the overriding consideration in this bond selective chemistry. While clear evidence was found by the authors for bond selective chemistry in the reactions of $NH_3^+$, the extent of the reaction control was much less than in the cases where H or Cl reacts with HOD.

## The Role of Lasers in the Design of the Experiment

How are these experiments carried out? Basically both teams of experimenters used three laser systems. One system prepares the vibrational excited reactant: HOH, HOD, or $NH_3^+$. A second laser system prepares the other reactant, H or Cl, from HI or $Cl_2$. The third system provides the probe laser beam to generate the laser-induced fluorescence signal. The overall apparatus is illustrated in Figure 15-1. Table 15-2 provides details of the specific lasers used for each function. Note that the Nd:YAG laser plays an important role in both systems, providing high-energy pulses that can be frequency-doubled, -tripled, and even -quadrupled to drive dye lasers or an OPO

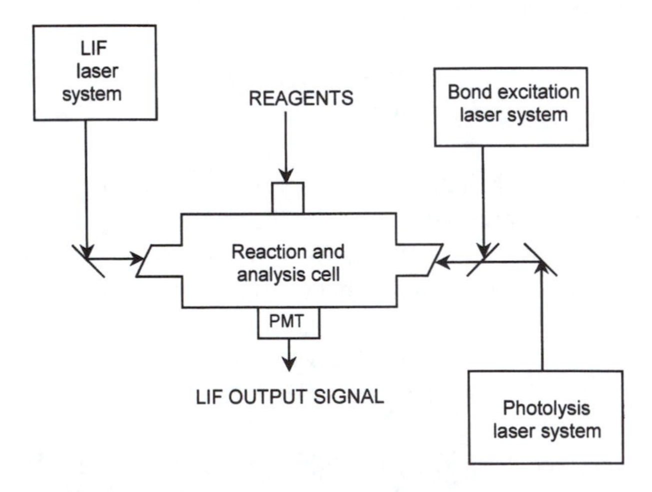


**Figure 15-1**
Schematic of the total laser apparatus used for the initiation and detection of bond selective chemistry. Three laser systems serve separate functions. The bond excitation laser system selectively excites one of the bonds of a reactant molecule and thus prepares a vibrationally excited reactant. The photolysis laser system breaks a specific bond in the second reactant to generate an atom for reaction with the vibrationally excited reactant. The LIF (laser-induced fluorescence) laser system provides excitation to induce a fluorescence signal in one of the products.

**Table 15-2**
Laser System Components for HOD + H or Cl Studies

| *Laser Function* | *Stanford (Zare Group)* | *Wisconsin (Crim Group)* |
|---|---|---|
| Bond excitation | Frequency-doubled Nd:YAG pumping an OPO | Frequency-doubled Nd:YAG pumping a dye laser which might be Raman-shifted using $H_2$ |
| Photolysis | Frequency-quadrupled Nd:YAG pumping a dye laser to dissociate HI | $H_2$ dissociated using a microwave discharge |
| Laser-induced fluorescence | Frequency-doubled Nd:YAG pumping a dye which is frequency-doubled | Excimer laser pumping a dye laser which is frequency-doubled |

with tuning capacities. In each investigation dye lasers pumped with tunable UV dyes were used to induce the fluorescence signals from the $A^2\Sigma^+ - X^2\Pi$ ultraviolet bands of OH or OD. The laser systems are complex in total, but each laser was chosen for a specific task using the special characteristics of that laser. This combination of laser systems is a powerful means to achieve the desired result.

## SUMMARY

Significant advances have been made in the quest for selectively controlling the course of a chemical reaction by the addition, in effect, of laser light as a catalyst. Clearly, additional exciting advances in the area of bond selective chemistry can be anticipated.

## FOR FURTHER EXPLORATION

The observation of mode selective or bond selective chemistry in polyatomic reaction systems will be an increasing occurrence. Two such complementary studies reveal fascinating comparisons between the reactions of the chlorine atom with the methane molecule in both the vibrational ground (v = 0) and excited (v = 1) states.[5,6] Frequency tripling of the output of a $Nd^{3+}$:YAG laser generates 355 nm light to photolyze molecular chlorine to 98% ground-state chlorine atoms. The 1064 nm fundamental output of the $Nd^{3+}$:YAG laser pumps a lithium niobate optical parametric oscillator that emits infrared excitation to vibrationally excited methane. In essence, by providing one quantum of asymmetric C–H stretch excitation, the two investigations probe the impact of the C–H stretch excitation on the overall reaction. The reaction under both sets of conditions generates HCl and $CH_3$, and the population distributions of the HCl product are determined by (2 + 1) resonance-enhanced multiphoton ionization (REMPI). The results reveal a marked contrast that clearly indicates that vibrational enhancement dramatically changes the energetics of the reaction. The ground-state reaction occurs only for head-on collisions between reactants, as a consequence of the requirement of a high degree of C–H–Cl collinearity in the transition state. Vibrational excitation of the C–H stretch lowers the barrier to reaction and allows reactivity to extend to the periphery of the methane molecule. These and future studies of mode selective chemistry will deepen our understanding of the mechanisms of chemical reactions.

## DISCUSSION QUESTION

1. Suppose bonds of the type XH can be considered good candidates for selective excitation. Would a reasonable criterion to pick a particular XH bond as a selective reactant be the fact that the XH stretch is relatively long or short?

## LITERATURE CITED

1. Sinha, A., Hsiaso, M. C., and Crim, F. F., "Controlling bimolecular reactions: Mode and bond selected

reaction of water with hydrogen atoms," *J. Chem. Phys.*, 1991, 94, 4928–4935.

2. Sinha, A., Thoemke, J. D., and Crim, F. F., "Controlling bimolecular reactions: Mode and bond selected reaction of water with translationally excited chlorine atoms," *J. Chem. Phys.*, 1992, 96, 372–376.
3. Bronikowski, M. J., Simpson, W. R., Girard, B., and Zare, R. N., "Bond-specific chemistry: OD:OH product ratios for the reactions H+HOD(001) and H+HOD(100)," *J. Chem. Phys.*, 1991, 95, 8647–8648.
4. Guettler, R .D., Jones, Jr., G. C., Posey, L. A., and Zare, R. N., "Partial control of an ion-molecule reaction by selection of the internal motion of the polyatomic reagent ion," *Science*, 1994, 266, 259–261.
5. Simpson, W. R., Rakitzis, T. P., Kandel, S. A., Orr-Ewing, A. J., and Zare, R. N., "Reaction of Cl with vibrationally excited $CH_4$ and $CHD_3$: State-to-state differential cross sections and steric effects for the HCl product," *J. Chem. Phys.* 1995, 103, 7313–7335.
6. Simpson, W. R., Rakitzis, T. P., Kandel, S. A., Lev-On, T., and Zare, R. N., "Picturing the transition-state region and understanding vibrational enhancement for the $Cl + CH_4 \rightarrow HCl + CH_3$ reaction," *J. Phys. Chem.*, 1996, 100, 7938–7947.

APPENDIX I

# The History of Lasers in Chemistry

1917 A. Einstein publishes a paper detailing the concept of stimulated emission.

1954 Charles H. Townsend constructs a microwave amplifier device called the maser using the phenomenon of stimulated emission; A. M. Prochorov and Nicolai G. Basov propose a similar device at the same time.

1958 Townsend and Arthur L. Schawlow publish a paper discussing the extension of maser principles to the optical region of the electromagnetic spectrum.

1960 The acronym "laser" is coined by Gordon Gould, a pioneer of laser development.

Theodore H. Maiman of the Hughes Aircraft Corporation Research Laboratories constructs the first laser—a solid-state ruby laser operated on a pulsed basis.

1961 Ali Javan and co-workers at Bell Telephone Laboratories fabricate the first CW laser—a helium-neon laser.

S. Jacobs, P. Rabinowitz, and Gordon Gould at Columbia University develop the first metal vapor laser (a cesium vapor laser).

W. Kaiser and C. G. Garrett perform the first demonstration of two-photon absorption.

Researchers at the University of Michigan demonstrate the process of frequency doubling or second-harmonic generation by focusing 694.3 nm light from a ruby laser onto a quartz crystal.

Zaret and colleagues are the first to employ lasers in ophthalmology in the treatment of retinal detachment (later published by M. M. Zaret, H. Ripps, L. M. Siegel, and G. M. Breinin, *Arch. Ophthalmol.,* 1963, 69, 97).

1962 The helium-neon laser is the first laser made available to a commercial market.

The occurrence of laser action in semiconductor *p–n*–junctions is found in several American laboratories (IBM, General Electric, and Lincoln Laboratory at MIT).

Britton Chance and Heinz Schleyer of the Johnson Research Foundation at the University of Pennsylvania report on the first application of lasers to molecular biology—the use of a liquid-nitrogen-cooled ruby laser to study electron transfer in photosynthetic algae.

1964 Townes, Prochorov, and Basov are awarded the Nobel Prize for Physics for their inventions of laser devices.

Infrared laser emission from $CO_2$ is first reported by C. K. N. Patel (*Phys. Rev. Lett.,* 1964, 12, 588).

1965 Gigawatt pulses are demonstrated using mode-locking techniques.

The cobalt-doped magnesium fluoride solid-state laser (Co:$MgF_2$) is discovered at AT&T Bell Laboratories.

1966 Peter P. Sorokin and John R. Lankard at IBM's Thomas J. Watson Research Center, Yorktown Heights, are the first to obtain stimulated emission from an organic compound—chloro-aluminum-phthalocyanine—thereby discovering the first dye laser (P. P. Sorokin, and J. R. Lankard, *IBM J. Res. Develop.,* 1966, 10, 162).

1970 Alfano and Shapiro discover the phenomenon of producing a broad continuum of light with ultra-short pulse duration—the ultrafast supercontinuum laser source, or picosecond continuum (R. R. Alfano, and S. L. Shapiro, *Phys. Rev. Lett.,* 1970, 24, 584).

1974 The separation of isotopes of an element (barium) is first reported by Anthony F. Bernhardt, Donald E. Duerre, Joe R. Simpson, and Lowell L. Wood of the Lawrence Livermore Laboratory.

1977 The first operation of a free-electron laser (wavelength of 3.417 $\mu$m) is reported by a Stanford research group (D. A. G. Deacon, L. R. Elias, J. M. J. Madey, G. J. Ramian, H. A. Schwettman, and T. I. Smith, *Phys. Rev. Lett.,* 1977, 38, 892–894).

Hurst and colleagues at Oak Ridge National Laboratory employ the technique of laser resonance ionization to detect single atoms of an element (cesium) for the first time.

1980 The shortest recorded laser pulse is of 150 fs duration.

1981 Arthur L. Schawlow, Nicolass Boembergen, and Kai M. Siegbahn are awarded the Nobel Prize for Physics for their contributions to the development of laser spectroscopy.

1982 Peter F. Moulton first observes lasing in the Ti:sapphire crystal at MIT Lincoln Laboratory, Lexington, Mass.

1985 The soft X-ray laser is developed with wavelengths 100 times shorter than that of visible light.

1987 The shortest laser pulse created to date is recorded at 6 fs on a dye laser system by R. L. Fork and associates at Bell Laboratories (R. L. Fork et al., *Opt. Lett.,* 1987, 12, 483).

1988 The Ti:sapphire laser is first offered commercially.

1991 Richard N. Zare and co-workers at Stanford University demonstrate the first use of lasers to selectively control the products of a chemical reaction—the reaction of energetic H atoms and deuterated water (HOD).

1992 Ahmed Zewail and researchers at Caltech use two 50 femtosecond pulses to control the reaction of xenon and iodine to produce xenon iodide.

1993 SOPRA develops a 1 kW XeCl excimer laser (very large excimer laser—VEL) which is capable of 10-J/pulse nominal energy at 308 nm.

1994 A 2 $\mu m$ InGaAsP semiconductor laser is available commercially from SDL, Inc.

1995 Researchers at the Center for UltraFast Optical Science at the University of Michigan have developed an ultrafast ($\leq$ 1 ps) extreme ultraviolet (XUV = 45–70 Å) source for time-resolved dynamical studies.

A high-intensity Nd:glass laser produces 13 terawatts of peak power (6 J in 450 fs) at 1.06 $\mu$m.

APPENDIX II

# Laser Types

Solid-State Lasers: These lasers are based on optical transitions in transition metal ions embedded in a transparent host ionic crystal or glass.

| *Laser* | *Transition Metal/ Host Lattice* | *Tuning Range or Principal Emission λ* | *Principal Features and Applications* |
|---|---|---|---|
| Ruby | Chromium ($Cr_2O_3$)/ corundum $Al_2O_3$ | 694 nm | Pulsed output (CW only with cryogenic cooling); pumped by flashlamp; used in holography and for plasma diagnostics in large fusion reactors. |
| Alexandrite | Chromium/ $BeAl_2O_4$ | 700–815 nm | Pulsed output; pumped by flashlamp; used in medical research and military applications. |
| Titanium: sapphire | Titanium ($Ti^{3+}$)/ corundum $Al_2O_3$ | 650–1100 nm | CW (2.5 W) or pulsed (10–50 Hz, 10–200 mJ/pulse) output; unusually wide tuning range; pumped by a variety of lasers; used in IR spectroscopy, particularly of semiconductors; photodynamic therapy research; LIDAR. |
| Nd:YAG | Neodymium ($Nd^{3+}$)/ yttrium aluminum garnet crystal ($Y_3Al_5O_{12}$) | 1.064 μm | CW (up to 100 W) and pulsed (20–100 ps with mode-locking); pumped by flashlamp or external lasers; numerous diverse applications; useful for harmonic generation of λs of 532, 355, and 266 nm. |
| Nd:glass | Neodymium ($Nd^{3+}$)/ amorphous glass | 1.064 μm | Pulsed output; extremely suitable for production of high-intensity ultrashort pulses via mode-locking; pumped by flashlamp. |
| $Co{:}MgF_2$ | cobalt/magnesium fluoride | 1.7–2.5 μm | Pulsed output (CW only with cryogenic cooling); pumped by Nd:YAG laser; medical research, remote sensing applications. |
| Rare-earth lasers | Erbium ($Er^{3+}$), holmium ($Ho^{3+}$), or thulium ($Tm^{3+}$)/yttrium aluminum garnet crystal | 2.94 μm (Er)<br>2.089, 2.127 μm (Ho)<br>2.01 μm (Tm) | Pumped by flashlamp or external laser, CW or pulsed output; 2 μm output useful for precision surgery on cartilage and eye-safe LIDAR; useful for cutting bone and drilling teeth. |

Dye Lasers: These lasers use fluorescent dyes as the gain medium. The dyes are typically large polyatomic organic molecules with absorption bands in the visible region and broad fluorescence bands for wide tunability.

| *Pump Laser* | *Pump λ (nm)* | *Output Mode* | *Principal Features* |
|---|---|---|---|
| Argon ion | Typically 488.0 or 514.5 | CW | Passively mode-locked output of < 100 fs pulsewidth and ≈ 100 MHz repetition rate is possible with 0.1 nJ energy per pulse. |
| Frequency-doubled Nd:YAG | 532 | Pulsed | Very high energy per pulse (up to 150 mJ), up to 50 Hz repetition rate; mode-locked output of < 250 fs. |
| Excimer laser | 308 (XeCl)<br>248 (KrF) | Pulsed | High pulse energy up to 80 mJ with high repetition rates up to 400 Hz possible. |
| Nitrogen | 337 | Pulsed | Economical with low-energy pulses (up to 70 μJ) and high repetition rates (up to 100 Hz). |

Monatomic Gas Lasers: Monatomic gases provide the active medium in these lasers.

| *Monatomic Gas(es)* | *Emission Wavelengths* | *Characteristics and Applications* |
|---|---|---|
| Argon | Numerous discrete lines between 350 and 530 nm; strongest lines at 488.0 and 514.5 nm | Fairly expensive, comparatively fragile, limited plasma tube lifetime; numerous spectroscopic applications, often used to pump dye lasers. |
| Helium–Neon | 632.8 nm; 1.152 μm; 3.391 μm | Relatively inexpensive, low output power, CW output; used in optical scanning devices, also for electronic printing and optical alignment. |
| Helium–Cadmium | 441.6 nm; 325 nm; 353.6 nm | Low power output, CW operation, sensitive to adequate Cd vapor pressure; numerous applications requiring blue or UV light of low power, including mastering of compact discs and CD-ROMS, laser-induced fluorescence, stereolithography, and biomedical diagnostics. |
| Krypton | Numerous discrete lines between 350 and 800 nm; strongest at 647.1 nm | Often mixed with argon gas to provide a wide range of discrete wavelengths; especially useful in biomedical applications. |

Molecular Gas Lasers: In these lasers the transitions responsible for stimulated emission occur in free gaseous molecules.

| *Molecular Gas* | *Principal Emission Lines* | *Output Characteristics* | *Principal Features and Applications* |
|---|---|---|---|
| $CO_2$ | 9.6 and 10.6 μm | Output power up to 10 kW when operated CW | Lasing action involves rotation-vibration energy levels; used for laser-induced chemical reactions, materials processing, surgical procedures. |
| $N_2$ | 337 nm | Pulsed output with 120 Hz maximum repetition rate | Operates on electronic transitions; inexpensive source of UV radiation; used in photochemical studies and as a pump for dye lasers. |
| $I_2$ | 1.315 μm | Microsecond pulses | Lasing transition occurs in free atomic iodine; used in temperature-jump experiments in aqueous solution and in LIDAR applications. |
| Excimer-inert gas halides | $F_2$:157 nm<br>ArF:193 nm<br>KrCl:222 nm<br>KrF:248 nm<br>XeCl:308 nm<br>XeF:351 nm | 10–20 ns pulses with average power of 20–100 W | Laser action achieved via excimers—combinations of two atoms that survive only in the excited state; useful for breaking chemical bonds and for vaporization (photoablation). |
| HF, HCl | 2.6–3.0 μm | Continuous wave power up to 150 W | Example of direct conversion of chemical energy into electromagnetic radiation, i.e., chemical reactions used to generate molecules with nonequilibrium inverted populations. |

Metal Vapor Lasers: These lasers are high-power, efficient sources of visible light in which the monatomic metal vapor lasing medium is created by using an electric discharge to heat a plasma tube containing a small amount of metal.

| *Metal Vapor Laser* | *Emission Line (nm)* | *Principal Features/Applications* |
|---|---|---|
| Copper | 510.6, 578.2 | Green and yellow lines present in 2:1 ratio; used in uranium isotope enrichment processes; high-speed photography. |
| Gold | 312.2, 627.8 | Red line useful in applications of cancer photodynamic therapy. |
| Strontium | 430.0 | A blue light source. |
| Manganese | 534.1, 1289.9 | No current commercial applications. |
| Barium | 1130.0, 1500.0 | No current commercial applications. |
| Lead | 722.9 | No current commercial applications. |

Semiconductor Lasers: Semiconductor lasers are solid-state lasers in which the lasing action occurs at a junction between p- and n-doped crystals (i.e., crystals with impurity atoms with fewer or more valence electrons than the atoms replaced, respectively). Pulsed operation at room temperature is possible with those lasers having such a junction (a heterojunction). Continuous wave operation at room temperature occurs when the active layers are of slightly lower refractive index, thus generating a double heterojunction.

| *Diode Laser Family* | *Example and Principal Emission Wavelengths* | *Applications* |
|---|---|---|
| Gallium arsenide | Binary and nonstoichiometric ternary and quaternary compounds of gallium and arsenic with aluminum, indium, and/or phosphorus:<br>GaAs: 0.904 μm<br>$Ga_{1-x}Al_xAs$: 0.8–1.3 μm<br>$In_{1-y}Ga_yAs$: 1.5 μm<br>$In_{1-x}Ga_xAs_{1-y}P_y$: 1.31–1.55 μm and 2 μm | Compact disc players, laser printers, and fiber optic communications; solid-state laser pumps (especially for Nd lasers). |
| Lead salt | Nonstoichiometric ternary and quaternary compounds, typically of lead, cadmium and/or tin, and tellurium, selenium, and/or sulfur:<br>$Pb_{1-x}Cd_xS$: 2.0–3.0 μm<br>$Pb_{1-x}Sn_xSe$: 8.0–29.0 μm | CW output (0.1 mW); wavelength tunability achieved by varying temperature—useful in high-resolution spectroscopy; space research. |

APPENDIX III

# Energy Scales and Intensity Units

Many units are used to measure the energy and wavelength of a photon. Moreover, many other units are used to describe beams of photons, usually referred to as beam power or intensity. Even though the scientific and technological world has in principle adopted System Internationale (SI) units, many practitioners in various subdisciplines persist in the use of older, now historical, units. We will first explore the units used to measure photon energy and then those used to characterize photon beams.

## Photon Energies

All units used to measure a photon's energy must start with Planck's law:

photon energy = constant × photon frequency

or $$\varepsilon = h\nu \qquad \text{(III-1)}$$

where $h$ is Planck's constant. Since Einstein asserted that the speed of a photon (in a vacuum) is a constant $c$, the wave properties of a photon allow us to write:

$$\varepsilon = h\left(\frac{c}{\lambda}\right) = \text{constant}\left(\frac{1}{\lambda}\right) \qquad \text{(III - 2)}$$

These two expressions set photon energy directly proportional to the frequency of the photon and directly proportional to the inverse of the wavelength of the photon. The first energy scale is natural and obvious. The second is less obvious and often the source of confusion. Inverse wavelength as an energy scale is very common, however, particularly in infrared (IR) spectroscopy and in visible (VIS) spectroscopy when compounds containing transition metals are involved (crystal or ligand field spectroscopy). Inverse wavelength is so commonly used that a special term has been developed for $(1/\lambda)$. The expression $(1/\lambda)$ is called a *wavenumber* and denoted symbolically most often by $\bar{\nu}$ or $\sigma$. The magnitudes of wavenumbers are most often presented as numbers of $cm^{-1}$.

Figure 1-2 in Chapter 1 presents a schematic comparison of energy units across the "entire" electromagnetic energy spectrum. The row $E$ should be the units in current usage, but these units are not commonly employed. The commonly used units are wavelength and frequency. However, in the IR range, energies are most often referred to in $cm^{-1}$, with typical values on the order of 200–8000 $cm^{-1}$. In the VIS and UV regions, "energies" are more often referred to by wavelength. Thus a UV/VIS spectral line is commonly referred to as an *xxx* nm line rather than a *yyyyy* $cm^{-1}$ line. For energies in the UV/VIS, and less frequently the IR, a unit called the Kaiser or Kayser is also reasonably common and is based on the wavenumber energy concept. Definitions are:

$$1 \text{ Kayser} = 1 \text{ cm}^{-1}$$
$$1 \text{ kiloKayser} = 1000 \text{ cm}^{-1}$$

As implied above, common practice often refers to the energy of a photon in terms of the wavelength of the particle. The units of wavelength used also vary with the spectral region; the common units are noted in Figure 1-2 of Chapter 1. In the IR region, microns $(10^{-6})$ m is the common unit. In the UV/VIS region, Angstroms Å = $10^{-8}$ cm and millimicron were customary units. Now the proper SI unit nanometer (nm) is in common usage, perhaps because of the easy conversions:

1 millimicron = 1 nanometer
1 Angstrom = 0.1 nanometer
thus 4000 Å = 400 nm

In Figure 1-2 of Chapter 1 the $E$ row is expressed in units of J $mol^{-1}$. While not explicitly stated, the properties $\nu$, $\bar{\nu}$, $\lambda$, and $E'$ refer to a single photon and not a mole of photons. What is the energy scale per mole of photons? Let's work through an example. All of the calculations will depend on the units used for Planck's constant:

$$h = 6.626 \times 10^{-34} \text{J s particle}^{-1}$$

where we have explicitly noted that Planck's constant refers to a single particle, not a mole of particles. Let $\nu = 10^{15}$ Hz (1 Hz = 1 $s^{-1}$). What is the energy $\varepsilon$ per mole?

$$\varepsilon = N_A\, h\, \nu$$
$$\varepsilon = 6.023 \times 10^{23} \text{ part mol}^{-1} \times 6.626 \times 10^{-34} \text{J s part}^{-1} \times 10^{15}\ \text{s}^{-1}$$
$$\varepsilon = 3.991 \times 10^{-10} \text{ J s mol}^{-1} \times 10^{15}\ \text{s}^{-1}$$
$$\varepsilon = 399.084 \text{ kJ mol}^{-1}$$
$$\varepsilon \approx 400 \text{ kJ mol}^{-1}$$

Moreover, convert this energy to electron volts:

$$\varepsilon \approx 400,000 \frac{\text{J}}{\text{mol}} \times \frac{1 \text{ eV}}{1.6022 \times 10^{-19} \text{ J}}$$
$$= 2.491 \times 10^{24} \frac{\text{eV}}{\text{mol}}$$
$$\varepsilon \approx 2.491 \times 10^{24} \text{eV} \frac{\text{mol}^{-1}}{6.023 \times 10^{23} \text{part mol}^{-1}}$$
$$= 4.13 \frac{\text{eV}}{\text{part}}$$
$$\varepsilon \approx 4 \frac{\text{eV}}{\text{part}}$$

By the way, the common X-ray energy unit is electron-volts (eV). While masers, not lasers, operate in the microwave region, we note that in the microwave portion of the electromagnetic spectrum frequency, generally gigahertz (GHz), is used to designate energy and spectral lines or features.

To conclude these remarks on units of photon energies, we briefly discuss energies in Raman spectroscopy. Without going into the detail of the experiment or theory, the critical measurements in Raman spectroscopy are the energies and the intensities of scattered incident photons. The response of the sample to incident photons is measured by the difference between the energy of the incident photon and the energy of the emitted photon. If an instrument or experimental apparatus measured the energy of the detected photons in frequency, Hz, calculating the effect of the incident photon on the target molecule would be simple:

$$\Delta E = h\Delta\nu$$
$$\text{sample energy change} = h\,|\,\text{detected } \nu - \text{incident } \nu| \qquad \text{(III-3)}$$

However, most laser setups detect the photon signals as a function of wavelength $\lambda$, and, to make the discussion even more complicated, common practice reports the energy changes in wavenumbers. Let's calculate the energy changes in wavenumbers and in Hz for a Raman signal observed at 508 nm using an incident 488 nm argon ion laser photon:

$$E_{\text{inc}} = \frac{hc}{\lambda_{\text{inc}}}$$
$$E_{\text{obs}} = \frac{hc}{\lambda_{\text{obs}}}$$
$$\Delta E = h\nu = E_{\text{obs}} - E_{\text{inc}}$$
$$= hc\left|\frac{1}{\lambda_{\text{obs}}} - \frac{1}{\lambda_{\text{inc}}}\right| = hc|\bar{\nu}_{\text{obs}} - \bar{\nu}_{\text{inc}}|$$
$$\nu = c\left|\frac{\lambda_{\text{inc}} - \lambda_{\text{obs}}}{\lambda_{\text{inc}}\lambda_{\text{obs}}}\right| = c|\bar{\nu}_{\text{obs}} - \bar{\nu}_{\text{inc}}| \qquad \text{(III - 4)}$$

Specifically from Equation III-4:

$$\Delta E = 6.626 \times 10^{-34} \frac{\text{Js}}{\text{part}} \times 3 \times 10^{8} \text{ms}^{-1}$$

$$\left|\frac{1}{508 \text{ nm}} - \frac{1}{488 \text{ nm}}\right| \times 10^{9} \frac{\text{nm}}{\text{m}}$$

$$= 1.998 \times 10^{-16} \frac{\text{J}}{\text{part}} \left|\frac{1}{508} - \frac{1}{488}\right|$$

$$= 1.604 \times 10^{-20} \frac{\text{J}}{\text{part}}$$

$$\text{or} \quad E = 9.66 \frac{\text{kJ}}{\text{mol}} \qquad \text{[near IR region]}$$

This 9.66 kJ mol$^{-1}$ should be the energy change expressed in SI units when a 488 nm photon induces the emission of a 508 nm photon by the Raman effect. In general, then, we have a formula:

$$\Delta E = \left|\frac{1}{\lambda_{\text{obs}}} - \frac{1}{\lambda_{\text{inc}}}\right| 1.197 \times 10^{8} \text{ J mol}^{-1}$$

To convert the energies to wavenumbers we note from Equation III-4:

$$\frac{\Delta E}{hc} = \left|\frac{1}{\lambda_{\text{obs}}} - \frac{1}{\lambda_{\text{inc}}}\right|$$

where $\Delta E/hc$ will have the units of reciprocal length in the units of the reported wavelength. If wavenumbers are desired and $\lambda$ is measured in nm:

$$\frac{\Delta E}{hc} = \left|\frac{1}{508 \text{ nm}} - \frac{1}{488 \text{ nm}}\right| 10^{9} \text{ nm m}^{-1} \times \frac{1 \text{ m}}{100 \text{ cm}}$$

$$= 807 \text{ cm}^{-1}$$

In general, for $\lambda$ measured in nm, we have the formula:

$$\frac{\Delta E}{hc} = 10^{7} \text{ cm}^{-1} \left|\frac{1}{\lambda_{\text{obs}}} - \frac{1}{\lambda_{\text{inc}}}\right|$$

where it should be noted that the energy change calculated here is per particle, that is, 807 cm$^{-1}$ part$^{-1}$, not 807 cm$^{-1}$ per mole.

Another variation in reporting Raman spectra is to report the energies of each photon line in wavenumbers while still experimentally measuring the signals in wavelength units. This is the less involved calculation. In terms of wavenumbers:

$$\frac{\varepsilon}{hc} = \frac{10^{7} \text{ cm}^{-1} \text{ part}^{-1}}{\lambda}$$

$$\left(\frac{\varepsilon}{hc}\right)_{\text{inc}} = \frac{10^{7} \text{ cm}^{-1} \text{ part}^{-1}}{488} = 20,491 \text{ cm}^{-1} \text{ part}^{-1}$$

$$\left(\frac{\varepsilon}{hc}\right)_{\text{obs}} = \frac{10^{7} \text{ cm}^{-1} \text{ part}^{-1}}{508} = 19,685 \text{ cm}^{-1} \text{ part}^{-1}$$

$$\Delta E = \Delta\left(\frac{\varepsilon}{hc}\right) = \left|19,685 \text{ cm}^{-1} - 20,491 \text{ cm}^{-1}\right|$$

$$= 806 \text{ cm}^{-1} \text{ part}^{-1}$$

which is the same result as before when significant figures are properly taken into account. To summarize the calculation, an incident photon whose energy is proportional to 20,491 cm$^{-1}$ causes by the Raman effect the emission of a photon whose energy is proportional to 19,685 cm$^{-1}$. Note here again these energies are per particle, but the per particle (photon) is customarily omitted.

## Photon Beams

A light source rarely emits one photon at a time, but in fact emits many. The number of photons in a unit volume or the number striking a unit area is the basis for quantifying beams of photons. If a beam of photons with energy $E$ passes a unit area in unit time, then we can write this number with units of energy: area$^{-1}$ time$^{-1}$. Since power is energy time$^{-1}$, this becomes power area$^{-1}$, where watts m$^{-2}$ are typical SI units. This concept is commonly known as the intensity of the photon beam or the intensity of a flux of photons. However, it is worthwhile noting that other terms are also used to discuss intensity and we will introduce these next.

Radiometry is the study of the energy and power of photons for all wavelengths. When only visible light is studied, the subject is called photometry. Since lasers operate at wavelengths outside of the VIS region, we will only mention some terms pertinent to laser power drawn from the radiometry literature.

The definition of intensity given above is properly called the irradiance, with typical units of watts $m^{-2}$, where the *area refers to the area illuminated by the photon beam.* The simple term power is replaced by radiant power. Two other terms should be introduced: radiant intensity and radiance. The concepts of radiant intensity and radiance involve the solid angle in which the beam of photons travels. If the photon source were a point source, photons would travel in all directions, traversing the solid angle of a sphere which is $4\pi$ steradians. That the angle subtended by the photon source has an effect on the radiant intensity or radiance can be easily visualized in the following manner. Suppose a flux of photons emitting from a source in all directions, that is, in a solid angle of sphere $4\pi$ steradians, is somehow directed to travel in the direction of the positive octant of a sphere. The radiant intensity and radiance of this beam are now 8 times greater than before. Obviously then if the solid angle were made very small, the radiant intensity would greatly increase. The very small divergence of a laser beam, hence its very small solid angle, is why laser beams are so "bright" or "intense." Note that the two concepts of radiant intensity and radiance are not the same, however. Radiant intensity is the power of the photon beam per solid angle, usually watts $\text{sterad}^{-1}$. Radiance is radiant intensity per unit area of the radiation source, watts $\text{sterad}^{-1}$ $m^{-2}$. Note the important distinction in the *areas* involved in the concepts of irradiance: ordinary intensity = watt $m^{-2}$, and radiance = watt $\text{sterad}^{-1}$ $m^{-2}$. Irradiance involves the area on which the beam of photons falls. Radiance involves the actual physical area of the photon source. Thus radiance relates the power of the photons emitted from some finite area to the solid angle subtended by those photons. When the photons studied are in the visible range of the electromagnetic spectrum, the concept of radiance is usually called brightness. In Chapter 2, the "brightness" of the sun is compared to that of a simple He/Ne laser.

At this time we should introduce the extremely important concept of spectral irradiance or spectral radiant intensity or spectral radiance. With the modifier *spectral* we are interested in the number of photons from a source at specified energies or wavelengths. In general, the energies of the photons emitted by a source are not equally distributed. For example, the number of photons of $\lambda = 500$ nm generated by the sun per unit time is not the same as the number of $\lambda = 250$ nm photons. Similarly, an argon ion laser produces many photons at 488 nm but essentially none at 489 nm. Thus, for practical purposes, the terms spectral irradiance, spectral radiant intensity, and spectral radiance specify the term per unit frequency or per unit wavelength. For example:

$(\text{spectral irradiance})_{\nu} = \text{power m}^{-2}\ \text{Hz}^{-1}$
(a frequency distribution)
$(\text{spectral irradiance})_{\lambda} = \text{power m}^{-2}\ \text{m}^{-1}$
(a wavelength distribution)

Spectral irradiance and related terms reflect much more accurately the actual electromagnetic radiation properties of photon sources.

APPENDIX IV

# Laser Safety

Those characteristics of laser light that are highly valued for scientific purposes often pose a significant hazard to various parts of the human body. To ensure the safety of all personnel, recommended procedures for the safe use of laser devices should be followed in the development and execution of all experiments.

## Health Hazards of Lasers

The intensity and directionality of laser light can cause significant and often irreparable damage to the retina. The lens of the eye can efficiently collect laser light and focus it to a very small focal spot on the photosensitive surface of the retina. Blind spots on the retina can develop because of the high power per unit area that is produced as a result of the eye's focusing power. The nature of the retinal damage depends on numerous factors, including the wavelength and intensity of the laser radiation, the exposure time, and the position and diameter of the focal spot. In particular, visible and near infrared light in the range of 400–1400 nm is most likely to cause retinal damage, as it is this radiation that penetrates the eye tissue and is focused on the retina.

Laser radiation in the ultraviolet and far-infrared region is not focused on the retina, but nevertheless poses an appreciable risk of eye damage. These wavelengths are absorbed by the cornea and lens, and serious burns can result to these parts of the eye. The risk of accidental injury is compounded by the invisibility of ultraviolet and far-infrared light to the observer.

It should be noted that eye damage can result both from direct exposure to the laser beam as well as from reflection of laser radiation into the eye. To reduce the exposure of the eye, protective eyewear should be worn. Manufacturers of laser safety eyewear typically specify both the wavelengths at which the filters in the eyewear offer protection, and the attenuation (optical density) at these wavelengths. To ensure adequate protection, eyewear should be matched to the particular laser in use.

The risk of serious skin burns is associated with high-power lasers. The extent of damage is dependent on the time of exposure, the power per unit area, and the wavelength of radiation. The skin is most susceptible to laser emission in the 200–320 nm range where skin reflectivity is low. Both direct and scattered exposure to pulses and CW radiation can be hazardous, although the limited time for heat dissipation with ultrafast pulses reduces the susceptibility to injury. High-powered laser sources are also potential fire hazards.

Additional considerations of laser safety extend beyond the dangerous aspects of the optical beam. In particular, the high-voltage power supply of most laser systems poses a serious electrical hazard. Adequate safety measures should also be employed when working with the organic dyes and solvents of dye lasers, with the cryogenic materials for cooling high-powered lasers, and with the vapors produced from laser-induced vaporization processes.

## Laser Safety Practices

Proper procedures for safe laser operation and viewing have been developed for general laser use. These basic safety practices may be summarized as follows:

1. Direct viewing of any laser beam (whether classified as hazardous or not) should be avoided at all times.

2. Only experienced personnel should be permitted to operate the lasers. All users should be instructed in the safe operation of all devices.
3. To avoid unauthorized use, laser devices should be used within a marked, controlled area, and no laser should be left unattended.
4. The lowest possible laser power should be used in all instances, either by the choice of laser or by the use of laser output neutral density filters.
5. Lasers should be firmly mounted on a breadboard or optical table to assure that the beam travels along the intended path. As much of the laser beam path as possible should be enclosed, and all primary and secondary beams must be terminated at the end of their useful paths. Laser beam paths should be kept above or well below eye level. All unnecessary mirror-like surfaces should be eliminated from the vicinity of the laser beam path to avoid added reflections.
6. Caution or warning signs required by federal product performance standards and supplied with each laser should remain attached to the laser to alert all users of the laser radiation contained therein.

## Laser Safety Standards

In addition to basic laser safety practices, specific safety measures commensurate with particular laser types have been devised by the American National Standards Institute (ANSI). The ANSI classification scheme considers the intensity of the emitted beam and its capability of producing a hazardous diffuse reflection in order to dictate appropriate safety measures. Lasers are organized into one of four classes, with increasing class number reflecting a higher level of hazard. Low-power diode or semiconductor lasers fall into Class I, i.e., those lasers systems that cannot under normal operating conditions produce a hazard. Low-power helium-neon lasers with outputs less than 1 mW constitute Class II lasers, for which a hazard exists only if an individual overcomes his natural aversion to bright light and stares directly into the laser beam. All of the above general measures will ensure the safe use of Class I and II lasers. Class III lasers under normal operating conditions do not generally cause skin damage but do present an eye hazard. Eye damage can occur during exposure to Class III radiation within a time period shorter that a blink response. Class III lasers do not produce a hazardous diffuse reflection but could produce a hazard if viewed directly. Measures should be taken to preclude the possibility of intrabeam viewing. This class is further subdivided into two levels according to output power: Class IIIa denotes laser systems with output power between 1 and 5 mW, as is present in some high-power helium-neon lasers; and Class IIIb corresponds to laser systems not exceeding 25 mW of output power, as observed in some argon ion-pumped and nitrogen-pumped dye laser systems. The final class of lasers, Class IV, can cause serious eye damage from either direct exposure or reflection, as well as severe skin injury. Lasers with high power, particularly those operated in continuous-wave mode, fall into this category, including the carbon dioxide, HF, HCl, and Nd:YAG lasers.

The design of experiments involving laser sources is not complete until appropriate laser safety standards have been recognized and adopted.

APPENDIX V

# Answers to Discussion Questions

## Chapter 5

**5-1.** With the use of multiple beam splitters and carefully controlled optical paths, the probe pulses could be precisely separated in time to probe the sample at multiple time intervals. Recall that a difference in optical path length of ≈ 0.3 mm translates to a delay of 1 ps, while an additional 30 cm in path length is equivalent to an approximate 1 ns time delay.

**5-2.** The rate of an ultrafast reaction could be decreased for measurement via a conventional technique through the use of lower reactant concentrations (for those reactants with positive reaction orders in the rate law). For elementary reactions (reactions that describe what happens on the molecular level), lower temperatures will also decrease the reaction rate. (In contrast, an overall reaction with a multistep reaction mechanism can increase in reaction rate with lower temperature if one or more of the elementary steps is an exothermic equilibrium reaction.) For those reactions in solution catalyzed by acid or base, adjustments in pH levels will also modify reaction rate.

**5-3.** With the availability of femtosecond laser pulses and femtosecond detectors, several additional events associated with fluorescence emission could be probed. For example, typical fluorescence measurements on a picosecond or nanosecond timescale reflect fluorescent signals originating from the ground vibrational state of the excited electronic state. While absorption has promoted a molecule to an excited vibrational state within the first excited electronic state, collisional and lattice relaxation nonradiative processes occur on a femtosecond timescale to lower the energy of the excited-state species. Thus, with techniques capable of femtosecond resolution, these nonradiative processes could be observed and characterized. Furthermore, with short detection times the orientation of an emitted photon is more likely to be preserved because the molecule is not as likely to have moved (translated) or rotated. Hence, with femtosecond capabilities the evolution of fluorescence anisotropy (i.e., the depolarization of the fluorescence signal) could be studied. In addition, fluorescence signals with very short lifetimes could be more accurately detected with instrumentation capable of femtosecond resolution. Finally, as the decay of a resonance Raman signal is generally faster than the process of fluorescence, the kinetics of resonance Raman decay could be monitored when femtosecond resolution is possible.

## Chapter 6

**6-1.** For single-photon spectroscopy, light is absorbed by a target molecule if the energy of the radiation matches the energy difference between two energy states. Furthermore, one photon is absorbed per activated molecule (Law of Photochemical Equivalence). Refer to Figure 6-1 to describe multiphoton spectroscopy in the case of sequential absorption (where the first photon possesses energy equivalent to the energy difference between an initial state and an intermediate state and the second photon matches the energy difference between the intermediate and final states) and in the case of concerted absorption (where the first photon's energy does not match an energy difference between the initial state and an intermediate state, but the energy of both photons is required to reach the final energy state).

**6-2.** Since only a single photon is required in CARS to reach the intermediate $E_k$ state prior to the emission of a second photon, the technique should not properly be called multiphoton spectroscopy.

## Chapter 7

**7-1.** Fluorescence experiments must demonstrate that excitation of $A$ leads to emission initially at $B$ and subsequently at $D$ (i.e., $A \rightarrow B \rightarrow D$) and that excitation of $B$ leads to emission at $D$ (i.e., $B \rightarrow D$). Similarly, excitation of $C$ should generate emission initially at $B$ and finally at $D$. Furthermore, excitation of any of $A$, $B$, or $D$ should not yield emission at $C$, nor should excitation of $C$ yield emission at $A$ or $D$.

**7-2.** Time-resolved absorption studies could be used to monitor the formation of the excited state and thus reveal the "donor" in energy transfer among pigment molecules. Any laser used to induce fluorescence must necessarily induce the requisite absorption. Rather than using a monochromator on the emission side of the sample, the monochromator should be placed between the excitation source and the sample.

**7-3.** Different relative amounts of light-harvesting pigments will alter the fluorescence intensities observed from these fluorophores as well as the lifetimes of the fluorescence. For example, for $A \rightarrow B$ transfer, increased amounts of $B$ shorten the fluorescence lifetime of $A$. No changes in the sequence of energy transfer would necessarily be expected.

**7-4.** In addition to examining vast and distant environments, remote sensing of laser-induced fluorescence is also convenient for examining "hostile" regions that are either unsafe for experimenters or damaging to instrumentation. Investigations involving systems at high temperatures (e.g., a flame), with potential hazardous substances (e.g., emissions generated from fires, nuclear explosions, or biological warfare), or in inconvenient locations (e.g., atmospheres of planets) are examples of studies that are ideally suited for remote sensing techniques.

**7-5.** Temporal studies to evaluate long-range changes in environmental conditions might investigate the impact on vegetation and/or the atmosphere of such factors as acid rain, atmospheric ozone depletion, wildfires, hazardous substance spills, and volcanic or meteor activity.

## Chapter 8

**8-1.** As a consequence of the varying R groups, the net charge on the three amino acids varies. Fluorescein-labelled arginine differs in net charge from fluorescein-labelled glycine by +1; similarly, the net charge of the fluorescent glycine derivative differs by +1 from the net charge on the tagged glutamic acid. Thus, separation of the amino acids by CE results in the initial detection of arginine, followed by glycine, and finally glutamic acid.

**8-2.** (*a*) Substitution of Gln-25 with lysine in Mutants I and III changes the net charge by +1; substitution of Glu-58 with alanine in Mutants II and III similarly changes the net charge by +1. Thus, Mutants I and II differ in net charge from the wild-type enzyme by +1, while Mutant III, with both mutations, differs in net charge from wild-type by +2. As these increases in net charge reduce the electrophoretic mobility toward the anode, the order of appearance should be Mutant III, followed by both Mutants I and II, and then followed by the wild-type protein.

(*b*) If Mutants I and II are further resolved, then Gln-25 and Glu-58 differ in solvent accessibility as a consequence of the tertiary structure of the protein. The greater the solvent accessibility, the more extensive the solvation of the amino acid residue, and the more impact mutation can have on the electrophoretic mobility of the enzyme.

(*c*) One possible explanation for the greater net positive charge on Mutant I is that the lysine substitution for glutamine leads to nearly a +1 change in net charge while the alanine substitution for glutamate leads to a change in net charge between 0 and +1. This situation could result if Gln-25 exhibits greater solvent accessibility than Glu-58. The glutamate carboxyl side chain is not as extensively deprotonated in the wild-type enzyme as anticipated. Hence, alanine substitution results in little change in overall charge.

**8-3.** Advantages:

1. The intrinsic fluorescence may be weak and require a sensitive detector and/or a powerful excitation source.

2. To observe the intrinsic fluorescence of several components, several laser emission wavelengths (and possibly several lasers) would be required.

Disadvantages:

1. Derivatization may lead to changes in the purity, stability, and electrophoretic mobility of the individual components.

2. Derivatization may significantly reduce the concentrations of the components so that detection is limited.

## Chapter 9

**9-1.** CARS is a useful spectroscopic technique to characterize even those samples with significant fluorescence emission. The wave vector addition of the input laser beams results in a highly directional CARS output signal that can be easily separated from fluorescence or scattered radiation.

**9-2.** The CARS technique requires the mixing of at least two laser input beams and the analysis of the resultant anti-Stokes beam. With the appropriate optical devices to direct and focus incident laser sources and emitted radiation, the intersection of the input laser beams and the detection of the highly coherent anti-Stokes output beam can be accomplished at some distance from the sample. As a consequence, hostile environments are more easily probed without risking the safety of either the experimentalist or the laser equipment.

## Chapter 10

**10-1.** At low laser powers a small extent of fragmentation is anticipated. Thus, the peak with the maximum intensity should correspond to the parent molecular ion $M^+$. The low-intensity peaks at double and triple the *m/z* ratio of the parent ion can then be assigned to $2M^+$ and $3M^+$, respectively. The low-intensity signal at one-half the *m/z* value of the parent ion could then result from an ion with the same mass but double the charge, i.e., $M^{2+}$.

If the peak at the highest *m/z* ratio were assigned the identity of the parent molecular ion $M^+$ (instead of the correct $3M^+$ assignment), then the signal at the actual *m/z* ratio of the parent molecular ion (as well as other peaks) would be assigned a nonintegral charge.

**10-2.** $t_2/t_1 = (MW_2/MW_1)^{1/2}$

**10-3.** In the case of optically selective ionization (also known as resonant ionization), a particular species is ionized by using a specific wavelength that will accomplish ionization of that substance. Thus, for every analyte of interest, the ionizing laser must be tuned to a different absorption wavelength. However, when probing for a specific substance, optically selective ionization is the most efficient approach. For nonselective (i.e., nonresonant) photoionization, a high-intensity ultraviolet laser beam is used to ionize uniformly any desorbed species, with the subsequent separation in the mass spectrometer as the means of chemically differentiating the desorbed sample. When surveying an extensive number of adsorbed species or when the identities of the adsorbed species (and hence their preferred absorption wavelengths) are unknown, nonselective ionization is the most practical strategy. A more accurate representation of the surface composition is efficiently obtained with nonselective photoionization. However, the possibility of a high degree of fragmentation is a drawback of this method.

**10-4.** As combustion reactions generally generate highly reactive radical intermediates, the high sensitivity and resolution of laser-assisted mass spectroscopy allows for an accurate analysis of the formation and consumption of these radicals. Furthermore, the laser-assisted approach permits the study of chemical reactions taking place in a "hostile" region with negligible perturbation of the flame or its environment. *In situ* analysis within the flame is a powerful advantage of laser-assisted mass spectrometry.

## Chapter 11

**11-1.** The photoacoustic signal was measured in the presence and absence of a magnetic field to test whether the thermal properties were affected by the increased order imposed on the liquid crystalline phase by the magnetic field.

**11-2.** The energy of the infrared wavelength used almost exactly matches the energy required for the C-H stretching motion observed in most organic materials.

## Chapter 12

**12-1.** Some possible experimental variables that may be altered include: (1) those parameters specifically related to the excitation source—wavelength of excitation, intensity of excitation, duration of irradiation, pulsed vs. CW nature of irradiation, variation of laser selection during synthesis, and (2) the conventional parameters that affect product identity and composition, including choice of solvent, concentrations of reagents, temperature, isolation or separation of product during synthesis, etc.

**12-2.** Multiphoton absorption would have the advantage of yielding a faster synthesis of carvone-camphor, since $k_2 > k_1$ as a consequence of the lifetime of state $^3[I']$ being shorter than the lifetime of state $^3[I]$, $\tau^3[I'] < \tau^3[I]$. However, one disadvantage of the multiphoton pathway is the possibility of a lower yield of product. High excitation intensities could promote $^3[I']$ to an ionization and dissociation continuum, lowering the production of II.

**12-3.** A switch to a laser-driven synthesis would require some cost-saving benefits such as a greater conversion yield (and thus lower cost of reagents per unit product), a faster rate of production, evidence of fewer competing side reactions (thereby increasing yield and minimizing separation of products), and/or lower energy costs.

## Chapter 13

**13-1.** In the two-stage photolysis of 7-dehydrocholesterol (7-DHC), the second photolyzing lasers are the nitrogen laser with emission at 337 nm and the YAG laser with third-harmonic emission at 354 nm. Each of the wavelengths is suitable to induce the photoisomerization of $T_3$ to $P_3$. Furthermore, $P_3$ preferentially absorbs the 337 nm radiation to undergo a subsequent cycloaddition reaction to generate a small additional amount of $L_3$ and regenerate some 7-DHC. This further conversion of $P_3$ must subsequently induce further isomerization of $T_3$ to $P_3$. Thus, the nitrogen laser converts more $T_3$ to other species than the frequency-tripled YAG (1.5% $T_3$ remaining vs. 11.0% $T_3$ remaining, respectively). While the final amount of $P_3$

is about the same for either laser, the nitrogen laser generates a small additional amount of $L_3$ over the YAG laser (9.8% vs. 8.7%, respectively) and a considerable additional amount of starting material 7-DHC (8.8% vs. 0.1%, respectively).

**13-2.** We have already hinted at the application of photochemical purification with lasers to remove hydrides of arsenic ($AsH_3$), phosphorus ($PH_3$), and boron ($B_2H_6$) from gaseous $SiH_4$. Highly pure silane serves as the starting material for the preparation of pure silicon for semiconductor production and for the manufacture of solar cells. The three hydride contaminants absorb 2-4 orders of magnitude more readily than $SiH_4$ at $\lambda = 193$ nm to yield easily separable polymeric solids or gaseous compounds.

Photochemical purification with lasers is ideally suited to any system where trace contaminants could readily undergo photolysis to yield separable products.

**13-3.** At 248 nm the absorption cross section of $H_2S$ is approximately $3 \times 10^{-19}$ cm$^2$ and that of syngas is approximately $7 \times 10^{-24}$ cm$^2$, for an absorption cross-section ratio of about $4 \times 10^4$.

## Chapter 14

**14-1.** No, photosensitizers should not be highly fluorescent species. A high likelihood of this additional primary process would lower the quantum yield of intersystem crossing, a primary process required to form the triplet state of the photosensitizer. The mechanisms of action of photosensitizers require either a direct or an indirect interaction of the triplet state of the photosensitizer with the biological target (type I mechanism vs. type II mechanism, respectively).

**14-2.** Species that, in a photogenerated triplet state, act either as electron donors or acceptors, or as proton donors or acceptors, are required for efficient photochemical action with tissues or chromophores via a type I mechanism.

**14-3.** Photosensitizers should: (1) exhibit low toxicity, (2) preferentially locate in a malignant tumor, (3) absorb in the red or near-infrared regions of the electromagnetic spectrum with a high absorptivity, (4) exhibit a large quantum yield of triplet formation for photochemical action, and (5) fluoresce weakly to enable detection of partitioning into tissues but without hindering triplet formation.

**14-4.** Incident light may be *reflected* by the surface of the tissue; be *absorbed* by a chromophore in the tissue; be *transmitted* by the tissue; or permeate the tissue, be *scattered* internally, and then be absorbed at a location distinct from the site at which the radiation entered the tissue.

**14-5.** All currently available optical fibers readily absorb the 10.6 $\mu$m radiation of $CO_2$ lasers and thus cannot transmit the light to the necessary tissues or chromophores within the human body.

**14-6.** The two primary mechanisms for the action of photosensitizers in biological systems are the free-radical mechanism (Type I) and the energy transfer mechanism (Type II). In each mechanism the photosensitizer absorbs light of the requisite wavelength to generate an excited singlet that undergoes intersystem crossing to produce a triplet state. In the free-radical mechanism, the triplet state reacts directly with the substrate or tissue of the biological system via an electron-transfer or hydrogen atom-transfer process to generate free radicals of both the substrate and the photosensitizer. Subsequent reactions, particularly involving oxygen, lead to cell damage. In the energy transfer mechanism, the triplet state of the photosensitizer reacts with oxygen dissolved in the biological tissue to generate a reactive and long-lived excited state of singlet oxygen and to regenerate the ground-state of the photosensitizer. Subsequent oxidation of the biological tissue by the singlet oxygen damages essential cellular components.

## Chapter 15

**15-1.** A criterion of a relatively large amount of X-H stretching (i.e., a long X-H stretch) would favor a significantly localized excitation and hence enhance selective excitation.

# Glossary

**acousto-optical modulator** *See* **optical modulator**.

**active medium** A substance in either the solid, liquid, or gas phase in which stimulated emission is induced. The characteristic quantized energy levels of each substance, in conjunction with external energy supplied by the pumping source, determine the laser output obtained. Active media range from simple monatomic gases to solutions of organic dyes to solid-state materials containing transition metal ions embedded in a host lattice.

**amplification** The process by which more than one photon of light can be emitted from a substance through stimulation by a single photon.

**anisotropic** Describes a physical property (of a substance or medium) that is dependent upon the direction in which it is measured.

**anti-Stokes transition** An inelastic scattering of light such that the scattered photons are characterized by a higher frequency than the incident photons. See also **inelastic scattering**.

**apparent mobility** The observed velocity of a particle under the influence of an electric field due to the combined effects of electrophoresis and electroosmosis. For cations in an aqueous medium, these two electrokinetic actions work in the same direction, and for anions, the two phenomena work in opposite directions. For neutral molecules, only the electroosmotic component contributes to the observed mobility.

**axial (longitudinal) mode** *See* **longitudinal mode**.

**beam diameter** The diameter of a mode of radiation at which the intensity has fallen to $1/e^2$ of the peak intensity.

**beam splitter** An optical device used to divide a laser beam into two or more beams. Operating on several different principles of reflection, beam splitters exist in many forms with varying characteristics. For example, cube beam splitters use dielectric coatings on the front surface of a crown glass or a fused silica substrate to reflect a portion of the beam by 90° without displacement. However, the reflectance and transmission properties of dielectric thin film coatings depend on wavelength, polarization, and angle of incidence, so cube beam splitters must be selected with these sensitivities in mind.

**birefringence** A phenomenon observed for some substances whereby the extent of refraction of light through the medium is dependent on the direction of propagation and the direction of polarization of the incident light.

**birefringent crystal** A crystal characterized by the property that different polarizations of light propagate through the medium at different velocities.

**Brewster's angle** The angle θ, given by the equation $\theta = \tan^{-1} n$, at which a material of refractive index $n$ minimizes the reflection loss for one component of polarization. Brewster angle windows are used as end windows in gas lasers to obtain polarized emission by enabling only one polarization component (parallel or perpendicular) to be amplified within the laser cavity.

**brightness** Radiance; radiant intensity per unit area of the source of radiation.

**Brillouin scattering** A form of inelastic scattering of light caused by the coupling of radiation with the acoustic modes of vibration of a sample.

**broadband operation** Simultaneous emission of several discrete wavelengths by a laser. A series of discrete emission lines results from the sustainment of population inversions between more than one pair of energy states in the active medium.

**CARS spectroscopy** Coherent anti-Stokes Raman spectroscopy involving a four-wave interaction to produce emission at $\nu_4 = 2\nu_1 - \nu_2$, where $\nu_1$ and $\nu_2$ are the frequencies of radiation of two input lasers. See also ***n*-wave mixing**.

**cavity** A region, generally bounded between two mirrors, which confines the lasing medium. A resonant cavity of length $L$ can sustain radiation of wavelength $\lambda = 2L/N$, where $N$ is an integer.

**cavity dumping** A method of achieving pulsed operation of a laser by rapidly emptying the stored energy within the laser cavity. An acousto-optic modulating device replaces the output mirror of a laser to produce off-axis diffraction of the laser beam. Intracavity power is allowed to build up by the device acting as a totally reflective mirror and is periodically "dumped" as an output pulse by reflecting the beam out of the cavity. Cavity dumping is a form of Q-switching in which the $Q$ factor of the cavity is changed from high to low. Cavity-dumped pulses appear with a frequency on the order of MHz and are typically $10^{-9}$ s in duration with peak powers of 10–100 W.

**chemical lasers** Lasers in which a pumping mechanism involving an exothermic irreversible chemical reaction is used to achieve a population inversion between two energy states.

**chirp** A variation or modulation in the frequency of light within a pulse of radiation, often appearing in ultrashort light pulses. Such modulation is characterized as negative chirp if the higher frequencies of light lead—i.e., appear earlier in time than—the lower frequencies within the pulse. A positively chirped pulse is characterized by longer wavelength components of light leading shorter wavelength components. Pulse compression to eliminate the frequency modulation of the pulse is accomplished by: (1) passage of a negatively-chirped pulse through a medium with normal dispersion or (2) reflection off a pair of gratings that shortens the path length for the time-delayed high frequencies in a positively chirped pulse.

**chromophore** A molecule, or generally a portion of a molecule, that absorbs light.

**circular polarization** Characteristic whereby the electric field vector of light describes a circular helix about the direction of propagation of the light.

**coherence** Condition in which electromagnetic waves are characterized as existing in phase with one another. Phase correlation has both a spatial (transverse) and temporal (longitudinal) component. Thus, a property of light in which light waves of the same wavelength travel in the same direction and have the same phase or polarization of the electric vectors in space.

**coherence length** The distance travelled by electromagnetic radiation over which the wave front remains in phase, calculated as $l_c = c \cdot t_c$, where $c$ is the speed of light and $t_c$ is the coherence time of the radiation.

**coherence time** The length of time over which the wave front of electromagnetic radiation remains in phase, calculated as $t_c = 1/\Delta\nu$, where $\Delta\nu$ is the range of frequencies of light emitted (i.e., the frequency width or the emission linewidth).

**collimated light** Radiation characterized by a nondivergent beam.

**collimation** The parallel propagation of light.

**collisional broadening** The appearance of a narrow range of frequencies for a particular lasing transition in an active medium as a consequence of momentum exchange during the random collision of particles in the lasing medium. Collisional broadening is an example of a homogeneous broadening mechanism.

**collisional quenching** A process by which an excited-state species is deactivated to the ground energy state through a collision with another species (either of similar or distinct chemical identity). The deactivation process involves no emission of radiation nor any transfer of energy to the colliding species. While collisional quenching is a bimolecular process, pseudo first-order kinetics is observed as the quencher concentration is generally significantly larger than the concentration of the excited-state species.

**conduction band** The unoccupied band of electronic orbitals for a crystalline solid.

**continuous wave (CW) laser** A device designed to sustain a population inversion between an upper energy state and a lower energy state. Such a continuous operation is maintained if the time interval between successive stimulated emissions of photons is comparable to the lifetime of the photon within the cavity. Thus, the kinetics of the pumping and stimulated emission processes, the physical dimensions of the cavity, the reflectivities of the mirrors, and the refractive index of the active medium all govern the CW condition.

**degeneracy** The number of states of a system with the same energy.

**dichroic mirror** A mirror with wavelength-dependent reflection properties, characterized by the reflection of light of a particular range of wavelengths and the transmission of all other wavelengths.

**diffraction** The change in the direction of propagation of a wave by the presence of an obstacle.

**diode laser** *See* **semiconductor laser**.

**directionality** A property of laser light that is a consequence of the low divergence of the beam.

**dispersion** The variation in the speed of light in a given medium as a function of wavelength.

**dispersive element** An optical device (e.g., prism, grating, etalon) capable of separating radiation into monochromatic components.

**divergence** A measure of the extent to which a beam of light travels in the same direction; a two-dimensional measure of the extent of spreading of the diameter of a light beam.

**Doppler broadening** The appearance of a narrow range of frequencies for a particular lasing transition in an active medium as a consequence of the random thermal motions of emitting excited-state species. Doppler broadening is an example of an inhomogeneous broadening mechanism.

**dye lasers** Lasers in which a fluorescent dye serves as the gain medium and another laser serves as the pumping source. Wide tunability is achieved for these lasers because the dye, typically a large polyatomic organic molecule, has broad spectral characteristics, including an absorption band in the visible region and a fluorescence band of significant width.

**dynamic scattering** An experiment that monitors the evolution of a scattered light wave as a function of time.

**efficiency of a laser** Ratio of the power of the output laser emission to the power of the input pumping action. A high efficiency, on the order of 5% to 35%, is characteristic of $CO_2$ lasers, while a low efficiency of 0.001% to 0.1% is common for argon lasers.

**Einstein** A mole of photons; Avogadro's number of photons.

**Einstein coefficients** For the three radiative processes that are possible in a system of two energy levels—absorption, spontaneous emission, and stimulated emission—the constants that relate that transition rate and the number of molecules in the appropriate energy level that can undergo that transition. The rate equations are $R_{\text{absorption}} = N_1 \cdot \rho_\nu \cdot B_{12}$; $R_{\text{spontaneous emission}} = N_2 \cdot A_{21}$; and $R_{\text{stimulated emission}} = N_2 \cdot \rho_\nu \cdot B_{21}$, where $R$ is the transition rate; $N_1$ and $N_2$ are the number of atoms or molecules with energies $E_1$ and $E_2$, respectively ($E_1 < E_2$); $\rho_\nu$ is the spectral density of radiation energy at frequency $\nu$ (in units of joule-sec/m$^3$); and $A_{21}$, $B_{12}$, and $B_{21}$ are the Einstein coefficients. The coefficients are constants determined by the energy levels and give the probability of the specified transition.

**Einstein relations** The mathematical relationships between the Einstein coefficients. The first Einstein relation states that the coefficients of stimulated absorption and stimulated emission are equal: $B_{12} = B_{21}$. The second Einstein relation states that the coefficient of spontaneous emission is equal to the coefficient of stimulated emission times the factor $8\pi \cdot h \cdot \nu^3/c^3$. Thus, the ratio of the probabilities of spontaneous and stimulated emission is proportional to the third power of the frequency of the emission.

**elastic scattering** A phenomenon in which incident photons interact with a target molecule and are not absorbed but reemitted with the same frequency. Elastic scattering, also known as **Rayleigh scattering**, involves the induction of an oscillating dipole in the target molecule by the oscillating electric field of the incident photons. The energy of the oscillating dipole radiates in all directions. The relative inefficiency of Rayleigh scattering—accounting for one photon in $10^3 – 10^4$—necessitates the use of intense light sources, such as lasers, to produce a sizable effect. The intensity of Rayleigh scattering shows a strong frequency dependence and is inversely proportional to the fourth power of the wavelength of the incident radiation. Rayleigh scattering experiments are useful in elucidating the size, shape, molecular weight, and intermolecular forces of target molecules.

**electromagnetic spectrum** The range of wavelengths, frequencies, or energies of light extending from the longest radio waves to the shortest known $\gamma$ rays.

**electro-optical modulator** *See* **optical modulator**.

**electroosmosis** In an electrophoretic separation process, the bulk flow of liquid due to the effect of an electric field on counterions adjacent to a negatively charged capillary wall.

**electrophoresis** A transport phenomenon describing the movement of particles through a solvent under the influence of an electrical field.

**elliptical polarization** Characteristic whereby the electric field vector of light rotates about the direction of propagation such that a projection of the vector on a plane perpendicular to the direction of propagation of light describes an ellipse.

**endogenous chromophore** A naturally occurring or native chromophore in a chemical or biological system.

**energy transfer** A radiationless process in which the excited-state energy of a donor species is transferred to an acceptor species. The interaction between donor and acceptor is critically dependent on several factors: (1) distance proximity, such that the rate of energy transfer varies as the inverse sixth power of the distance of separation; (2) overlap between the emission spectrum of the energy donor and the absorption spectrum of the energy acceptor; and (3) the relative orientation of the acceptor and donor transition dipoles. Energy transfer typically occurs on a timescale of $10^{-10} – 10^{-8}$ s.

**etalon** An optical element placed within a laser cavity to reduce the number of axial cavity modes output.

**excimer lasers** Lasers in which the excited energy state undergoing population inversion is a stable complex of two atoms (e.g., $ArF^*$, $KrF^*$, $XeCl^*$) which dissociates in the ground state to regenerate the separate atoms. Correctly speaking, an excimer refers to a homonuclear excited-state diatomic complex, while exciplex is the accurate term for a heteronuclear species. Since the laser transition produces a dissociative (i.e., unstable) state, a population inversion is guaranteed upon excimer (exciplex) formation.

**exogenous chromophore** A molecule added to a chemical or biological system to preferentially absorb a certain wavelength of light.

**extrinsic fluorescence** Emission emanating from an exogenous chromophore added to a chemical or biological system lacking native or naturally occurring fluorophores.

**extrinsic fluorophore** A molecule added to a chemical or biological system to preferentially fluoresce; a molecule naturally occurring in a chemical or biological system that is chemically modified in order to induce fluorescence emission upon the absorption of a specific wavelength of light.

**feedback** Amplification of laser light within a laser cavity using a configuration that provides for multiple passages of the radiation through the active medium. A pair of spherical mirrors is generally employed to amplify the light and to direct the passage of photons effectively through the entire volume of the active medium. Laser output is generally obtained by coupling one mirror that is essentially 100% reflective and a second mirror that is partially reflective and partially transmissive.

**femtosecond** A measure of time equal to $10^{-15}$ second; the time it takes light in a vacuum to travel a distance of $2.998 \times 10^{-7}$ m or 0.2998 $\mu$m.

**figure of merit** A parameter used to quantify the quality of the crystal in solid-state lasers, defined as the absorption by the crystal of the pump light divided by the absorption of the laser light.

**filled bands** The fully occupied bands of electronic orbitals for a crystalline solid.

**fluorescence** Radiative relaxation of a species in a singlet excited electronic state to the ground electronic state. The phenomenon of fluorescence is characterized by an emission wavelength that is longer than the wavelength absorbed to generate the excited state. This "red-shifting" of the emitted photon results from: (1) the promotion of a species through light absorption to the first excited electronic state and an excited vibrational state (as a consequence of the shifting of the equilibrium position of the excited-state potential energy function to a larger distance than observed in the ground state), and (2) a rapid relaxation to the lowest vibrational level of the excited state prior to fluorescence emission. The lifetime of fluorescence (i.e., the lifetime of the excited state) is defined as the reciprocal of the sum of the rate constants for all possible de-excitation pathways of the excited state. Fluorescence lifetimes are typically in the $10^{-12} - 10^{-9}$ s range.

**fluorescence decay curve** The time dependence of fluorescence emission at a fixed wavelength after light absorption at a select wavelength.

**fluorescence lifetime** The reciprocal of the sum of the rate constants for all possible de-excitation pathways of a singlet excited electronic state.

**fluorophore** A molecule, or a portion of a molecule, that fluoresces.

**Forster distance** The distance of separation between a donor and an acceptor species at which energy transfer is 50% efficient (i.e., the quantum yield of energy transfer is 0.50). Symbolized by the variable $R_o$, the Forster distance for a typical donor-acceptor pair is on the order of 20–50 Å.

**Forster transfer** The radiationless process by which the excited-state energy of a donor is transferred to an acceptor. The rate constant of energy transfer is characterized by a dependence on the inverse sixth power of the distance of separation between the donor-acceptor pair. See also **energy transfer**.

**four-level laser** A scheme consisting of a ground state, $S_1$, and three excited-state levels $S_2 < S_3 < S_4$, whereby the pumping action involves the transition from $S_1$ to $S_4$ and the laser emission results from the transition $S_3 \rightarrow S_2$. Fast, nonradiative transitions from $S_4 \rightarrow S_3$ and $S_2 \rightarrow S_1$ create a population inversion between states $S_2$ and $S_3$. As $S_2$ is initially unpopulated, population inversion is readily achieved with any atoms or molecules in state 3. Thus, lasing action can be obtained despite a significantly populated ground state, $S_1$. A neodymium-doped YAG laser is based on a four-level scheme.

**frequency doubling** Also known as **second-harmonic generation;** a method by which the output of a laser is modified to yield similar coherent radiation with twice its initial frequency. Frequency doubling is accomplished when two photons of laser light with frequency $\nu$ are pumped into and absorbed by a suitable transparent substance in its ground state which subsequently returns to the ground state via emission of a single photon of frequency $2\nu$. Such a multiphoton phenomenon requires the doubling medium to respond nonlinearly to the incident radiation. In other words, the induced electric dipole $\mu$ is a nonlinear function of the incident electric field $E = E_o \cos \nu t : \mu = \alpha E + \beta E^2 + \ldots$ Expansion of the nonlinear term leads to $\beta E^2 = \beta E_o^2 \cos^2 \nu t = \frac{1}{2} \beta E_o^2 (1 + \cos 2\nu t)$, indicating that a contribution to the induced electric dipole oscillates at a frequency of $2\nu$. Exact phase matching of the incident radiation and the second-harmonic-generated radiation ensures the buildup of the intensity of the frequency-doubled light. Thus, suitable frequency-doubling media include those materials lacking a center of symmetry (therefore, no liquids or gases) and also possessing the property of optical anisotropy or birefringence. Birefringent substances enable the matching of the refractive indices of the incident and frequency-doubled light, as the orientation of the material affects the direction of propagation and the direction of polarization of the light. Typical frequency-doubling crystals are inorganic salts, including potassium dihydrogen phosphate ($KH_2PO_4$ or KDP), lithium niobate ($Li_3NbO_4$), and barium sodium niobate ($Ba_2NaNb_5O_{15}$).

**gain** For a round-trip pass of electromagnetic radiation through the active medium of a laser with cavity length $L$, the ratio of the beam intensity at the end of the round trip to the beam intensity at the start of the round trip.

**gain curve** The frequency dependence of the small-signal gain coefficient, i.e., a plot of $\beta$ vs. $\nu$.

**gas lasers** Lasers in which the transition responsible for stimulated emission occurs in a monatomic gas (e.g., Ar, Kr, or Ne in combination with He) or in a simple molecular gas (e.g., $CO_2$, $N_2$, or $I_2$).

**gigawatt** A measure equal to $10^9$ watts, denoted GW.

**ground state** The lowest energy configuration of a system.

**harmonic generation** Technique in which specific-wavelength lasers are used to provide additional laser lines of higher energy. Harmonic generation depends on the nonlinear response of a laser medium to an applied electric field; that is, the polarization $P$ created in the medium by the applied electric field $E$ is represented by a series expansion in terms of powers of the electric field.

**hertz** Denoted by the symbol Hz, the number of cycles per second and the SI unit for frequency.

**hole burning** The appearance of frequency-dependent minima in the gain curve of a laser at those frequencies where the cavity is resonant. The value of the gain at the "holes" (minima) in the gain curve is the threshold value, $\beta_{th} = \alpha_e + \alpha_o$, the sum of the losses of radiation in the laser due to absorption and scattering in the medium and due to useful output, respectively.

**homogeneous broadening mechanism** A process to which all atoms or molecules in a lasing medium respond identically to produce a lasing emission line that is characterized by a range of frequencies. The effects of homogeneous broadening produce lasing transitions with the same central frequency (lasing line of peak intensity) and the same distribution of frequencies (Lorentzian profile or lineshape) for all atoms or molecules in the laser medium. Homogeneous broadening arises from the consequences of the uncertainty principle (natural broadening) and from the momentum exchange between colliding particles (collisional broadening).

**hyperpolarizability** The constant of proportionality, denoted by $\beta$, for the second-order term relating the magnitude of an induced dipole moment in a molecule to the strength of the applied electric field: $\mu = \alpha E + \beta E^2 + \ldots$ The hyperpolarizability has units of C $m^3$ $V^{-2}$ and a magnitude characteristic of the molecule interacting with the electric field.

**inelastic scattering** A phenomenon in which incident photons interact with a target molecule and are not absorbed but reemitted with a different frequency. Inelastic scattering, also known as **Raman scattering**, involves two types of processes: a Stokes transition in which the scattered photon has less energy than the incident photon, and an anti-Stokes transition in which the scattered photon is higher in energy than the incident photon. Although the Stokes and anti-Stokes photons are shifted in frequency by an equal magnitude from the Rayleigh (elastic) transition, the probabilities of the two types of inelastic scattering are not equal. The probability (and therefore intensity) of each transition is directly proportional to the population of the energy level of the target molecule with which the incident photon interacts. These initial energy levels of the target molecule, $E_1$ and $E_2$, typically differ by a small amount ($E_2 - E_1 = h\nu_k$) compared to the energy of the incident photon. A Stokes transition is observed for the interaction of a photon with frequency $\nu_o$ with a molecule in energy level $E_1$ to scatter a photon of frequency $\nu_s = \nu_o - \nu_k$. An anti-Stokes transition occurs as an incident photon of frequency $\nu_o$ interacts with the target molecule in energy level $E_2$ to scatter a photon of frequency $\nu_{AS} = \nu_o + \nu_k$. As a consequence of the Boltzmann distribution of molecules in the available energy states, the Stokes transition is stronger in intensity than the anti-Stokes transition. The intensities thus show a temperature dependence as well as a fourth-power dependence on the frequency of the scattered photon. Nevertheless, the Raman effect is an especially inefficient process—accounting for one photon in $10^6$–$10^7$—and thus requires the intensity of laser radiation sources.

**inhomogeneous broadening mechanism** A process to which atoms or molecules in a lasing medium respond in a nonuniform statistical fashion to produce lasing emission lines that are characterized by a range of frequencies. The central frequencies of the resulting lasing transitions (i.e., the lasing lines of peak intensity) differ from the collection of atoms or molecules. The distribution of frequencies for a particular lasing transition is characterized by a Gaussian function (Gaussian profile or lineshape). The asymmetric Maxwell-Boltzmann distribution of velocities exhibited by the lasing medium is one source of inhomogeneous broadening (Doppler broadening). Local variations in the temperature, pressure, and magnetic field exerted on a lasing medium also give rise to inhomogeneous broadening.

**insulators** Solids in which the highest occupied band of electronic orbitals (the valence band) is fully occupied by electrons and in which the energy difference between the valence band and the conduction band (the unoccupied quantum states) is large. Under the influence of an electric field, no movement of electrons occurs between the valence and conduction bands, and no measurable electrical conductivity is observed.

**intensity of light** The number of photons entering a unit volume per second, i.e., the rate at which photons are transferred per unit volume. Intensity has units of photons $V^{-1}$ $s^{-1}$ or Einsteins $V^{-1}$ $s^{-1}$.

**interference filter** A filter with a thin film optical coating characterized by a wavelength-dependent reflectance that is designed to pass radiation of a narrow wavelength range.

**intrinsic fluorescence** Emission from an endogenous or naturally occurring fluorophore in a system.

**intrinsic fluorophore** A molecule exhibiting native fluorescence without requiring chemical modification or labelling to observe fluorescence.

**isotropic** Characterized by a physical property (of a substance or medium) that is not dependent upon the direction in which it is measured.

**Kerr cell** An optical device used to change the polarization of incident light. Analogous to a Pockels cell, the Kerr cell operates by the application of an electric field to a dipolar liquid that exhibits the Kerr effect—a change in refractive index proportional to the *square* of the magnitude of the applied field.

**laser** Acronym for the process of light amplification by stimulated emission of radiation; a device generally consisting of the three main components of an active medium, a pumping source, and a cavity or optical resonator by which the phenomenon is promoted. The first successful laser device was constructed by Theodore H. Maiman of the Hughes Aircraft Corporation in 1960, consisting of a solid-state ruby laser operated on a pulsed basis. The acronym is attributed to Gordon Gould, a pioneer of laser development.

**laser cavity** The resonant cavity or space between the end mirrors of a laser including the intervening active medium.

**lasing** The phenomenon by which a population is established in a high-energy state in such a way as to be suitable for stimulated emission to yield nearly monochromatic light.

**Law of Photochemical Equivalence** A statement detailing a one-to-one correspondence between the number of photons absorbed and the number of molecules undergoing photoactivation.

**LIDAR** An acronym for the technique known as laser-induced detection and ranging; a method by which a high-power pulsed laser and a telescope receiver are combined for the detection of laser-induced radiation on land, in ocean waters, in the atmosphere, and in space.

**line broadening** Condition by which a laser emission line is characterized, not by a single discrete frequency, but by a distribution of frequencies defined by a mathematical function known as the lineshape. A frequency distribution is quantified by the range or "width" of frequency values at one-half the maximum intensity of the emission, i.e., the full width at half maximum (FWHM). Mechanisms contributing to line broadening induce small shifts in the energy levels of the states responsible for the lasing transition, creating a distribution of energy states about each energy level and leading to the observed frequency distribution for the laser emission. Homogeneous broadening is defined as the condition whereby every atom or molecule in the lasing medium possesses the same distribution of energy levels, while the term inhomogeneous broadening applies when the atoms or molecules of the lasing material experience distinct variations in the characteristic energy levels.

**lineshape** The profile or distribution of frequencies observed for a particular lasing transition in an active medium. Gaussian and Lorentzian distributions are the most typical lineshapes.

**linewidth of lasing transition** A measure of the distribution of frequencies exhibited by emitted photons generated by a single lasing transition. Both homogeneous and inhomogeneous line-broadening mechanisms contribute to the linewidth by producing a distribution of excited states with lasing transition frequencies spread about a central value. The profile of the frequency distribution is symmetrical with the central frequency as the most commonly observed. The number of emitted photons, i.e., the intensity of emission, decreases as the frequency increases or decreases about the central value. The linewidth is quantitatively defined as the full width of the frequency distribution at one-half the maximum intensity (FWHM). The frequency profile (i.e., the lineshape) is often mathematically described by either a Gaussian or Lorentzian function.

**longitudinal axis of a laser** A reference line in the direction of the laser cavity length.

**longitudinal (axial) mode** A frequency of radiation that gives rise to constructive interference within a laser cavity of length $L$, $\nu = N \cdot c/2L$, where $N$ is an integer and $c$ is the speed of light.

**longitudinal mode spacing** The frequency separation $\Delta\nu$ of adjacent longitudinal modes of a laser with cavity length $L$ whereby $\Delta\nu = c/2L$.

**maser** Acronym for the process of microwave amplification by stimulated emission of radiation; a device invented in 1954 by Charles H. Townsend and co-workers and independently by A. M. Prochorov and Nicolai G. Basov that operated on the principle of stimulated emission and served as the predecessor of the laser.

**metals** Solid materials possessing partially filled valence bands in which electrons move freely upon the application of an electrical field.

**metal vapor lasers** Lasers in which the monatomic metal vapor lasing medium is created by using an electrical discharge to heat a plasma tube containing a small amount of metal. Metals such as copper, gold, and strontium are used to construct metal vapor lasers which are high-power, efficient sources of light.

**mobility** In an electrophoretic separation process, the velocity of a molecule in a charged field per unit field.

**mode-locking** A method of achieving ultrashort, high-power pulsed operation of a laser by correlating the longitudinal modes present within the laser cavity so that their phases are matched to give rise to completely constructive interference at just one point in the cavity. Destructive interference of the modes occurs everywhere else. Thus, an intense pulse of laser output is obtained at a regular time interval corresponding to the time required for a complete round trip in the resonator cavity. Mode-locking can be achieved: (1) electro-optically or acousto-optically using a modulating device that acts as a shutter with a frequency of $c/2L$ (i.e., active mode-locking), and (2) chemically using a saturable absorber dye within the laser cavity. Pulses of 1–200 ps duration and, consequently, extremely high peak power are attained via mode-locked operation.

**mode of radiation** A form of radiation with defined temporal, spatial, and spectral characteristics, i.e., set values of frequency, direction of propagation, polarization, and power distribution. The modes of radiation with frequency $\nu = N \cdot c/2L$ ($N$ = integer) which give rise to lasing in a cavity of length $L$ are known as the longitudinal or axial modes. The distribution of power across the beam diameter is the characteristic spatial or transverse mode.

**molecular gas lasers** *See* **gas lasers**.

**monochromaticity** Condition in which electromagnetic radiation is characterized by a single wavelength as a consequence of transitions between only two levels. Monochromaticity is quantified by the $Q$ factor, the ratio of the frequency of radiation to the range of frequencies emitted (i.e., linewidth), $Q = \nu/\Delta\nu$.

**multiphoton process** A process involving the absorption of two or more photons per molecule or atom. If the frequencies of all absorbed photons match inherent absorption frequencies of the molecule, then the process is termed a resonant multiphoton process.

**natural broadening** The appearance of a narrow range of frequencies for a particular lasing transition in an active medium as a consequence of the uncertainty principle that limits the simultaneous determination of a particle's exact position and momentum. Natural broadening is an example of a homogeneous broadening mechanism.

**natural fluorescence lifetime** The reciprocal of the rate constant for fluorescence. The natural fluorescence lifetime is observed for an excited-state species when fluorescence is the only decay path available for the excited state.

**net migration velocity** Velocity of a particle in a charged field equal to the sum of the electrophoretic velocity and the electroosmotic flow velocity.

**neutral density filter** A filter designed to attenuate an incident beam of radiation without altering its spectral distribution. Metallic neutral density filters attenuate the beam by a combination of reflection and absorption processes; glass neutral density filters attenuate by absorption.

***n*-wave (or *n*-color) mixing** An experiment in which $n$-1 laser beams of specified frequency $\nu_1, \nu_2, ..., \nu_{n-1}$ and specified spatial direction are focused on a sample to produce the $n^{th}$ color (or wave) as output. An interesting example of $n$-wave mixing is the spectroscopic technique known as coherent anti-Stokes Raman spectroscopy (CARS), a 4-wave mixing experiment. In CARS, the output of two lasers at frequencies $\nu_1$ and $\nu_2$ are combined to generate radiation of frequency $\nu_4 = 2\nu_1 - \nu_2$ (note $\nu_1 = \nu_3$). In a CARS experiment, energy equal to $h\nu_1$ is absorbed by a molecule which is subsequently induced with the input of a second laser to undergo stimulated emission at $\nu_2$. This stimulated emission occurs to an energy level not quite at the ground state (i.e., $\nu_1 > \nu_2$). Absorption of a second photon of energy $h\nu_1$ is followed by emission of radiation of frequency $\nu_4$, returning the molecule to its ground state.

**optical modulator** A device that operates as a high-speed shutter to control the amount of light transmitted. An electro-optical modulator applies a voltage across a birefringent crystal to vary differentially the refractive indices along various crystal axes. The operational mode of the electro-optical modulator is voltage dependent and generally consists of either a conversion of linearly polarized light to circularly polarized light (i.e., a quarter-wave plate mode) or a conversion of the plane of polarization by 90° (i.e., a half-wave plate mode). An acousto-optical modulator uses a propagating sound wave to create spatially periodic variations in the refractive index of a medium of glass or water. A light beam incident upon the three-dimensional pattern of refractive indices is variably diffracted to create losses in the cavity and modulations in the intensity of the emitted beam.

**optical parametric oscillator** A device driven by a laser to output tunable coherent light by nonlinear optical processes.

**optical pumping** The use of a light source (e.g., a xenon flashlamp or a laser) to induce excitation of an organic dye for laser emission.

**photoablation** The light-induced removal of material on a solid surface. The ablation process is initiated by the focusing of light on a portion of a sample to cause heating and partial vaporization. The rapid expansion due to vaporization causes ejection of an even larger volume of the sample as a plume of vapor and liquid particles.

**photoacoustic effect** The phenomenon by which light that is modulated in time produces audible sounds when focused on an absorbing medium. The energy of a portion of the photons absorbed by a sample is generally converted into heat. Time modulation of the incident light will produce a periodic heating of the sample which in turn will generate pressure pulses in the atmosphere at the sample/atmosphere interface. These pressure waves will be heard as sound. Alexander Graham Bell discovered the photoacoustic phenomenon in the 1880s.

**photodynamic therapy (PDT)** A medical procedure using laser-activated exogenous photosensitizers as chromophores which preferentially localize in cancerous tissues and eradicate such tissue through a light-induced chemical reaction.

**photosensitizer** A molecule activated by the absorption of light. Photosensitizing drugs are used in photodynamic therapy to absorb light of a specific wavelength to yield fluorescence, thus permitting an estimate of the amount as well as the precise location of a tumor.

**photothermal process** A heating induced upon the absorption of light and resulting from the radiationless decay of the electronically excited states produced via absorption.

**picosecond** A measure of time equal to $10^{-12}$ second; the time it takes light in a vacuum to travel a distance of $2.998 \times 10^{-4}$ m or 0.2998 mm.

**picosecond continuum** *See* **ultrafast supercontinuum laser source**.

**plane (or linear) polarization** Characteristic whereby the electric field vector of light is constrained to a single plane perpendicular to the direction of propagation of light.

**plane waves** Waves travelling in perfect spatial and temporal coherence.

**Pockels cell** An optical device used to change the polarization of incident light. The cell operates by the application of an electric field to a crystalline material that exhibits the Pockels effect—a change in refractive index proportional to the magnitude of the applied field. Modulation of the electric field at a particular frequency leads to modulation of the incident light polarization.

**polarizability** The ease with which the electronic distribution of an atom or molecule can be distorted by an applied electric field. The polarizability is denoted by the symbol $\alpha$ and is a constant for a particular chemical species with units of C $m^2$ $V^{-1}$. In the absence of a strong electric field, $\alpha$ is the proportionality constant relating the magnitude of the induced dipole moment in a molecule to the electric field strength: $\mu = \alpha E$.

**polarization** A measure of the phase relationship between a photon's electric field vector and the perpendicular direction of the light's propagation. Common descriptions of the polarization of a beam of radiation include: (1) plane (or linear) polarization, (2) left- or right-handed circular polarization, and (3) elliptical polarization.

**polarizer** An optical device or element that absorbs light polarized in a given direction more strongly than light polarized at 90° to this direction. Polarizers include sheet polarizers made of long-chain polyvinyl alcohols, polarizing prisms typically made of calcite crystals, and beam splitter cubes made of high optical quality glass.

**population inversion** A condition existing between a pair of energy levels for an atomic or molecular species in which there are more atoms or molecules in the higher state than in the lower state. A population inversion may be achieved directly as a result of pumping action or indirectly through an intermediate state populated by the pumping pulse. The transition from the state of greater population to the lower-energy state is responsible for the laser emission. A population inversion cannot satisfy the condition of thermal equilibrium whereby the population of a state with a lower energy is always larger.

**primary processes** Competitive de-excitation pathways of an excited-state species produced as a consequence of the absorption of light. Some primary processes regenerate the ground state of the absorbing molecule; these reactions include fluorescence, collisional quenching, and energy transfer. Other primary processes, including dissociation and isomerization, produce new ground-state species.

**pulse compression** An experimental technique for generating ultrashort pulses. Pulses from a dye laser, optically pumped by a mode-locked laser, are focused into short-length optical fibers to spectrally broaden the light (as a consequence of the nonlinear response of the optical fibers to frequency). The broadened pulse is generally characterized by positive chirp, whereby lower frequencies of light lead higher frequencies (i.e., the longer wavelength components appear in the front part of the pulse with respect to the shorter wavelength components). The spectrally broadened pulse of light is recollimated and passed through a matched pair of gratings to compress the pulse in time by providing a shorter path length for the higher-frequency components of the pulse.

**pulsed laser** A device in which a population inversion between two energy states is maintained for short durations of time at regular intervals. A population inversion cannot be sustained because the rate of stimulated emission is faster than the rate of absorption (pumping). The pulse duration is shorter than the time interval between pulses. While some lasers are inherently pulsed, various techniques are available to operate CW lasers in a pulsed mode. The converse situation, continuous operation of an inherently pulsed laser, is not possible.

**pulse duration** The width of a pulse of laser radiation at half the maximum power.

**pumping** The action of generating excited states in an active medium in order to create or sustain a population inversion. If the rate of pumping is slower than the decay rate of the excited state giving rise to stimulated emission, then a population inversion cannot be sustained.

**pumping source** An external supply of energy provided to generate excited states in the active medium of a laser. Common pumping mechanisms include the use of an electrical discharge or electron impact, the application of electromagnetic radiation, and the initiation of an exothermic chemical reaction. Optimal pumping sources include broadband flashlamps as well as lasers.

**pump-probe technique** A specialized absorption technique developed to monitor ultrafast processes not suited to conventional methods. Central to the pump-probe technique is the use of two pulses, often originating from the same laser source, to both initiate a photochemical reaction (with the pump pulse) and examine the subsequent absorption of the sample (with the probe pulse). A time delay is established between pump and probe pulses to explore phenomena occurring after the initial excitation.

**pump rate** The rate at which energy (photons) is provided to promote atoms from the ground state to a higher energy state. The pump rate is usually specified in terms of the number of atoms per second undergoing the transition between energy states.

***Q* factor** The quality factor of a mode of radiation in a resonator or laser cavity. The $Q$ factor is a quantitative measure of the relative amount of energy remaining within a cavity after one complete cycle of oscillation of the mode. A complete cycle corresponds to two passes of the energy through the active medium via reflections involving both mirrors forming the resonant cavity. Factors contributing to energy loss include the finite reflectivities of the mirrors and both scattering and absorption by the lasing medium. One quantification of the $Q$ factor is the ratio of the peak emission frequency to the linewidth (FWHM) of the emission, $Q = \nu/\Delta\nu$.

**Q-switching** A method of achieving pulsed operation of a laser by alternately reducing then rapidly increasing the $Q$ factor of the laser cavity. A situation of low cavity quality ($Q$) corresponds to an arrangement whereby stimulated emission cannot be enhanced via the multiple passage of light through the active medium. The inability of a laser cavity to act as a resonator to trap emitted light (i.e., due to the presence of a partially reflective mirror) will establish an appreciable population inversion in the active medium without a significant amount of accompanying stimulated emission. Restoration of the proper functioning of the resonator by the insertion of a totally reflective element in the laser cavity will permit the buildup of stimulated emission. Rapid removal of the reflective device will release the stored energy in a single pulse of intense light. The technique of $Q$-switching is dependent upon a rate of pumping greater than the rate of spontaneous emission to ensure a buildup of population inversion and a rate of altering the cavity quality that is sufficiently fast to generate a short pulse of light. Q-switching can be achieved: (1) mechanically, through the rotation of one of the laser mirrors, (2) electro-optically, through the action of a shutter, Pockels cell, and/or polarizer, and (3) chemically through the use of a saturable absorber dye. Pulses of $10^{-9}$–$10^{-8}$ s duration are achieved via $Q$-switching operation.

**quantum yield of a primary process** A ratio of the number of excited-state species undergoing a particular primary process relative to the number of photons absorbed to generate the excited-state species. Except for situations involving very high intensity pulsed lasers, primary processes are characterized by the absorption of only one photon of light per molecule activated to an excited state.

**radiance** Brightness; radiant intensity per unit area of the source of radiation.

**radiant intensity** Power of the light source per unit solid angle.

**Raman effect** The spectral phenomenon of the inelastic scattering of light first observed by C. V. Raman. See also **inelastic scattering**.

**Raman scattering** *See* **inelastic scattering**.

**Rayleigh scattering** *See* **elastic scattering**.

**reflection** The 180° change in direction of propagation of a wave incident on a surface.

**refraction** The change of direction of propagation of a wave as a consequence of a change in wave velocity as the wave passes from one medium to another.

**refractive index** For a given medium, the ratio of the speed of light in a vacuum to its speed in the medium ($n = c/v$). By definition, the refractive index of a medium is always greater than or equal to 1. A slight wavelength dependence is observed for the refractive index such that the longer the wavelength, the larger the refractive index.

**remote sensing** A technique by which data characterizing terrain and ocean surfaces are obtained from a distance using visible, infrared, and/or microwave radiation. Remote sensing systems may be ground-based or airborne and either passive (utilizing sunlight for excitation) or active (using a laser excitation source). Airborne remote sensing techniques have wide-ranging applications, including the detection of oil spills in the ocean, the identification of forest fires, the analysis of the Martian atmosphere, and the characterization of terrestrial and oceanographic vegetation.

**REMPI** An acronym for the condition known as 1+1 resonance-enhanced multiphoton ionization. The REMPI process involves the absorption of two photons by a molecule desorbed from a solid surface—one photon serves to promote the desorbed species to an excited electronic state and the second photon acts to ionize the excited-state desorbed molecule.

**resonance Raman effect** The phenomenon whereby incident photons possess energy matching the energy required to raise a molecule to an excited but stable electronic state and lead to emission of a Stokes or anti-Stokes photon in an inelastic manner. The resonance Raman effect has an enhanced probability (on the order of $10^2 – 10^4$-fold greater) than the spontaneous Raman effect. See also **spontaneous Raman effect, inelastic scattering, Stokes transition**, and **anti-Stokes transition**.

**resonator** The cavity or region of the laser containing the active medium through which radiation is amplified, generally via multiple reflections off a pair of mirrors.

**ruby laser** A pulsed three-level, solid-state laser in which the transition responsible for stimulated emission occurs in $Cr^{3+}$ ions embedded in corundum. The ruby laser was the first type of laser ever constructed; the inventor of the ruby laser was Theodore H. Maiman of the Hughes Aircraft Corporation who built the device in 1960.

**saturable absorber** A dye whose transmission at a laser wavelength increases nonlinearly with increasing light intensity.

**second-harmonic generation (SHG)** *See* **frequency doubling**.

**second hyperpolarizability** The constant of proportionality, denoted by $\chi$, for the third-order term relating the magnitude of an induced dipole moment in a molecule to the strength of the applied electric field: $\mu = \alpha E + \beta E^2 + \chi E^3 + ...$ The second hyperpolarizability has units of C m$^4$ V$^{-3}$ and is important in the observation of nonlinear Raman spectroscopies at high $E$ values. The second hyperpolarizability is also referred to as the third-order susceptibility.

**selection rules** Theoretically derived rules based on quantum mechanics that define the allowable transitions between quantized energy states of chemical species.

**semiconductor laser** A solid-state laser in which the lasing action occurs at a junction between p- and n-doped crystals (i.e., crystals with impurity atoms that have fewer or more valence electrons, respectively, than the atoms replaced). A semiconductor laser operating at a fixed wavelength is typified by the gallium-arsenide laser, while wavelength tunability is achieved with a temperature variation of a lead salt laser composed of a nonstoichiometric tertiary compound of lead, cadmium, and sulfur ($Pb_{1-x}Cd_xS$).

**semiconductors** Solids in which thermal excitation can promote electrons from the filled valence level to the conductance bands to yield temperature-dependent electrical conductivities of intermediate magnitude.

**single-line operation** Production of laser emission at a single wavelength or frequency, either by sustaining a population inversion between a single pair of energy states of the active medium or by using an etalon or prism wavelength selector to isolate one emission line in a laser operating in broadband mode.

**single-mode operation** A means of obtaining laser output with a single longitudinal and single transverse mode. Apertures in the laser cavity can be used to support only a single transverse mode. Elimination of all axial modes except one can be achieved by either: (1) reducing the length of the cavity to create a rigid mirror spacing such that the cavity mode spacing, $c/2L$, is greater than the linewidth broadening, or (2) introducing a dispersive element in the laser cavity to vary randomly the losses of the laser cavity resonances and enhance a single mode.

**small-signal gain coefficient** A quantity that reflects a frequency-dependent property of the lasing medium that defines per unit cavity length the fractional gain in the intensity of radiation that passes through a laser cavity, $\gamma(\nu) = 1/I_\nu \cdot dI_\nu/dL$.

**solid-state lasers** Lasers in which the transition responsible for stimulated emission occurs in transition metal ions embedded in a transparent host ionic crystal or glass (e.g., ruby lasers with $Cr^{3+}$ in a host lattice of corundum [$Al_2O_3$] or Nd:YAG lasers with $Nd^{3+}$ in a yttrium aluminum garnet crystal [$Y_3Al_5O_{12}$]).

**spatial filter** A unit designed to improve the spatial coherence of a laser beam by reducing random variations in beam intensity arising from irregular scattering. Such scattering can arise from surface imperfections or dust particles in an optical system and is typically characterized by a high frequency. One design of a spatial filter consists of a pinhole placed at the focus of a microscope objective to block the high-frequency scattered light and transmit a more uniform energy distribution.

**spatial (transverse) mode** A characteristic distribution of power across the diameter of a laser beam. Propagation of radiation slightly off the longitudinal axis of the laser cavity gives rise to the observed variations in the cross-sectional intensity distribution patterns. Spatial modes are a function of mirror radii, cavity length, wavelength of operation, and the aperture (opening) at the end of the optical cavity that limits the light emitted from the laser. A spatial mode is denoted by the label $TEM_{mn}$, a transverse electromagnetic mode with integral numbers ($m$, $n$) of nodes (minima) in the intensity of the beam's cross section in directions perpendicular to the direction of propagation of the beam. The $TEM_{00}$ mode is the most desirable spatial mode for laser operation as: (1) the mode provides the smallest possible beam diameter and divergence angle to ensure focusing to a minimum spot size and (2) the cross section of the beam is characterized by a Gaussian power distribution that is maintained as the beam propagates through free space or optical elements.

**speckle** A phenomenon in which the scattering of light from a highly coherent source by a rough surface generates a random distribution of light intensity through destructive and constructive interference. Such a phenomenon gives the surface a granular appearance.

**spectral brightness** Brightness per unit of frequency.

**spectral distribution** The range of colors output by a source of light.

**spectroscopy** The study of the interaction of light with matter.

**spherical waves** Waves propagating from a point with surfaces of equal phase that form a series of concentric circles.

**spontaneous emission** The process by which a photon of light is emitted from an atom or molecule in an excited energy state in the absence of any external stimulus or radiation. The energy of the emitted radiation matches the energy difference between the initial excited state and the final state. The rate of spontaneous emission is solely dependent on the population of the excited state. Spontaneous emission is characterized by a random direction of propagation of the emitted photons.

**spontaneous emission rate** The number of atoms per second that undergo the process of stimulated emission; the rate constant for spontaneous emission is symbolized by $A_{mn}$ for emission from the energy state $E_m$ to the energy state $E_n$.

**spontaneous Raman effect** The phenomenon whereby incident photons of sufficient energy raise a molecule to a virtual state which exists long enough to emit a Stokes or anti-Stokes photon in an inelastic manner. See also **inelastic scattering, Stokes transition**, and **anti-Stokes transition**.

**static scattering** A time-independent scattered light wave or an experiment that is not concerned with the time evolution of a scattered light wave.

**stimulated absorption** The process by which an atom or molecule is promoted to an excited energy state by the absorption of photons. The rate constant for stimulated absorption is symbolized by $B_{nm}$ for absorption of a photon by the energy state $E_n$ to the energy state $E_m$.

**stimulated emission** The process by which the emission of a photon of light from an atom or molecule in an excited energy state is induced by the interaction of the excited-state species with a photon whose energy exactly matches the energy difference between the excited and lower-energy states. The preferred direction of stimulated emission is identical to the direction of the applied incident radiation. Furthermore, the incident and stimulated radiation constructively add to increase the amplitude of the stimulating radiation. Einstein proposed the existence of both spontaneous and stimulated emission in 1917. The rate constant for stimulated emission is symbolized by $B_{mn}$ for emission from the energy state $E_m$ to the energy state $E_n$.

**Stokes transition** An inelastic scattering of light such that the scattered photons are characterized by a lower frequency than the incident photons. See also **inelastic scattering**.

**supercontinuum laser** *See* **ultrafast supercontinuum laser source**.

**superradiance** An intense pulse of radiation generated by the production of an excited state via a single input of energy from the pumping source. A sufficient population inversion is achieved without the use of mirrors and is rapidly depleted via stimulated emission at the lasing frequency. Pulsed nitrogen lasers and excimer lasers are superradiant sources.

**synchronous pumping** A method of generating ultrashort laser pulses by matching the optical cavity length of a dye laser to that of a mode-locked pump laser. A continuous train of extremely narrow pulses is a consequence of the time interval between mode-locked pump pulses being exactly equal to the round-trip time of the pulses propagating within the dye cavity. The ultrashort nature of the output pulse train arises from the short time period in which light is amplified in the dye medium.

**terawatt** A measure equal to $10^{12}$ watts, denoted TW.

**tetrapyrroles** A class of macrocyclic compounds consisting of four pyrrole groups (heterocyclic rings of four carbon atoms and a nitrogen atom) linked by methene bridges.

**third-order susceptibility** The constant of proportionality, denoted by $\chi$, for the third-order term relating the magnitude of an induced dipole moment in a molecule to the strength of the applied electric field: $\mu = \alpha E + \beta E^2 + \chi E^3 + ...$ The third-order susceptibility (sometimes called the **second hyperpolarizability**) has units of C $m^4$ $V^{-3}$ and a magnitude characteristic of the molecule interacting with the electric field.

**Thomson scattering** Scattering of light through an interaction with free electrons.

**three-level laser** A device in which a population inversion in a state $S_2$ is achieved indirectly by the fast radiationless decay of a higher-energy intermediate state $S_3$ ($S_3 \rightarrow S_2$) populated directly by a pumping pulse ($S_1 \rightarrow S_3$). The lasing action is associated with the slower transition from $S_2$ to the ground state $S_1$ ($S_2 \rightarrow S_1$). In practice, $S_3$ represents a range of energy states that can be populated by the action of the pumping source. Molecules in these states relax nonradiatively to $S_2$. The advantage of a three-level system over a two-level laser is the enhanced number of opportunities to populate the $S_2$ state for a lasing transition. A disadvantage of both two- and three-level arrangements is that the population inversion relies on the input of energy to convert ground-state species to the highest energy level, i.e., population inversion is directly dependent on pump power. A ruby laser is a characteristic three-level laser.

**threshold condition for laser action** A statement of the extent of population inversion required for laser action in a laser of known loss and mirror reflectivity characteristics. Satisfying this condition achieves stable, steady-state operation of the laser by balancing the factors which enhance a population inversion (i.e., the small-signal gain of the lasing medium) and the factors which deplete a population inversion (scattering and absorption by the lasing medium, reflectivities of the mirrors, and length of the laser cavity).

**time-resolved fluorescence spectrum** The wavelength dependence of fluorescence emission at characteristic time intervals after light absorption at a particular wavelength.

**transverse (spatial) mode** *See* **spatial mode**.

**tunability** Capacity to obtain laser emission at discrete selected wavelengths over a broad region of the electromagnetic spectrum.

**tunable lasers** Lasers emitting highly monochromatic light throughout a region (e.g., visible) of the electromagnetic spectrum through a combination of broad spectral characteristics of the lasing medium and additional dispersive optical elements (e.g., etalon, diffraction grating) within the laser cavity.

**tuning curve** A plot of laser output (i.e., energy per pulse or power) as a function of wavelength for a particular laser system. For a nitrogen-pumped dye laser, for example, a tuning curve will consist of a series of output power curves for individual dyes. Depending on the number of dyes examined and the variable width (30–60 nm) of each dye's power curve, continuous tuning of the laser throughout the visible region (e.g., 360-800 nm) can be achieved.

**two-level laser** An unattainable system in which a molecule is promoted to a higher energy state ($S_1 \rightarrow S_2$) by a pumping source to induce a population inversion for the laser emission ($S_2 \rightarrow S_1$). The attainment of a population inversion is directly dependent on the pump power.

**ultrafast supercontinuum laser source** A broad continuum of light of ultrafast duration ($\approx$100 fs) produced by focusing the output of a picosecond laser through a transparent medium. Alfano and Shapiro discovered this phenomenon in 1970.

**vacuum ultraviolet region** Light in the wavelength range of 4 to 200 nm. This radiation, also known as far-UV radiation, is absorbed by air to effect electronic transitions in oxygen, nitrogen, and carbon dioxide. Thus, since a vacuum apparatus must be used to avoid interfering absorbances by air, this region is referred to as the vacuum region.

**valence band** The highest occupied band (either partially or fully occupied) of electronic orbitals for a crystalline solid.

**vibrational relaxation** The dissipation of excess vibrational energy to return a photoexcited molecule to the lowest vibrational level in an excited electronic state. Such a process occurs on a timescale of $10^{-12}$ s.

**vibronic lasers** Solid-state lasers in which the energy levels of a transition metal ion combine or couple with the vibrational frequencies of the host lattice to yield bands of closely spaced energy levels capable of lasing. Tunable emission over a range of wavelengths arises from such closely spaced, vibrationally broadened electronic energy states in the transition metal ion. Examples of vibronic lasers include titanium:sapphire ($Ti^{3+}$ in $Al_2O_3$), alexandrite ($Cr^{3+}$ in $BeAl_2O_4$), and $Ni^{2+}$- and $Co^{2+}$-doped $MgF_2$ lasers.

**wavenumber** The number of waves in a unit distance. Denoted by $\bar{\nu}$, the wavenumber is equal to the reciprocal of the wavelength of the radiation.

## Dictionary of Acronyms

**CARS** Coherent anti-Stokes Raman spectroscopy

**CSRS** Coherent Stokes Raman spectroscopy

**CW** Continuous wave

**FEL** Free electron laser

**FTS** Femtosecond transition-state spectroscopy

**FWHM** Full width at half maximum

**HORAS** Higher order anti-Stokes scattering

**HORSES** Higher order Stokes effect scattering

**IR** Infrared

**IR-LDI** Infrared laser desorption/ionization

**IRS** Inverse Raman scattering

**KDP** Potassium dihydrogen phosphate

**LAMS** Laser-assisted mass spectroscopy

**LASER** Light amplification by stimulated emission of radiation

**LED** Light-emitting diode

**LIDAR** Laser-induced detection and ranging

**LIF** Laser-induced fluorescence

**MASER** Microwave amplification by stimulated emission of radiation

**MCP** Multichannel plate detector

**MPI** Multiphoton ionization

**MVL** Metal vapor laser

**PARS** Photoacoustic Raman spectroscopy

**PDT** Photodynamic therapy

**PES** Potential energy surfaces

**PMMA** Poly(methyl methacrylate)

**PMT** Photomultiplier tube

**REMPI** 1+1 Resonance-enhanced multiphoton ionization

**RIKES** Raman-induced Kerr effect spectroscopy

**SHG** Second-harmonic generation

**SR** Spontaneous Raman spectroscopy

**SRG** Spontaneous Raman gain spectroscopy

**TAC** Time-to-amplitude converter

**TEM** Transverse electromagnetic mode

**TIRE** The inverse Raman effect

**TOF** Time-of-flight

**TPA** Two-photon absorption spectroscopy

**TRIKE** The Raman-induced Kerr effect

**UV** Ultraviolet

**UV-LDI** Ultraviolet laser desorption/ionization

**VUV** Vacuum ultraviolet

**YAG** Yttrium aluminum garnet

**YLF** Yttrium lithium fluoride

# Index

ST. JOHN FISHER COLLEGE LIBRARY
0 1220 0053395 2